M000293834

Related Kaplan Books

College Admissions and Financial Aid

Kaplan/Newsweek College Catalog

Parent's Guide to College Admissions

Scholarships

Yale Daily News Guide to Internships

Yale Daily News Guide to Succeeding in College

You Can Afford College

Test Preparation

SAT II Biology E/M

SAT or ACT? Test Your Best

SAT & PSAT

SAT & PSAT Essential Review

SAT Math Workbook

SAT Verbal Workbook

SAT II: Chemistry

SAT II: Mathematics

SAT II: Writing

ACT

ACT Essential Review

AP BIOLOGY

**By Glenn E. Croston, Ph.D. and
the Staff of Kaplan Educational Centers**

Simon & Schuster

SYDNEY · LONDON · SINGAPORE · NEW YORK

Kaplan Books
Published by Simon & Schuster
1230 Avenue of the Americas
New York, New York 10020

Copyright © 2000 by Kaplan Educational Centers
All rights reserved. No part of this book may be reproduced or transmitted in any form or by any means, electronic or mechanical, including photocopying, recording, or by information storage and retrieval system, without the written permission of the Publisher, except where permitted by law.

Kaplan® is a registered trademark of Kaplan Educational Centers.

For bulk sales to schools, colleges, and universities, please contact: Order Department, 100 Front Street, Riverside, NJ 08075. Phone: 800-223-2336. Fax: 800-943-9831.

Contributing Editor: Ida Delmendo
Project Editor: Larissa Shmailo
Production Editor: Maude Spekes
Interior Design: Michael Shevlin
Interior Production: Amparo Graf
Art: Vincent Jeffrey
Cover Design: Cheung Tai
Desktop Publishing Manager: Michael Shevlin
Managing Editor: David Chipps
Executive Editor: Del Franz

Special thanks are extended to Pauline Carrico, Laurel Douglas, Craig Dubois, Jan Gladish, Cooper McDonald, and Sara Pearl.

Advanced Placement test directions selected from 1994 *AP Biology: Free Response Scoring Guide with Multiple Choice Section*. Reprinted by permission of the College Board and Educational Test Service. Copyright © 1994. Permission to reprint AP test materials does not constitute review or endorsement by ETS or the College Board.

Manufactured in the United States of America
Published Simultaneously in Canada

March 2000

10 9 8 7 6 5 4 3 2

ISBN 0-684-87327-3

Table of Contents

DISCLAIMER

The material in this book is up-to-date at the time of publication. The College Board may have instituted changes after this book was published. Please read all the official materials you receive regarding the AP Biology Examination carefully.

KAPLAN

ABOUT THE AUTHOR

Glenn E. Croston, Ph.D. is head of biomolecular screening at ACADIA Pharmaceuticals in San Diego, CA. He has worked as a research scientist and published extensively since earning his Ph.D. in biology at the University of California, San Diego in 1992. He is a co-author of Kaplan's *SAT II Biology E/M*. Glenn has also written and edited a wide range of MCAT biology preparative materials.

ACKNOWLEDGMENTS

This book would not have been possible without the efforts and insights of the staff of Kaplan Educational Centers.

Ida Delmendo and Craig Dubois were among those who made indispensible contributions.

THE AP BIOLOGY EXAM

So, you've decided to take the Advanced Placement Biology exam. What exactly is the AP Biology exam? Why should you be interested in taking it? If you have taken Advanced Placement Biology in high school or have a good foundation in biology, taking the AP Biology exam could help you earn college credit and/or placement into advanced coursework.

This book is designed to help you prepare for the AP Biology exam. We've included information about the AP Biology exam, test-taking strategies, and an extensive review of essential topics in biology. Each chapter in the biology review includes review questions to help you identify your strengths and weaknesses. Also included are two practice tests with answers and explanations. With Kaplan's proven test-taking strategies and the targeted biology review in this book, you'll have everything you need to ace the AP Biology exam.

Test Information and Strategies

ABOUT THE AP BIOLOGY EXAM

What is the AP Biology exam?

The AP Biology exam is a three-hour exam designed to test knowledge of a year of introductory, college-level biology. The first section of the exam consists of 120 multiple-choice questions, while the second section of the exam consists of four essay questions. Each section of the exam is 90 minutes long.

What's covered on the AP Biology exam?

The AP Biology exam covers three major areas:

1. Molecules and Cells—25%
2. Heredity and Evolution—25%
3. Organisms and Populations—50%

The three major areas are further broken down into topics. The percentages are related to the number of questions on the AP exam. For example, about 10% of the multiple-choice questions will cover cells, or 12 out of 120 total questions.

I—Molecules and Cells (25%)

Chemistry of life (7%)

Water

Organic molecules in life

Free energy changes

Enzymes

Cells (10%)
Prokaryotic and Eukaryotic cells
Membranes
Subcellular organization
Cell cycle and regulation
Cellular Energetics (8%)
Coupled reactions
Fermentation and cellular respiration
Photosynthesis

II—Heredity and Evolution (25%)

Heredity (8%)
Meiosis and gametogenesis
Eukaryotic chromosomes
Inheritance patterns
Molecular Genetics (8%)
RNA and DNA structure and function
Gene regulation
Mutation
Viral structure and replication
Nucleic acid technology and applications
Evolutionary Biology (8%)
Early evolution of life
Evidence for evolution
Mechanisms of evolution

III—Organisms and Populations (50%)

Diversity of Organisms (8%)
Evolutionary patterns
Survey of the diversity of life
Phylogenetic classification
Evolutionary relationships
Structure and Function of Plants and Animals (32%)
Reproduction, growth, and development
Structural, physiological, and behavioral adaptation
Response to the environment
Ecology (10%)
Population dynamics
Communities and ecosystems
Global issues

Anatomy of the AP Biology Exam

Section I

This section consists of three types of multiple-choice questions. The first type is a straight-forward, multiple-choice question with five answer choices. Here is an example:

1. If the diploid number of a female organism is 48, then the number of chromosomes in each egg cell is

   (A) 12
   (B) 24
   (C) 48
   (D) 72
   (E) 96

   Answer: (B) If the diploid number of an organism is 48, then 2n = 48. The number of chromosomes in an egg cell would be half the diploid number, or 24.

The second type of multiple choice question asks you to match statements to a diagram or a list of choices. For example:

Neuron

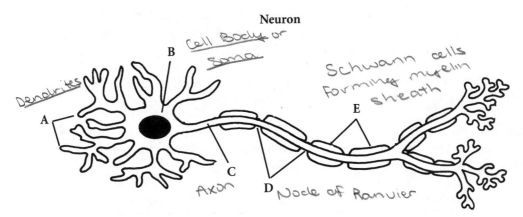

2. Which labeled structure receives input?

   Answer: (A) Dendrites are the neural structures which receive input signals and direct them toward the cell body. The structure labeled B is the cell body, which contains the nucleus. The structure labeled C is the axon. Structures labeled E are areas of the neuron covered in myelin, an insulating material. In saltatory conduction, an action potential jumps from one Node of Ranvier to another Node of Ranvier, skipping areas of the neuron covered by myelin. Structures labeled D are Nodes of Ranvier, uninsulated areas of the neuron.

Here are examples of questions that ask you to match statements to a list of choices:

> Questions 3–5:
>
> (A) Mesoderm
> (B) Endoderm
> (C) Ectoderm
> (D) Morula
> (E) Blastula

3. This germ layer produces the nervous system C

4. This germ layer produces the digestive tract B

5. This germ layer produces the circulatory system A

Answers:

3. **(C)** The ectoderm develops into the epidermis (outer layer of skin), nervous system, and sweat glands.

4. **(B)** The endoderm develops into the digestive and respiratory tracts, parts of the liver and pancreas, and the bladder lining.

5. **(A)** The mesoderm develops into the muscles and skeletal system, circulatory system, excretory system, gonads, and dermis, or inner layer of skin.

The third type of multiple-choice question asks you to interpret a set of data or experimental results. Here are some examples of the this type of multiple-choice question:

A couple suspects that their daughter may have been accidentally switched with another baby at the hospital. They send out samples of their blood as well as the blood of their "daughter" for testing. The antiserum tests gave the following results:

	Father	Mother	"Daughter"
anti-A	no agglutination	agglutination	agglutination
anti-B	no agglutination	no agglutination	agglutination

6. What is the blood type of the father?

(A) A
(B) B
(C) AB
(D) O
(E) It cannot be determined from the information given

7. What is the blood type of the mother?

   (A) A

   (B) B

   (C) AB

   (D) O

   (E) It cannot be determined from the information given

8. What is the blood type of the "daughter"?

   (A) A

   (B) B

   (C) AB

   (D) O

   (E) It cannot be determined from the information given

Answers:

6. **(D)** Anti-A serum contains antibodies to A antigens on red blood cells. If A antigens are present in a person's blood, the antiserum will show an agglutination reaction when mixed with the blood. Anti-B serum contains antibodies to B antigens on red blood cells. If B antigens are present in a person's blood, the antiserum will show an agglutination reaction when mixed with the blood. Since the father's blood did not show an agglutination reaction with either anti-A serum or anti-B serum, neither A antigens nor B antigens are present on his red blood cells. Therefore, the father has type O blood.

7. **(A)** The mother's blood showed an agglutination reaction with anti-A serum but did not show an agglutination reaction with anti-B serum. Therefore, A antigens are present on her red blood cells while B antigens are not. The mother has type A blood.

8. **(C)** The "daughter's" blood showed an agglutination reaction with both anti-A serum and anti-B serum. Therefore, both A and B antigens are present on her red blood cells. The "daughter" has type AB blood.

(None of the questions asked whether the "daughter's" blood type could have resulted from the father's and mother's blood type, but let's look at it anyway. Since the father is type O, he has neither A nor B antigens on his red blood cells. Since the mother is type A, she has only A antigens on her red blood cells. Therefore it is not possible for a child of theirs to have B antigens. Their children cannot have type B or AB blood.)

Section II

The second section of the AP Biology exam consists of four essays. The essays will cover the three major areas as follows:

Molecules and Cells—1 question
Heredity and Evolution—1 question
Organisms and Populations—2 questions

Scoring the AP Biology Exam

Section I (multiple-choice) accounts for 60% of the overall grade, and Section II (essays) accounts for 40% of the overall grade.

The maximum number of points you can earn on the AP Biology Exam is 150. Section I (multiple-choice) contributes 60% of those points, or 90 points $(0.60 \times 150 = 90)$. Section II (essays) contributes 40% of those points, or 60 points $(0.40 \times 150 = 60)$.

Scoring of Section I

The raw score for Section I (multiple-choice) is calculated by taking the number of questions answered correctly and subtracting 1/4 point for every question answered incorrectly. For example, if you answered all 120 questions correctly, your raw score would be calculated as follows:

Raw score = # correct–(1/4)(# incorrect)
Raw score = 120–(1/4)(0) = 120–0 = 120

This raw score is converted to a composite score by multiplying by 0.75:

Composite score = (0.75)(raw score)

So for our example, the composite score is (0.75)(120) = 90.

Scoring of Section II

Section II consists of four essay questions. Each essay question is worth a total of 10 points, so the maximum number of points in the raw score is 40. To convert the raw score to a composite score, multiply by 1.5. For example, if you received 10 points for each essay, your raw score would be 40. Your composite score would be calculated as follows:

Composite score = (1.5)(raw score)

So for our example, the composite score is (1.5)(40) = 60.

Total Composite Score

The composite score from Section I is added to the composite score from Section II to give the total composite score. For our example, the total composite score is $90 + 60 = 150$.

The composite scores are then converted to an AP score using the following conversion:

Composite score	AP grade	Comment
91–150	5	Extremely well qualified
70–90	4	Well qualified
52–69	3	Qualified
32–51	2	Possibly qualified
0–31	1	No recommendation

This scale applied to the 1994 AP Biology exam. Please note that this scale should be used as an estimate only, as the actual scale varies somewhat from year to year.

TEST-TAKING STRATEGIES

Section I: Multiple-Choice

Section I consists of 120 multiple-choice questions to be completed in 90 minutes. That works out to only 45 seconds per question to read, analyze, choose the best answer, then code it onto the answer grid. Here are some strategies to maximize your score on the multiple-choice section:

1. *Answer the easier questions first.* Easy questions are worth just as many points as hard questions. To maximize your score, you need to answer as many questions correctly as possible, but it doesn't matter if the questions are easy or hard. On your first pass through the multiple-choice section, answer all the easy questions first.

2. *Circle harder questions and come back to them later.* If you come across a harder question, circle it and move on. Don't waste valuable time on harder questions early in the exam. If you start to answer a question then find yourself confused, move on and come back to that question later. You're better off spending those extra few minutes answering 3 or 4 easier questions.

3. *Mark up your test booklet.* As a student, you may be used to having teachers tell you not to write in your books. But when taking the AP Biology exam, it is to your advantage to mark up your test booklet. Label diagrams, cross out incorrect answer choices, write down key mnemonics in the margins (for example, **PMAT** can help you remember the

order of the stages of mitosis: **P**rophase, **M**etaphase, **A**naphase, and **T**elophase). Just remember that no credit is given for anything you've done in your test booklet, so make sure to transfer your final answers to your answer sheet.

4. *Be careful with your answer sheet.* Speaking of your answer sheet, it's easy to forget to skip a row on your answer sheet when you skip a question. Be careful to skip spaces on your answer sheet when you skip questions. Otherwise, you may spend valuable time erasing and re-gridding your answers—time that you could be using to answer questions. Also, if you change an answer, erase your previous answer cleanly.

5. *Guess.* Yes, guess! There is a 1/4 point deduction for wrong answers, which cancels out the effect of random guessing; random guessing will not improve your score. However, if you can eliminate one or two answer choices, your odds of getting the correct answer improve considerably, and chances are you will gain more points than you lose.

6. *Pace yourself.* The AP Biology exam is three hours long. It's easy to get distracted and lose focus. Try to keep yourself on task and actively working throughout the exam. Don't get discouraged if you don't know the answers to some questions. You don't need to answer all of the questions correctly to get a good score.

Section II: Essays

Section II consists of 4 essay questions to be completed in 90 minutes. That's just over 22 minutes per essay. But don't worry, these aren't the same essays you would write for your English Composition class. Here are a couple pointers to help you get the highest score possible on Section II:

1. *Read each question carefully.* Read the question, then read it again. What is it asking about? Make sure your essay answers the question they ask.

2. *Prioritize the questions according to difficulty.* Again, your goal is to score as many points in this section as possible. Read all the questions first, then decide which ones you will be able to answer most effectively. Do those essays first, then go back to questions that are a little harder for you to answer.

3. *Prepare an outline and a list of key terms.* Graders are looking for main ideas, supporting details, and key terms. Think about what you will write and organize the ideas into outline form before you begin writing. Make a list of key terms to include in each paragraph.

4. *Label drawings.* If you use a drawing, graph, or diagram in your answer to an essay question, be sure to label it appropriately.

5. *Pace yourself.* In Section II, you have 90 minutes to write 4 essays. Plan to spend about 22 minutes on each essay question. Spend 2 minutes planning your essay, about 18 minutes writing, and about 2 minutes reviewing your essay. Make sure you've included all the key terms you listed when planning your essay and labeled any drawings or graphs.

The evening before the exam. . .

It's tempting to spend the last few hours before the AP exam cramming. If possible, spend the evening before the test relaxing. Get a good night's sleep, and eat something before the exam. There's nothing like having your stomach growling two hours into a three hour exam. Most of all, be confident! You have the knowledge and test-taking skills to conquer the AP Biology exam.

Additional Resources

For more information on the AP Program and the AP Biology Exam, please contact the College Board at their Web site at www.collegeboard.org/ap/biology or at their National office at 45 Columbus Avenue, New York, NY 10023-6992, telephone (212) 713-8600.

Stress Management

The countdown has begun. Your date with THE TEST is looming on the horizon. Anxiety is on the rise. The butterflies in your stomach have gone ballistic. Perhaps you feel as if the last thing you ate has turned into a lead ball. Your thinking is getting cloudy. Maybe you think you won't be ready. Maybe you already know your stuff, but you're going into panic mode anyway. Worst of all, you're not sure of what to do about it.

It is possible to tame that anxiety and stress—before and during the test. We'll show you how. You won't believe how quickly and easily you can deal with that killer anxiety.

MAKE THE MOST OF YOUR PREP TIME

Lack of control is one of the prime causes of stress. A ton of research shows that if you don't have a sense of control over what's happening in your life, you can easily end up feeling helpless and hopeless. So, just having concrete things to do and to think about—taking control—will help reduce your stress. This section shows you how to take control during the days leading up to taking the test.

IDENTIFY THE SOURCES OF STRESS

Jot down (in pencil) anything you identify as a source of your test-related stress. The idea is to pin down that free-floating anxiety so that you can take control of it. Here are some common examples to get you started:

- I always freeze up on tests.
- I need a good/great score to go to Acme College.

- My older brother/sister/best friend/girlfriend- or boyfriend did really well. I must match their scores or do better.
- My parents, who are paying for school, will be really disappointed if I don't score well.
- I'm afraid of losing my focus and concentration.
- I'm afraid I'm not spending enough time preparing.
- I study like crazy, but nothing seems to stick in my mind.
- I always run out of time and get panicky.
- I feel as though thinking is becoming like wading through thick mud.

Take a few minutes to think about the things you've just written down. Then rewrite them in some sort of order. List the statements you most associate with your stress and anxiety first, and put the least disturbing items last. Chances are, the top of the list is a fairly accurate description of exactly how you react to test anxiety, both physically and mentally. The later items usually describe your fears (disappointing Mom and Dad, looking bad, etcetera). As you write the list, you're forming a hierarchy of items so you can deal first with the anxiety provokers that bug you most. Very often, taking care of the major items from the top of the list goes a long way toward relieving overall testing anxiety. You probably won't have to bother with the stuff you placed last.

TAKE STOCK OF YOUR STRENGTHS AND WEAKNESSES

Take one minute to list the areas of the test that you are good at. They can be general ("algebra") or specific ("quadratic equations"). Put down as many as you can think of, and if possible, time yourself. Write for the entire time; don't stop writing until you've reached the one-minute stopping point.

Next, take one minute to list areas of the test you're not so good at, just plain bad at, have failed at, or keep failing at. Again, keep it to one minute, and continue writing until you reach the cutoff. Don't be afraid to identify and write down your weak spots! In all probability, as you do both lists, you'll find you are strong in some areas and not so strong in others. Taking stock of your assets and liabilities lets you know the areas you don't have to worry about, and the ones that will demand extra attention and effort.

Now, go back to the "good" list, and expand it for two minutes. Take the general items on that first list and make them more specific; take the specific items and expand them into more general conclusions. Naturally, if anything new comes to mind, jot it down. Focus all of your attention and effort on your strengths. Don't underestimate yourself or your abilities. Give yourself full credit. At the same time, don't list strengths you don't really have; you'll only be fooling yourself.

Every area of strength and confidence you can identify is much like having a reserve of solid gold at Fort Knox. You'll be able to draw on your reserves as you need them. You can use your reserves to solve difficult questions, maintain confidence, and keep test stress and anxiety at a distance. The encouraging thing is that every time you recognize another area of strength, succeed at coming up with a solution, or get a good score on a test, you increase your reserves. And, there is absolutely no limit to how much self-confidence you can have or how good you can feel about yourself.

WHAT DO YOU WANT TO ACCOMPLISH IN THE TIME REMAINING?

The whole point to this next exercise is sort of like checking out a used car you might want to buy. You'd want to know up front what the car's weak points are, right? Knowing that influences your whole shopping-for-a-used-car campaign. So it is with your conquering-test-stress campaign: Knowing what your weak points are ahead of time helps you prepare.

So let's get back to the list of your weak points. Take two minutes to expand it just as you did with your "good" list. Be honest with yourself without going overboard. It's an accurate appraisal of the test areas that give you trouble. So, pick up your pencil, check the clock, and start writing.

Facing your weak spots gives you some distinct advantages. It helps a lot to find out where you need to spend extra effort. Increased exposure to tough material makes it more familiar and less intimidating. (After all, we mostly fear what we don't know and are probably afraid to face.) You'll feel better about yourself because you're dealing directly with areas of the test that bring on your anxiety. You can't help feeling more confident when you know you're actively strengthening your chances of earning a higher overall test score.

IMAGINE YOURSELF SUCCEEDING

This next little group of exercises is both physical and mental. It's a natural follow-up to what you've just accomplished with your lists.

First, get yourself into a comfortable sitting position in a quiet setting. Wear loose clothes. If you wear glasses, take them off. Then, close your eyes and breathe in a deep, satisfying breath of air. Really fill your lungs until your rib cage is fully expanded and you can't take in any more. Then, exhale the air completely. Imagine you're blowing out a candle with your last little puff of air. Do this two or three more times, filling your lungs to their maximum and emptying them totally. Keep your eyes closed, comfortably but not tightly. Let your body sink deeper into the chair as you become even more comfortable.

With your eyes shut you can notice something very interesting. You're no longer dealing with the worrisome stuff going on in the world outside of you. Now you can concentrate on what happens inside you. The more you recognize your own physical reactions to stress and anxiety, the more you can do about them. You might not realize it, but you've begun to regain a sense of being in control.

Let images begin to form on the "viewing screens" on the back of your eyelids. You're experiencing visualizations from the place in your mind that makes pictures. Allow the images to come easily and naturally; don't force them. Imagine yourself in a relaxing situation. It might be in a special place you've visited before or one you've read about. It can be a fictional location that you create in your imagination, but a real-life memory of a place or situation you know is usually better. Make it as detailed as possible, and notice as much as you can.

If you don't see this relaxing place sharply or in living color, it doesn't mean the exercise won't work for you. Some people can visualize in great detail, while others get only a sense of an image. What's important is not how sharp the details or colors, but how well you're able to manipulate the images. If you have only a faint sense of the images, that's okay—you'll still experience all the benefits of the exercise.

Think about the sights, the sounds, the smells, even the tastes and textures associated with your relaxing situation. See and feel yourself in this special place. Stay focused on the images as you sink farther back into your chair. Breathe easily and naturally. You might have the sensations of any stress or tension draining from your muscles and flowing downward, out your feet and away from you.

Take a moment to check how you're feeling. Notice how comfortable you've become. Imagine how much easier it would be if you could take the test feeling this relaxed and in this state of ease. You've coupled the images of your special place with sensations of comfort and relaxation. You've also found a way to become relaxed simply by visualizing your own safe, special place.

Now, close your eyes and start remembering a real-life situation in which you did well on a test. If you can't come up with one, remember a situation in which you did something (academic or otherwise) that you were really proud of—a genuine accomplishment. Make the memory as detailed as possible. Remember how confident you felt as you accomplished your goal. Now start thinking about the upcoming test. Keep your thoughts and feelings in line with that successful experience. Don't make comparisons between them. Just imagine taking the upcoming test with the same feelings of confidence and relaxed control.

This exercise is a great way to bring the test down to Earth. You should practice this exercise often, especially when the prospect of taking the exam starts to bum you out. The more you practice it, the more effective the exercise will be for you.

EXERCISE YOUR FRUSTRATIONS AWAY

Whether it is jogging, walking, biking, mild aerobics, pushups, or a pickup basketball game, physical exercise is a very effective way to stimulate both your mind and body and to improve your ability to think and concentrate. A surprising number of students get out of the habit of regular exercise, ironically because they're spending so much time prepping for exams. Also, sedentary people—this is a medical fact—get less oxygen to the blood and hence to the head than active people. You can live fine with a little less oxygen; you just can't think as well.

Any big test is a bit like a race. Thinking clearly at the end is just as important as having a quick mind early on. If you can't sustain your energy level in the last sections of the exam, there's too good a chance you could blow it. You need a fit body that can weather the demands any big exam puts on you. Along with a good diet and adequate sleep, exercise is an important part of keeping yourself in fighting shape and thinking clearly for the long haul.

There's another thing that happens when students don't make exercise an integral part of their test preparation. Like any organism in nature, you operate best if all your "energy systems" are in balance. Studying uses a lot of energy, but it's all mental. When you take a study break, do something active instead of raiding the fridge or vegging out in front of the TV. Take a 5- to 10-minute activity break for every 50 or 60 minutes that you study. The physical exertion gets your body into the act, which helps to keep your mind and body in sync. Then, when you finish studying for the night and hit the sack, you won't lie there, tense and unable to sleep because your head is overtired and your body wants to pump iron or run a marathon.

A warning about exercise, however: It's not a good idea to exercise vigorously right before you go to bed. This could easily cause sleep onset problems. For the same reason, it's also not a good idea to study right up to bedtime. Make time for a "buffer period" before you go to bed: For 30 to 60 minutes, just take a hot shower, meditate, simply veg out.

GET HIGH . . . NATURALLY

Exercise can give you a natural high, which is the only kind of high you should be aiming for. Using drugs (prescription or recreational) specifically to prepare for and take a big test is definitely self-defeating. (And if they're illegal drugs, you can end up with a bigger problem than the AP Biology test on your hands.) Except for the drugs that occur naturally in your brain, every drug has major drawbacks—and a false sense of security is only one of them.

You may have heard that popping uppers helps you study by keeping you alert. If they're illegal, definitely forget about it. They wouldn't really work anyway, since amphetamines make it hard to retain information. Mild stimulants, such as coffee, cola, or over-the-counter caffeine pills can sometimes help as you study, since they keep you alert. On the down side, they can also lead to agitation, restlessness, and insomnia. Some people can drink a pot of high-octane coffee and sleep like a baby. Others have one cup and start to vibrate. It all depends on your tolerance for caffeine. Remember, a little anxiety is a good thing. The adrenaline that gets pumped into your bloodstream helps you stay alert and think more clearly. But, too much anxiety and you can't think straight at all.

Alcohol and other depressants are out, too. Again, if it's illegal, forget about it. Depressants wouldn't work anyway, since they lead to the inevitable hangover/crash, fuzzy thinking, and lousy sense of judgment. These would not help you ace the test.

Instead, go for endorphins—the "natural morphine." Endorphins have no side effects and they're free—you've already got them in your brain. It just takes some exercise to release them. Running around on the basketball court, bicycling, swimming, aerobics, power walking—these activities cause endorphins to occupy certain spots in your brain's neural synapses. In addition, exercise develops staying power and increases the oxygen transfer to your brain. Go into the test naturally.

TAKE A DEEP BREATH . . .

Here's another natural route to relaxation and invigoration. It's a classic isometric exercise that you can do whenever you get stressed out—just before the test begins, even during the test. It's very simple and takes just a few minutes.

Close your eyes. Starting with your eyes and—without holding your breath—gradually tighten every muscle in your body (but not to the point of pain) in the following sequence:

1. Close your eyes tightly.
2. Squeeze your nose and mouth together so that your whole face is scrunched up. (If it makes you self-conscious to do this in the test room, skip the face-scrunching part.)
3. Pull your chin into your chest, and pull your shoulders together.
4. Tighten your arms to your body, then clench your hands into tight fists.
5. Pull in your stomach.
6. Squeeze your thighs and buttocks together, and tighten your calves.
7. Stretch your feet, then curl your toes (watch out for cramping in this part).

At this point, every muscle should be tightened. Now, relax your body, one part at a time, in reverse order, starting with your toes. Let the tension drop out of each muscle. The entire process might take five minutes from start to finish (maybe a couple of minutes during the test). This clenching and unclenching exercise should help you to feel very relaxed.

. . . AND KEEP BREATHING

Conscious attention to breathing is an excellent way of managing test stress (or any stress, for that matter). The majority of people who get into trouble during tests take shallow breaths. They breathe using only their upper chests and shoulder muscles, and may even hold their breath for long periods of time. Conversely, the test taker who by accident or design keeps breathing normally and rhythmically is likely to be more relaxed and in better control during the entire test experience.

So, now is the time to get into the habit of relaxed breathing. Do the next exercise to learn to breathe in a natural, easy rhythm. By the way, this is another technique you can use during the test to collect your thoughts and ward off excess stress. The entire exercise should take no more than three to five minutes.

With your eyes still closed, breathe in slowly and deeply through your nose. Hold the breath for a bit, and then release it through your mouth. The key is to breathe slowly and deeply by using your diaphragm (the big band of muscle that spans your body just above your waist) to draw air in and out naturally and effortlessly. Breathing with your diaphragm encourages relaxation and helps minimize tension.

As you breathe, imagine that colored air is flowing into your lungs. Choose any color you like, from a single color to a rainbow. With each breath, the air fills your body from the top of your head to the tips of your toes. Continue inhaling the colored air until it occupies every part of you, bones and muscles included. Once you have completely filled yourself with the colored air, picture an opening somewhere on your body, either natural or imagined. Now, with each breath you exhale, some of the colored air will pass out the opening and leave your body. The level of the air (much like the water in a glass as it is emptied) will begin to drop. It will descend progressively lower, from your head down to your feet. As you continue to exhale the colored air, watch the level go lower and lower, farther and farther down your body. As the last of the colored air passes out of the opening, the level will drop down to your toes and disappear. Stay quiet for just a moment. Then notice how relaxed and comfortable you feel.

THUMBS UP FOR MEDITATION

Once relegated to the fringes of the medical world, meditation, biofeedback, and hypnosis are increasingly recommended by medical researchers to reduce pain from headaches, back problems—even cancer. Think of what these powerful techniques could do for your test-related stress and anxiety.

Effective meditation is based primarily on two relaxation methods you've already learned: body awareness and breathing. A couple of different meditation techniques follow. Experience them both, and choose the one that works best for you.

BREATH MEDITATION

Make yourself comfortable, either sitting or lying down. For this meditation you can keep your eyes opened or closed. You're going to concentrate on your breathing. The goal of the meditation is to notice everything you can about your breath as it enters and leaves your body. Take three to five breaths each time you practice the meditation; this set of breaths should take about a minute to complete.

Take a deep breath and hold it for 5 to 10 seconds. When you exhale, let the breath out very slowly. Feel the tension flowing out of you along with the breath that leaves your body. Pay close attention to the air as it flows in and out of your nostrils. Observe how cool it is as you inhale and how warm your breath is when you exhale. As you expel the air, say to yourself a cue word such as calm or relax. Once you've exhaled all the air from your lungs, start the next long, slow inhale. Notice how relaxed feelings increase as you slowly exhale and again hear your cue words.

MANTRA MEDITATION

For this type of meditation experience you'll need a mental device (a mantra), a passive attitude (don't try to do anything), and a position in which you can be comfortable. You're going to focus your total attention on a mantra you create. It should be emotionally neutral, repetitive, and monotonous, and your aim is to fully occupy your mind with it. Furthermore, you want to do the meditation passively, with no goal in your head of how relaxed you're supposed to be. This is a great way to prepare for studying or taking the test. It clears your head of extraneous thoughts and gets you focused and at ease.

Sit comfortably and close your eyes. Begin to relax by letting your body go limp. Create a relaxed mental attitude and know there's no need for you to force anything. You're simply going to let something happen. Breathe through your nose. Take calm, easy breaths and as you exhale, say your mantra (one, ohhm, aah, soup—whatever is emotionally neutral for you) to yourself. Repeat the mantra each time you breathe out. Let feelings of relaxation grow as you focus on the mantra and your slow breathing. Don't worry if your mind wanders. Simply return to the mantra and continue letting go. Experience this meditation for 10 to 15 minutes.

Quick Tips for the Days Just Before the Exam

- The best test takers do less and less as the test approaches. Taper off your study schedule and take it easy on yourself. You want to be relaxed and ready on the day of the test. Give yourself time off, especially the evening before the exam. By then, if you've studied well, everything you need to know is firmly stored in your memory banks.

- Positive self-talk can be extremely liberating and invigorating, especially as the test looms closer. Tell yourself things such as, "I choose to take this test" rather than "I have to"; "I will do well" rather than "I hope things go well"; "I can" rather than "I cannot." Be aware of negative, self-defeating thoughts and images and immediately counter any you become aware of. Replace them with affirming statements that encourage your self-esteem and confidence. Create and practice visualizations that build on your positive statements.

- Get your act together sooner rather than later. Have everything (including choice of clothing) laid out days in advance. Most important, know where the test will be held and the easiest, quickest way to get there. You will gain great peace of mind if you know that all the little details—gas in the car, directions, etcetera—are firmly in your control before the day of the test.

- Experience the test site a few days in advance. This is very helpful if you are especially anxious. If at all possible, find out what room you are assigned to, and try to sit there (by yourself) for a while. Better yet, bring some practice material and do at least a section or two, if not an entire practice test, in that room. In this situation, familiarity doesn't breed contempt, it generates comfort and confidence.

- Forego any practice on the day before the test. It's in your best interest to marshal your physical and psychological resources for 24 hours or so. Even race horses are kept in the paddock and treated like princes the day before a race. Keep the upcoming test out of your consciousness; go to a movie, take a pleasant hike, or just relax. Don't eat junk food or tons of sugar. And—of course—get plenty of rest the night before. Just don't go to bed too early. It's hard to fall asleep earlier than you're used to, and you don't want to lie there thinking about the test.

Handling Stress During the Test

The biggest stress monster will be the test itself. Fear not; there are methods of quelling your stress during the test.

- Keep moving forward instead of getting bogged down in a difficult question. You don't have to get everything right to achieve a fine score. The best test takers skip difficult material temporarily in search of the easier stuff. They mark the ones that require extra time and thought. This strategy buys time and builds confidence so you can handle the tough stuff later.

- Don't be thrown if other test takers seem to be working more furiously than you are. Continue to spend your time patiently thinking through your answers; it's going to lead to better results. Don't mistake the other people's sheer activity as signs of progress and higher scores.

- Keep breathing! Weak test takers tend to forget to breathe properly as the test proceeds. They start holding their breath without realizing it, or they breathe erratically or arrhythmically. Improper breathing interferes with clear thinking.

- Some quick isometrics during the test—especially if concentration is wandering or energy is waning—can help. Try this: Put your palms together and press intensely for a few seconds. Concentrate on the tension you feel through your palms, wrists, forearms, and up into your biceps and shoulders. Then, quickly release the pressure.

Feel the difference as you let go. Focus on the warm relaxation that floods through the muscles. Now you're ready to return to the task.

- Here's another isometric that will relieve tension in both your neck and eye muscles. Slowly rotate your head from side to side, turning your head and eyes to look as far back over each shoulder as you can. Feel the muscles stretch on one side of your neck as they contract on the other. Repeat five times in each direction.

With what you've just learned here, you're armed and ready to do battle with the test. This book and your studies will give you the information you'll need to answer the questions. It's all firmly planted in your mind. You also know how to deal with any excess tension that might come along, both when you're studying for and taking the exam. You've experienced everything you need to tame your test anxiety and stress. You're going to get a great score.

MOLECULES AND CELLS

Living organisms display amazing diversity, ranging from the simplest bacteria to blue whales, but all living organisms share basic unifying principles starting with the chemistry of life. For example, water is essential to all forms of life, no matter how simple or complex. A second principle is that all living organisms, and the molecules and reactions they are composed of, must obey the same physical laws of chemistry and energy that rule the rest of the universe.

All life shares common biological molecules including carbohydrates, lipids, proteins, and nucleic acids. Organisms from fungi to man even share many reactions and metabolic pathways. These basic features of all life will play a role in the more complex life activities of cells, organs, organisms, and ecosystems presented later.

At a molecular level the human cell shares a great deal in common with the single-cell yeasts that make bread, a fact that may not be immediately evident to the baker. Simple organisms like yeasts, worms, and fruit flies have proven invaluable to biologists in discerning the complexities of human biology since there are so many features that they share despite their differences. These common traits contribute to the interdependence of all living organisms on earth, another important trait common to all life, including man.

The Chemistry of Life

1.1   THE PROPERTIES OF WATER

Water is essential to all life on Earth. Water covers a majority of the surface of the earth and is the site of some of the key ecosystems of the earth. Each cell contains water that bathes the reactions of life and is indispensable to all life. The presence of liquid water on Earth is one of the features of the planet that probably allowed life to originate and persist. It is the physical properties of water that allow it to play this key role for life. The properties of water that make it ideally suited to play this unique role are:

1. Water molecules are polar.
2. Water expands when it freezes.
3. Water absorbs a great deal of heat when it evaporates.
4. Water absorbs a large amount of heat when it is heated.
5. Water is cohesive and has a high surface tension.
6. Water is an excellent solvent for a large variety of molecules.
7. Water dissociates to form protons and hydroxyls in solution.

Water molecules are polar.

Water molecules have one atom of oxygen with two atoms of hydrogen at an angle from each other. Each water molecule as a whole lacks a net charge, but within each molecule the oxygen atom pulls electrons toward itself more than the hydrogens, causing the oxygen to have a partial negative charge and the hydrogens to have a partial positive charge (see figure). This unequal distribution of charges is what makes water a polar substance.

Water is a polar molecule

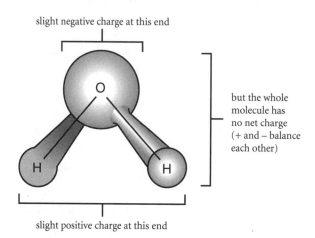

slight negative charge at this end

but the whole molecule has no net charge (+ and – balance each other)

slight positive charge at this end

The polarity of water allows water molecules to readily form hydrogen bonds with each other. A hydrogen bond is formed when a hydrogen atom with a partial positive charge interacts with a negatively charged atom such as oxygen in another molecule. The polarity of a water molecule allows it to form hydrogen bonds with other polar molecules as well, such as sugars or proteins, allowing water to dissolve these substances. The polar nature of water molecules is involved in most of the exceptional properties of water.

Water expands when it freezes.

One of the consequences of the polarity of water is that water molecules interact with each other in a network of hydrogen bonds. In liquid water, these bonds form and break very rapidly as the individual water molecules move. When water freezes, the individual water molecules stop moving and the hydrogen bonds between molecules are frozen in place in a rigid crystalline structure. The positions of the water molecules are further apart in frozen water than in liquid water, leading to one of the unusual properties of water. In most substances, the solid occupies less space than the liquid as molecules fall into the crystalline lattice. Since solid water occupies more space than liquid water, water expands as it freezes and ice is less dense than liquid water. This property of water affects life on earth. When the temperature of the environment falls below the freezing point for water, a lake or ocean will freeze on the surface, with ice floating on top of the denser liquid water beneath. The ice on top insulates the water beneath it, and slows further freezing, allowing life to continue beneath the surface ice. If, like most substances, water became denser as it froze, then a lake or ocean would freeze from the bottom up and would freeze solid, fish, plankton and all. Freezing of lakes and oceans would be far more extensive and destructive to life in this scenario and it is possible that if ice were denser than liquid water, a past period of glaciation would have frozen the oceans solid, perhaps forever.

Water absorbs a great deal of heat when it evaporates.

Water molecules in liquid water interact with each other through a large number of hydrogen bonds. When water molecules are converted into a gas in the process called evaporation, the water molecules are separated from each other in space and no longer have these hydrogen bonds. When water is heated on a stove, the molecules have more kinetic energy and are able to break the hydrogen bonds more readily. Breaking the hydrogen bonds to cause evaporation takes a large amount of energy called the **molar heat of evaporation**.

The heat of evaporation is used by terrestrial organisms as a cooling mechanism. Since heat energy is required for evaporation, evaporating water will absorb heat (see figure). Water on the skin when it evaporates on a hot day will draw heat from the skin. The sweating or panting of mammals on a hot day uses the absorption of heat by evaporating water to draw heat out of the body and make it possible to maintain a cooler interior temperature than the external environment. In the absence of this, the body temperature would equilibrate with the exterior temperature on a hot day, causing harm or even death.

Water absorbs heat when it evaporates

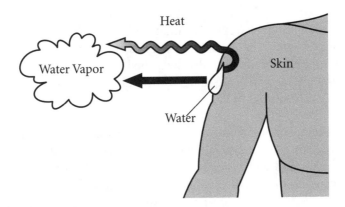

Water absorbs a large amount of heat when it is heated.

The temperature of a liquid is a measure of the kinetic energy of its molecules. The quicker molecules move, the greater the temperature. When heat energy is added to water, a great deal of the heat goes to breaking hydrogen bonds between molecules and not directly to making the molecules move faster. As a result, water can absorb a large amount of heat energy while its temperature changes little. If this were not the case, it would be much more difficult for the body to maintain a constant internal temperature. It also means that the temperature of aquatic environments does not fluctuate dramatically or rapidly. Fish do not need to adapt to sudden temperature changes the same way that a terrestrial mammal does. In fact, the oceans of the earth have a strong moderating influence on the climate of the planet, absorbing and redistributing heat around the globe and moderating the larger changes in temperature found on land. The role of the oceans in absorbing heat is likely to play an important role in the effects of global warming on the world's weather.

Water is cohesive and has a high surface tension.

In solution, water molecules have many hydrogen bonds with each other, and therefore stick readily to each other. The hydrogen bonds make water molecules near the surface stick to each other, pulling inward causing a force called **surface tension**. Surface tension causes water to pull together into round droplets on wax paper rather than spreading out flat and allows insects like water striders to float on the surface of water instead of sinking. Surface tension plays a role at any interface between air and liquid, such as in the lungs. Detergents tend to break up surface tension. Detergents play an essential role in the human lungs where they are secreted, and a harmful role in the environment when dumped as pollutants in the environment. The cohesiveness between water molecules is also what draws water up from roots into trees in an unbroken column.

Water is an excellent solvent for a large variety of molecules.

The polar nature of water makes it an excellent solvent for a wide variety of polar and charged substances, including salts, sugars, amino acids, and other molecules essential to life. Water hydrogen bonds with these substances in the same manner it does to itself, drawing these substances into the water, surrounded by a shell of interacting water molecules. Polar substances that water interacts with are called hydrophilic (water loving). The ability of water to dissolve substances is essential to life.

Water does not dissolve nonpolar molecules well. Hydrocarbons such as benzene or long-chain alkyl groups do not have any polar groups that water can hydrogen bond with and so are repelled from water and will not mix with it. These molecules are called hydrophobic (water hating). The repulsion of hydrophobic groups from water causes hydrophobic groups to draw together to present the smallest possible surface to water. An example of this occurs when oil is stirred with water in salad dressing—the hydrophobic oil separates, unable to dissolve or mix with the water. This repulsion is what causes membranes to form spontaneously from lipids mixed with water, allowing one region of the cell to be separated from another by the lipid bilayer membrane. Hydrophobic interactions also cause proteins to fold with hydrophobic regions on the inside of the protein, hidden from water.

Water dissociates to form protons and hydroxyls in solution.

The aqueous portion of a cell contains a large variety of solutes, including salt ions, sugars and macromolecules such as nucleic acids. Some of the most important solutes in water are acids and bases. At a small but predictable frequency, water molecules in solution will break down into a hydrogen ion (H^+) and a hydroxide (OH^-). The hydrogen ion is not really a free proton in solution, as it is often referred to, but complexes with another water molecule to make a hydronium ion: H_3O^+. For the sake of simplicity though it will be referred to as the

H^+ ion. The concentration of H^+ ions is the acidity of a solution and is given by a term called pH, the negative log of the concentration of H^+ ions.

$$pH = -\log[H^+]$$

A concentration of 10^{-7} M H^+ ions translates into a pH of 7, for example. At a pH of 7, the concentration of H^+ ions is equal to the concentration of hydroxide ions and the pH is said to be neutral, neither acidic or basic. Pure water has a pH of seven, with equal concentrations of H^+ ions and hydroxide ions. The inside of the body and the cytoplasm of the cell have a pH of 7.4, close to neutral pH. At acidic pHs, in which the pH is lower than 7, the concentration of H^+ ions is greater than 10^{-7}M, and the concentration of hydroxide ions is lower.

If a substance donates H+ ions in water, it is called an acid, and if it accepts H^+ ions in water, it is called a base. A base will decrease the H^+ ion concentration in water, increase the hydroxide ion concentration and increase the pH. Depending on how strongly a molecule donates or accepts H^+ ions, it will be called a weak acid or a strong acid. HCl ions completely dissociate in water for example, making HCl a strong acid. If one mole of HCl is placed in water, at equilibrium virtually all of it will dissociate to create one mole of H^+ ions, as well as one mole of Cl^- ions. A weak acid has more affinity for the H^+ ions and dissociates more weakly in water, leaving less than a mole of H^+ ions for every mole of acid added to water.

A measure of how strong an acid binds protons (H^+ ions) is the pK_a of an acid. The pK_a is the negative log of the equilibrium constant for the dissociation of an acid. For example, for acetic acid, the dissociation of the acid is given by the equation:

$$HA \leftrightarrow H^+ + A^-$$

For this equation, the equilibrium constant is K_a, where

$$K_a = \frac{[H^+][A^-]}{[HA]}$$

The smaller the value of pK_a, the stronger the acid.

Some materials called **buffers** have properties that allow them to act as either an acid or a base, minimizing changes in pH. Biological molecules and reactions can be quite sensitive to changes in pH, making buffers important for life. A change in the pH of blood from 7.4 to 7 can be enough of a change to cause a coma in humans. In the absence of a buffer, the addition of a very small amount of acid, 1×10^{-6} micrometer acid, will cause a large change in pH, a whole pH unit, while with sufficient buffer present, the hydrogen ions will be neutralized and the change in pH will be negligible.

A common laboratory procedure is to add acid (or base) to a solution, note the volume of acid added, and measure the change in pH that occurs. The resulting plot from this experiment is termed a **titration curve** (see figure). When a buffer is present in the solution being titrated, the pH changes very rapidly until the pH nears the pK_a of the buffering material. For example, in the figure shown, a base is added to the solution, so that the pH is basic. As acid is added to the base, the pH changes rapidly at first, with most + hydrogen ions remaining in solution. When the concentration of hydrogen ions in solution nears the pK_a for the buffer, the hydrogen ions start to bind to the buffer, driving it into the acid form. As the buffer binds the hydrogen ions, the pH changes little. When the buffer is fully protonated, additional acid causes a large change in pH. When the pH equals the pK_a for the buffer, half of the buffer is protonated and the other half is not.

Acid/Base Titration

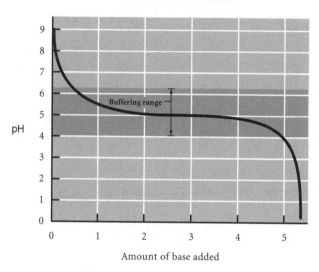

Buffers minimize changes in pH

There are many different buffers in the body, but the most important buffer in blood in humans is carbonic acid. Carbon dioxide dissolved in water can react with water to form carbonic acid. Carbonic acid can dissociate to release H^+ ions and form bicarbonate ions in a reversible reaction. If hydrogen ions are added to blood, some of them will combine with the bicarbonate ion to reform carbonic acid. The body actively controls the pH of the blood to ensure the maintenance of the pH in the range fit for life. One means to control pH is by changing the rate of breathing to alter the rate of carbon dioxide removal from the body.

Carbonic acid as a buffer

$$H_2CO_3 \longrightarrow HCO_3^- + H^+$$

CARBONIC ACID   BICARBONATE

When blood becomes more acidic, more H+ becomes bound to the base, thus forming the partner acid:

$$HCO_3^- + H^+ \longrightarrow H_2CO_3$$

BICARBONATE   CARBONIC ACID

1.2 ORGANIC MOLECULES OF LIFE

All living organisms use the same basic molecules to form the structures and perform the activities of life. The primary types of biological molecules are carbohydrates, lipids, proteins, and nucleic acids. All of these molecules are organic, meaning that they are based on carbon skeletons.

Carbon is extremely flexible in its chemistry, with four valence electrons that can form covalent bonds with many other molecules. The simplest organic molecules are hydrocarbons, containing chains or rings of carbons with hydrogens attached. Carbon molecules are reactive enough to be useful, since inert molecules are of little use to perform reactions that will support life, but carbon-based biochemistries are also stable enough that biological molecules do not rapidly degrade. Methane, ethane and butane are examples of hydrocarbons with 1, 2 or 3 carbon atoms. Hydrocarbons are very nonpolar, and hydrophobic as a result, repelling water.

While all biological molecules are organic and based on carbon backbones, functional groups containing other atoms are often responsible for the unique form and function of each type of biological molecule. Functional groups are attached to carbon skeletons and give compounds their functionality. These groups are commonly involved in chemical reactions. Examples of important functional groups found in biological molecules include the following:

Hydroxyl (OH). Hydroxyl groups are contained in polar compounds, such as ethanol or glucose, making them more soluble in water. Hydroxyl groups are commonly involved in hydrogen bonds.

Carbonyl ($C=O$). Carbonyl groups are polar groups with a double-bond between C and O. Carbonyl groups are contained in aldehydes and ketones, including formaldehyde and chain-form monosaccharides.

Carboxyl (COOH). Carboxylic groups are polar groups contained in carboxylic acids. They donate their H^+ ions in solution to dissociate as acids such as acetic acid (vinegar), producing a negative charge (COO^-). Fatty acids and amino acids have carboxylic acid groups.

Amino (NH_2). Amines have a polar nature and depending on the pH and the pK_a of the amine may be charged. Primary amines such as those in amino acids can often act as bases to accept protons to have a positive charge (NH_3^+) at a physiological pH. Secondary amines are also found in many biological molecules.

Sulfhydryl (SH). Sulfhydryl groups play a specific role in proteins. The amino acid cysteine has a sulfhydryl functional group that can form a disulfide bond with another cysteine to stabilize protein structure. Sulfhydryl-containing compounds such as b-mercaptoethanol are also sometimes used as reducing agents.

Phosphates (PO_4^{2-}). Found in organic phosphates like glycerol phosphate, these groups store energy that can be passed from one molecule to another by the transfer of a phosphate group. Nucleic acids and ATP contain phosphate.

Biological Macromolecules

There are four main types of biological macromolecules: carbohydrates, lipids, nucleic acids, and proteins. Carbohydrates, lipids, and proteins will be discussed in this section, but the structure of nucleic acids will primarily be presented in a later chapter. Carbohydrates, proteins and nucleic acids are all polymers, composed of subunits that are covalently joined together into long chains. Polymers allow the subunits to be joined in specific sequences along the chain in proteins and nucleic acids. Different sequences of subunits allows proteins to assume a multitude of forms and functions and allow nucleic acids to contain and transmit information.

Carbohydrates (Simple Sugars and Polysaccharides).

Monosaccharides are simple sugars that contain carbon, hydrogen, and oxygen, and have the general formula $(CH_2O)_n$ where n indicates the number of carbons. The most ubiquitous monosaccharide is the sugar glucose (see figure), found in all living cells as an energy source. Glucose has six carbons, making it a hexose with the molecular formula $C_6H_{12}O_6$. In water, the hydroxyl on the last carbon reacts with the aldehyde in glucose to make a cyclic molecule. At equilibrium in water, most glucose is found in ring form. Since the hydroxyl can attack the aldehyde from either the top or the bottom, there are two different ring forms of glucose, α-glucose and β-glucose. The different ring forms of a sugar are called **anomers**.

Simple Sugars

GLUCOSE
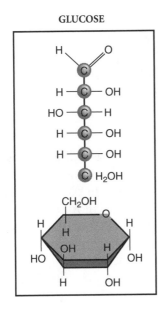

FRUCTOSE
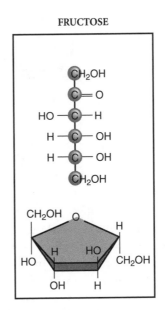

There are other common monosaccharides in addition to glucose. Fructose is also a hexose, with six carbons, and has the same molecular formula as glucose, $C_6H_{12}O_6$, but has a different structure, making it a structural isomer of glucose (see figure). Fructose is commonly found in fruit. Since fructose has a ketone group rather than an aldehyde, ring formation results in a five-membered ring instead of a ring with six members. Three carbon sugars such as glyceraldhyde are key metabolic intermediates when glucose is burned for energy, and the five-carbon sugars ribose and deoxyribose are key components of nucleic acids.

Monosaccharides are all chiral molecules, with complex stereochemistry. In the straight chain drawings of the sugars above, the last carbon (drawn at the bottom, farthest from the carbonyl-carbon) describes the stereochemistry of a sugar. In the drawings of straight chain version of these sugars, if the hydroxyl on the last carbon is to the right, the sugar is the D-stereoisomer. Almost all natural sugars are the D-form, and only D-forms are usually metabolized. Glucose is normally found as D-glucose stereoisomer and the enzymes that catalyze reactions involving glucose will recognize only the D-form, providing an important example of the specificity of enzymes in catalyzing reactions.

Disaccharides contain two monosaccharides joined by a **glycosidic** bond. Common disaccharides include sucrose, table sugar, and lactose, from milk. Sucrose (see figure) is composed of a glucose unit joined to a fructose unit, with the removal of a water molecule during the synthesis. The orientation of the glycosidic linkage can be either alpha or beta, changing the orientation of the two monosaccharides to each other.

Sucrose

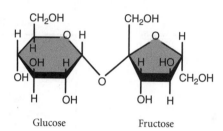

Glucose   Fructose

Polysaccharides such as starch, cellulose, and glycogen are chains of repeating monosaccharides of a particular type, usually glucose, joined one to another by glycosidic bonds. Glucose is a commonly used source of energy for both plants and animals, and it is highly soluble, but takes up a lot of space, making it inefficient for storage of energy. Plants such as potatoes store glucose as starch, a polysaccharide containing large numbers of glucose monomers joined together. Glycogen is the storage form of glucose found in animals, mostly the liver and skeletal muscle in humans, and also consists of polymerized glucose as a branched polymer, rather than simply a straight chain linear structure (see figure).

Glycogen Structure

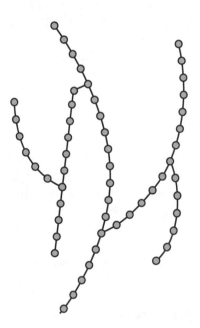

Specific enzymes are responsible for the storage of glucose in these forms and the release of glucose from these forms when energy is needed. For example, when glucose is needed, liver enzymes will break down glycogen and release the glucose into the blood. After a meal, however, a different set of liver enzymes will respond to hormonal control by synthesizing glycogen for storage until the energy is needed later on, between meals or during exercise.

Another function of polysaccharides is as a structural component. An abundant biopolymer is the structural polysaccharide cellulose, found in the cell walls of plants. Cellulose fibers are very strong, providing strength to trees and other plants. Cellulose is also composed of glucose, like starch and glycogen, but with beta-1,4 glycosidic bonds unlike the alpha-1,4 linkages in glycogen. This difference not only makes cellulose strong, but also makes it indigestible to humans since humans do not make an enzyme that can digest the beta-1,4 glycosidic bonds in cellulose. Chitin, found in the exoskeletons of arthropods and crustaceans, is another structural polysaccharide that is very similar to cellulose.

Lipids (Fats and Oils)

The next major class of biological molecules is the lipids. Like sugars, lipids are also composed of carbon, hydrogen, and oxygen, but their hydrogen-to-oxygen ratio is much greater than carbohydrates. There are several different types of lipids that are very different structurally, but they are all very hydrophobic and do not dissolve well in water. The high degree of hydrophobicity of lipids is caused by the large number of nonpolar bonds they contain, primarily C-H and C-C bonds. The hydrophobic portion of lipids tend to group together to stay away from water, such as when oil is mixed with water.

One of the roles of lipids is as an energy source. Fat tissue in animals (known as adipose tissue) is composed largely of lipids called **tryglycerides**. Triglycerides are a much more efficient form of energy storage than polysaccharides. Triglycerides have more energy that can be burned for metabolism per gram of weight since they have more high-energy (reduced) C-C and C-H bonds than sugars, and they exclude water so they take up less space than glycogen to store. Most animals store only a temporary small amount of energy as glucose or glycogen, and most of their energy as fats. Storage of energy in triglycerides occurs in animals while plants use starch.

Triglycerides

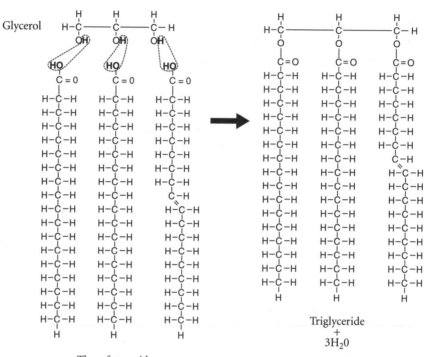

Glycerol

Three fatty acids

Triglyceride
+
3H$_2$0

The structure of triglycerides includes two components: a glycerol chain and three fatty acid molecules joined to the glycerol. **Fatty acids** are a component in many lipids, including triglycerides (see figure). Fatty acids each have a long carbon chain, with a carboxylic acid group at one end. In triglycerides, fatty acids form ester linkages with the glycerol and are no longer charged. Since triglycerides have no charge they are sometimes called neutral fats. Fatty acids differ in their length, how many carbons long the carbon chain is, and in the number of double bonds contained in the carbon/chain. Fatty acids that lack double bonds are **saturated** and fatty acids with one or more double bonds are called **unsaturated**. The charged carboxylate in fatty acids is hydrophilic while the carbon tail is very hydrophobic. This dual nature of fatty acids tends to make the tails cluster together, with the heads sticking out when they are mixed with water. When mixed with a hydrophobic greasy substance, fatty acids will surround it with their tails and bring it into solution with their hydrophilic heads oriented outward. In this way, fatty acids can act as detergents. Soap was commonly derived in the past from animal fat, by hydrolyzing triglycerides to release fatty acids that could be used as soap.

Lipids also play an important role in membranes. The predominant class of lipid in membranes is **phospholipids**, with a glycerol and two fatty acids, similar to triglycerides, but with a phosphate group and an additional polar group joined to the glycerol (see figure). The phosphate and the additional polar group make the glycerol end of phospholipids very polar, referred to sometimes as the polar head compared to the nonpolar fatty acid tails. The fatty acid chains in phospholipids are very hydrophobic while the phosphate-alcohol end is very hydrophilic. As a result, phospholipids are like detergents and when mixed with water will spontaneously form structures with the fatty acids gathered together to keep out water and the phosphate group pointing out toward water. This is the basic structure of the lipid bilayer formed by phospholipids in cell membranes.

Phospholipids

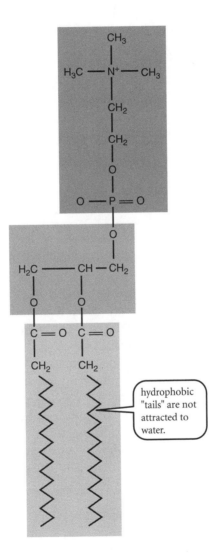

hydrophobic "tails" are not attracted to water.

Waxes are esters of fatty acids and alcohols. Waxes form protective coatings to repel water on skin, fur, leaves of higher plants, and the exoskeleton of many insects. Bees use waxes to form their hive. Waxes are solid but malleable, making them useful to humans.

Steroids have three fused cyclohexane rings and one fused cyclopentane ring. Examples of this lipid derivative include **cholesterol** (see figure), the sex hormones estrogen and testosterone, and corticosteroids. Cholesterol is present in the membranes of eukaryotes, and modifies the fluidity of membranes.

Cholesterol

Proteins

Proteins make up 50 percent of the dry weight of a cell and are vital to almost everything a cell does. Proteins are formed from building blocks called amino acids, of which there are 20 common types. Amino acids each contain an amino group and a carboxyl group, and each amino acid has a unique R group that is distinct in each of the 20 amino acids (see figure). There are several classes of amino acids, including some that are very hydrophobic (phenylalanine, isoleucine, leucine, alanine, tryptophan), some that are polar with a hydroxyl group (serine, threonine, tyrosine), some that have polar amide side chains (glutamine, asparagine), some charged basic amines (lysine, arginine), and some charged carboxylic acids (glutamate, aspartate). Proline is a unique amino acid since it is cyclic, and affects the structure of proteins containing it. The sulfhydryl of cysteine also plays an important role in protein structure. The chemical nature of the amino acid side groups in a protein determines the structure and function of the protein.

Generalized Amino Acid Structure

The ® could be a hydrogen atom, a carbon, or a chain of carbons or other groups.

Proteins are formed by the covalent linking of amino acids. When two amino acids are joined, they form a peptide bond (see figure) between the carboxyl group of one amino acid and the amino group of the next. A peptide bond is formed with the input of energy and the loss of water; this process is called **dehydration synthesis**. Amino acids are joined together by peptide bonds during **translation** (see chapter 5, Molecular Genetics). The biosynthesis of two amino acids together forms a **dipeptide**, in which each amino acid is called a **residue**. Many amino acids joined together form a **polypeptide**. With twenty different possible amino acids at each position in a polypeptide, there is an enormous number of possible proteins that could be made, far more than are formed in nature. Even for a relatively small polypeptide with only four amino acid residues, there are $20 \times 20 \times 20 \times 20 = 160{,}000$ possible proteins that could be formed.

Two amino acids joined by a peptide bond

peptide
bond

The primary structure of a protein is the linear sequence of amino acids.

The sequence of amino acids and their R groups in a protein determine how the protein folds its polymer chain into a 3D shape and what the protein does. If a protein is made with the wrong amino acids or folds incorrectly, it cannot do its job properly. Globular proteins are typically functional proteins, such as carriers or enzymes. Long fibrous proteins are structural proteins; one example is collagen.

Functions

Various types of proteins and their functions are listed in the table below.

Type of Protein	Function	Examples
Hormonal	Chemical messengers	Insulin, glucagon
Transport	Transport of other substances	Hemoglobin, carrier proteins
Structural	Physical support	Collagen
Contractile	Movement	Actin, myosin
Antibodies	Immune defense	Immunoglobulins, interferons
Enzymes	Biological catalysts	Amylase, lipase, ATPase

Levels of Protein Structure

To do their job, proteins must have a specific shape or structure. An enzyme that has the wrong shape will not catalyze its reaction, for example. If an enzyme is heated, subjected to changes in pH or other environmental changes, it may lose its structure as well as its function. Such a protein is **denatured**. The denaturation process is reversible in some cases, but not always. An example of protein denaturation can be observed when lemon juice is mixed with milk. The acid in the lemon juice will cause the milk proteins to unfold and denature, exposing hydrophobic amino acids from their interior that are normally hidden from water. When exposed, the hydrophobic regions of one denatured protein can interact with hydrophobic regions from other denatured proteins, causing a visible mass of tangled denatured proteins; the milk looks curdled.

Protein structure has four levels: primary, secondary, tertiary, and quaternary.

Primary structure is the sequence of amino acids joined by peptide bonds in the linear polypeptide chain. When proteins are synthesized, the first amino acid is left with the amino end free, not in a peptide bond. The last amino acid in a protein has a free carboxyl group. When the amino acid sequence of a protein is written on paper it is written from the free amino terminus on the left to the carboxyl terminus on the right.

The **secondary level** of protein structure involves coils or folds of the local polypeptide chain in patterns that contribute to the overall conformation of the protein. Secondary structure is caused by hydrogen bonds at regular intervals along the local region of the polypeptide backbone; **alpha-helices** and **beta-pleated sheets** are the most common types of secondary structure. The structure of proline does not allow it to fit within the normal alpha-helix structure, making it a helix-breaker. The small size of glycine, with only a hydrogen side chain, makes it fit well within tight turns in secondary structure, such as between stretches of beta-sheet.

The **tertiary level** of structure can be characterized as folding due to bonding between side chains of the various amino acids. Tertiary structure can be reinforced by strong covalent bonds, called disulfide bridges, between cysteines in different parts of the primary structure of the protein. A protein with disulfide bridges in its structure will be more resistant to denaturation. Tertiary structure can also be modular, with two or more globular regions called domains that fold independently which are connected by relatively flexible regions of the polypeptide chain. Secondary structure involves bonds between nearby amino acids in the linear sequence. Tertiary structure can bring amino acids that are far apart in the linear polymer very near each other in the 3D protein structure. Amino acids with hydrophobic side chains tend to fold in the interior of proteins, away from water, while charged or polar amino acids fold near the surface.

The final **quaternary level** of protein structure of the protein results from the relationship between different polypeptides, called subunits. For example, some proteins, especially cell surface receptors, are made up of two proteins that interact to produce large protein complexes. Such interactions between subunits can be involved in the regulation of the process in which the protein is involved.

Proteins fold spontaneously as they are synthesized to form the lowest energy folded structure. Sometimes proteins called chaperonins are required for a protein to fold correctly. When heated, a protein will often lose activity, then refold as it is cooled and regain activity. The process of protein folding probably proceeds from one level of organization to the next, starting with the linear primary protein sequence, followed by interactions between nearby amino acids to form secondary structure, then folding of the alpha-helices and beta-sheets to form tertiary structure. Finally, subunits come together to form tertiary structure and the complete protein.

Scientists often wish to study the function of individual proteins, a difficult task when proteins are present in a dense mixture of other proteins in the cell. To study the function of proteins, biochemists in the laboratory will purify proteins from the cell, isolating a given protein for study. Different proteins have different physical properties, including size, shape, and charge, that the biochemist can use to separate proteins. If the protein of interest is an enzyme, the location of the enzyme during purification can be determined by following the activity of the enzyme in various fractions of proteins produced during the purification.

A means of examining the proteins in a mixture is **SDS-polyacrylamide gel electrophoresis**. When placed in an electric field, a charged molecule will move according to its charge. SDS is a charged detergent that denatures proteins, giving them all the same denatured globular shape, and binds to them, giving them a charge in addition to other charges. Gel electrophoresis involves the migration of the proteins treated with SDS in an electric field through a matrix of cross-linked acrylamide polymers. If the proteins are placed in the gel with SDS and a charge is then applied to the gel, the proteins will separate in the gel according to their molecular weight. Small proteins move through the gel more quickly since they can fit more easily through the holes in the porous gel, while large proteins move more slowly. After the proteins have moved through the gel, their position can be visualized by staining the gel with a dye that binds to proteins and not to the gel (see figure). The molecular weight of proteins in the sample analyzed can be determined by comparison to standard proteins in the same gel that have a known molecular weight.

SDS gel electrophoresis

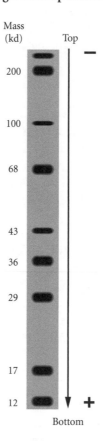

SDS combines with proteins, giving them a net negative charge. The negatively charged SDS–protein complexes travel through the gel toward the positive anode.

Nucleic Acids

Nucleic acids are the information molecules of the cell. DNA is a polymer that carries genetic information from one generation to another and encodes the information for producing proteins. RNA is the nucleic acid that helps to take the protein-coding information in DNA and produce the proteins of the cell. DNA and RNA are both polymers composed of **nucleotide** building blocks. Each nucleotide consists of a nitrogenous base, a five-carbon sugar (either ribose or deoxyribose) and phosphate groups. There are four types of nitrogenous bases in both DNA and RNA. The nitrogenous bases in DNA are adenine, thymine, cytosine and guanine. RNA differs in having uracil instead of thymine. The bases of RNA and DNA possess the information content of the nucleic acids, since it is these that vary from one subunit in the polymer to the next. The backbone of DNA and RNA is composed of the alternating pentose sugar and phosphate groups. The structure and function of DNA and RNA will be discussed in detail in chapter 5, Molecular Genetics.

ATP, a nucleotide

Some nucleotides play roles other than as building blocks for nucleic acids. ATP, for example, is a nucleotide that also acts as a key energy currency in all cells (see figure). Other nucleotides, such as NADH, NADPH and $FADH_2$, are coenzymes, playing a key role in enzymatic reactions and energy transfer.

1.3   FREE ENERGY CHANGES

Energy is continually flowing in the universe in many different forms, including heat, vibrational energy, chemical energy, kinetic energy, potential energy, and light. The reactions of life involve the flow of energy within organisms and the environment, using the energy to drive the activities of life. Whether the energy is used to drive formation of proteins or to move a muscle, it must still follow rules of thermodynamics that govern all energy in the universe. Thermodynamics is the study of the relationship between different forms of energy.

Free Energy Change: Like water flowing

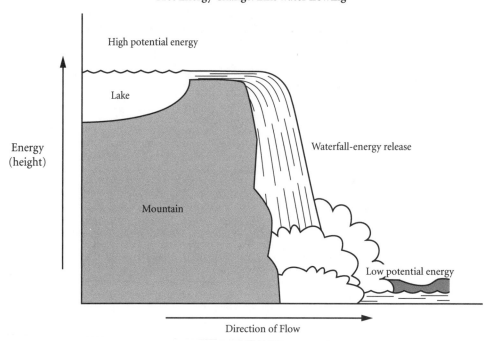

The laws of thermodynamics can be used to predict if a reaction such as a metabolic pathway in a cell will occur or not. If the products have less free energy (ΔG) than the reactants, then the reaction has a negative ΔG and will occur spontaneously, without putting additional energy into the reaction. The energy in a reaction is like water flowing down a river—water will only flow downhill on its own, with a release of energy (see figure). Thermodynamics does not describe the rate of reactions however—this is left to the study of kinetics. For example, thermodynamics predicts that a piece of sugar in the presence of oxygen will be oxidized, releasing energy, but it does not state the rate or the route by which this will take place. In a room without a catalyst, the reaction will take place slowly, but with either a flame or a cell to release the energy, the reaction will be rapid. Thermodynamics states where things start and end up while kinetics states the path used to get from the start to the end.

The first law of thermodynamics is the **conservation of energy**. This law states that energy is never lost from the universe, or any closed system, but that it is simply converted from one form into another. For life, some energy is useful, such as the energy found in certain high-energy chemical bonds like those in glucose. Other forms of energy such as heat energy are not useful since they cannot be captured to drive essential activities. Heat energy when it is produced is lost to the exterior system and is no longer useful to drive biological reactions. Biological reactions are never 100% efficient in the transfer of energy, so that some energy is lost to the environment as heat during the reaction. Every living organism generates heat as a result of the reactions of life. Organisms extract as much energy as they can, losing a percentage to the environment as heat, with the total amount of energy at the end unchanged but in different forms.

The **second law of thermodynamics** states that (since no reaction or any other process in the universe can ever be 100% efficient in the conversion of energy) systems move over time toward a more disordered state with less useful energy. **Entropy** is the term used to describe the disorder of a system. An example of entropy is that molecules in solution favor a less-ordered state. Having many molecules of solute in one location in a solution is an ordered state. According to entropy, molecules will spontaneously assume a less ordered state, and so diffusion occurs, with molecules dispersing into a less ordered randomly spaced solution. Entropy dictates that over time large complex molecules will become smaller, less complex molecules. In any closed system, the amount of energy will remain the same over time (although it will change form), but the entropy will increase.

Living organisms at first glance may appear to contradict the second law, since life perpetuates a high degree of complexity in the face of entropy. Living organisms derive energy from an outside source however—the sun. Living organisms use the energy from the sun to maintain the complex ordered state of the molecules of life. Considering the broader system of the sun and earth along with living organisms, the second law is upheld.

Entropy is denoted by the symbol "S" in thermodynamic equations. Another thermodynamic measure is **enthalpy**, "H." **Enthalpy** is the total thermodynamic heat content of a system. Changes in entropy are denoted by "ΔS" and changes in enthalpy are denoted by "ΔH." If $\Delta S > 0$ in a chemical reaction, then the product of a reaction has more entropy than the starting materials. If $\Delta H < 0$, then the products have less heat energy than the reactants, having lost it to the environment as an **exothermic** reaction. Both ΔS and ΔH play a role in calculating whether a chemical reaction will proceed or not, but neither of these factors alone can determine this.

A factor that can be used to predict the spontaneity of a reaction is called the **Gibbs Free Energy**, denoted by the symbol "G." ΔG is the change in free energy of a reaction between reactants and products. If $\Delta G < 0$ for a reaction, the reaction is favorable and will proceed spontaneously. If $\Delta G = 0$ for a reaction, it is at equilibrium, with the forward and backward reactions occurring at the same rate and no net change in the concentrations of reactants or products occurring.

The free energy change (ΔG) for a reaction can be calculated from the change in enthalpy (ΔH) and the change in entropy (ΔS) during the reaction. The relationship between these values is given by the following equation, in which T is the temperature:

$$\Delta G = \Delta H - T\Delta S$$

A decrease in enthalpy, releasing heat from the reaction into the environment, makes the reaction more favorable. Increases in entropy or temperature also make ΔG more negative and indicate a more favorable reaction. A reaction with a negative ΔG can be used by an organism to do work. Reactions with $\Delta G < 0$ are called **exergonic** and reactions with $\Delta G > 0$ are called **endergonic**. These terms should not be confused with endothermic and exothermic, which refer to the enthalpy of a reaction.

If $\Delta G < 0$, then a reaction will spontaneously move forward to form product until equilibrium is achieved. At equilibrium the forward and backward reactions occur at the same rate, and $\Delta G = 0$. It is important to remember that the free energy change of a reaction predicts only whether the reaction will spontaneously move forward; it says nothing of the rate. In biology, almost all reactions require enzymes as catalysts to allow the reaction to move forward at an appreciable rate. All that an enzyme (or any catalyst) can do is to increase the rate of a reaction that is favorable. A biological reaction catalyzed by an enzyme must still have a negative ΔG for it to move forward and a reaction with a positive ΔG will not occur even if an enzyme is present.

1.4   ENZYMES

The chemical reactions required for life may be thermodynamically spontaneous, with a $\Delta G < 0$, but still occur at an extremely slow rate if the reactants are simply mixed together without a catalyst. Thermodynamics says nothing of the rate at which a reaction will occur. Even if a reaction is determined by thermodynamics to be favorable, it can take thousands of years to reach equilibrium without a catalyst. A factor that governs the rate of reactions is the **activation energy**.

In a reaction proceeding from reactants to products, with products at a lower free energy level than the reactants, the reaction is thermodynamically favorable and will occur spontaneously. The path of the reaction determines how rapidly it will occur. To get from reactants to products generally requires the reactants to go through a transient high-energy reaction intermediate, and the formation of the intermediate is the rate-limiting step in the reaction. The intermediate is chemically unstable, thus very short-lived, and will rapidly either fall back to the original reactants or go forward to form the products. The chemical instability of the intermediate also accounts for its high-energy state. Since the reaction intermediate is at a higher energy level than the reactants, energy must be put into the reactants for them to form the intermediate. Increasing the kinetic energy of molecules in solution by adding heat increases their energy, supplying the activation energy to form the intermediate, and increasing the reaction rate. This is one reason why heating a reaction will often make it go faster.

Another way to make a reaction go faster is to use a catalyst. A catalyst is a substance that increases the rate of a chemical reaction but is not itself consumed or altered in the reaction. A catalyst decreases the activation energy for the reaction by making it easier for molecules to form the transient reaction intermediate. The biological catalysts used by organisms to speed reactions are called **enzymes**.

Enzymes Lower Activation Energy

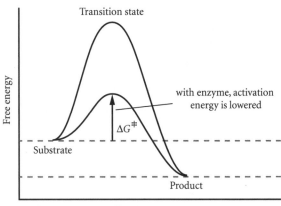

Enzymes are specialized proteins that act as biological catalysts (although a few examples of RNAs that act as catalysts are known as well), speeding the rate of reactions by many orders of magnitude. By bringing reactants together in the right orientation to form the reaction intermediate and by stabilizing the reaction intermediate, enzymes reduce the activation energy and make biological reactions occur rapidly (see figure). Since enzymes are not catalysts, they are not altered or consumed at the end of each reaction cycle, and one enzyme can catalyze the reaction of many molecules of reactant, termed **substrate** in enzyme-catalyzed reactions.

Enzymes bring reactants together in the correct orientation in a region of the enzyme called the **active site**. The active site is generally a cleft or pocket in the enzyme structure that binds substrate in the correct orientation to catalyze the reaction. Specific amino acid residues in the enzyme's 3D structure fold together into precise locations in the active site to stabilize substrate binding, to interact chemically in the reaction process, or to stabilize the reaction intermediate. The interactions of substrate with the active site can be hydrophobic, ionic, and hydrogen bonds. The amino acids involved in catalysis at the active site are brought near each other in space in the tertiary structure of the folded enzyme, but are not necessarily near each other in the primary, linear structure of the enzyme. Since the active site requires specific residues in precise positions for the enzyme to catalyze its reaction, enzymes must fold correctly to catalyze a reaction (to have activity). Denaturation of an enzyme, which destroys the 3D structure of the enzyme, disrupts the active site structure along with the rest of the structure and eliminates enzyme activity.

An example of the importance of correct folding for enzyme activity is an experiment performed with the enzyme ribonuclease, which hydrolyzes RNA, breaking the RNA polymer into smaller pieces. The enzyme has 124 amino acids including 8 cysteines that form 4 disulfide bonds to stabilize the protein structure. Purified ribonuclease protein was subjected to several treatments, and then analyzed for enzyme activity. The treatments included reduction with b-mercaptoethanol (b-ME), or treatment with urea, a denaturing reagent that disrupts hydrogen bonds required for secondary and tertiary protein structure. The table below summarizes the findings of these experiments:

	Step 1	**Step 2**	**Observation**
Condition 1	No reagents added	No reagents added	100% activity
Condition 2	Urea and b-ME	Nothing removed	0 % activity
Condition 3	Urea and b-ME	b-ME removed, then urea removed	1% activity
Condition 4	Urea and b-ME	urea removed, then b-ME removed	100% activity

What conclusions can be made based on these results? With urea and b-ME present, the secondary and tertiary structure of the enzyme are disrupted, the active site is not folded correctly, and the enzyme has lost all activity (Condition 2 compared to Condition 1). If the b-ME is removed first with urea still present, the enzyme regains only 1% of the original activity (Condition 3). In this condition, urea blocks correct protein folding during disulfide bridge formation, and disulfide bridges reform randomly. With 8 cysteines, the correct combination of disulfide bonds will form randomly about 1% of the time, correlating with the amount of active enzyme observed with condition. When the urea is removed, only those enzyme molecules that had randomly formed the correct disulfides are able to form the correct secondary and tertiary structure and regain enzyme activity. If the urea is removed first followed by the mercaptoethanol, however (Condition 4), then full enzyme activity is regained. In this circumstance, the enzyme will spontaneously form the correct secondary and tertiary structure when urea is removed. When b-ME is removed, the enzyme will be locked into the correct structure by the correct disulfides formed when the enzyme is folded correctly. This experiment demonstrates the importance of the correct enzyme folded structure for enzyme activity. It also demonstrates that the protein sequence alone is required for the protein to spontaneously fold into the active enzyme structure, since the enzyme correctly refolds after denaturation with no extra energy or information.

The three dimensional nature of the active site at which substrates bind leads to one of the key features of enzymatic catalysis—the tremendous specificity of enzymes for a given reaction. The shape of the active site restricts the substrate that can bind, increasing the specificity of the reaction.

One model of substrate binding to the active site is termed the **lock and key** model, in which the substrate is like a key that binds in an active site with an exactly complimentary shape (the lock). It appears that a slightly more complicated model, the **induced fit** model, may more accurately represent enzyme function (see figure). In this model, the binding of substrate changes the structure of the enzyme and the active site. Changes in enzyme structure caused by substrate binding have been physically measured, supporting this model.

Enzymes - Lock and Key Model　　　　**Enzymes - Induced Fit Model**

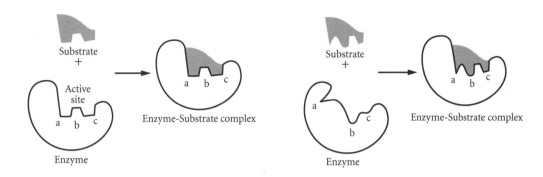

Enzymes do not alter the final energy of the reactants or products, and so do not change the equilibrium or change in free energy of the reaction, only the rate at which equilibrium is reached. In fact, enzymes will catalyze both the forward and backward reaction rates. In a model of enzyme action known as the **Michaelis-Menten model**, the activity of an enzyme is assumed to progress in several general steps. In the first step, the enzyme binds the substrate at the active site. This step is reversible, since the substrate can fall out of the active site in its original form. For enzymes with substrate bound to the active site, the other option is that the substrate is converted to the high energy reaction intermediate, from where it can go on to form product. Each step in the reaction model can be described by a rate constant:

Michaelis Menten Equation

$$\text{Enzyme} + \text{Substrate} \underset{k_2}{\overset{k_1}{\rightleftharpoons}} \text{Enzyme-Substrate Complex} \xrightarrow{k_3} \text{Enzyme} + \text{Product}$$

Enzymes in this equation do not refer to enzyme molecules, but to the number of enzyme active sites, not exactly the same thing. As more substrate is added to a solution of enzyme, more of the active sites have substrate bound to them in ES (enzyme-substrate) complexes. The more ES complexes formed, the more product formed. When a high concentration of substrate is added, all of the active sites will have substrate bound and adding more substrate cannot increase the reaction rate any further, since the ES concentration is already maximal. This is another hallmark of enzyme catalyzed reactions—they are saturable with increasing reactant (substrate) concentration. For an enzyme, once the active sites are full of substrate, adding more does not increase the rate further. For a non-enzymatic reaction, increasing the reactant concentration will continue to increase the rate since the formation of the reaction intermediate does not require a specific site and happens randomly in solution.

Based on the Michaelis-Menten model, an equation can be derived that describes the relationship of the reaction rate on the substrate concentration and the affinity of the enzyme for the substrate. This equation is called the Michaelis-Menten equation.

Michaelis Menten Equation

$$V = V_{max} \frac{[S]}{[S] + K_M}$$

In this equation, V_{max} is the maximal reaction rate that is achieved when 100% of the enzyme molecules in solution are in the ES form, occupied with substrate. K_m is a measure of the affinity of the enzyme for the substrate. V is the reaction rate at a specific substrate concentration [S]. Practically speaking, K_m is given in units of concentration and K_m is the concentration of substrate requires to achieve 50% of the maximal possible reaction rate (1/2 V_{max}).

K_m and V_{max} are often determined experimentally for an enzyme in experiments in which the concentration of enzyme is held constant while the substrate concentration is varied and the reaction rate measured. The reaction rate is measured by waiting a period of time, stopping the reaction, and then determining how much product has formed in a unit of time. When this is done, and the resulting data plotted with the substrate concentration on the *x*-axis and the measured reaction rate (V) is plotted on the *y*-axis, several of the features of enzymes can be observed (see figure), as predicted from the Michaelis-Menten model and equation.

Reaction Velocity as a Function of Substrate Concentration

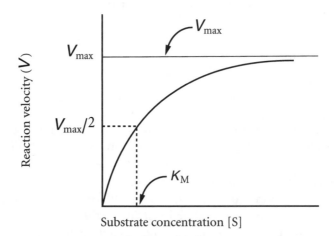

At low substrate concentrations, adding more substrate rapidly increases the reaction rate. At very high substrate concentrations, as the active sites become saturated, adding more substrate has only a very small increase in the reaction rate. In these experiments, it is assumed that the reaction rate is measured before a significant amount of product has accumulated, since the product can alter or even inhibit the initial reaction. The curve approaches the maximal reaction rate V_{max}, which can be estimated on the plot. K_m can also be determined from the plot by first finding V_{max}, finding the reaction rate (V) that is 1/2 V_{max} and then determine the substrate concentration when the curve achieves 1/2 V_{max}. This substrate concentration is K_m.

A common way to plot the data from the above graph is by plotting the reciprocal of the reaction rate (1/V) on the *y*-axis and the reciprocal of the substrate concentration (1/[S]) on the *x*-axis. In this format, if an enzyme obeys the Michaelis-Menten model, a linear relationship will be observed (see figure).

Lineweaver-Burk Plot

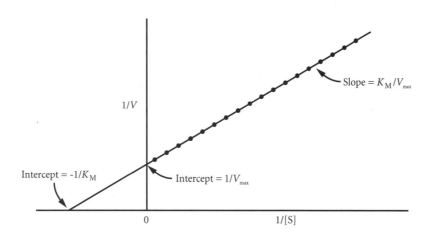

This graph is called the **Lineweaver-Burk plot** and can be used to determine the K_m and V_{max} more accurately. In this plot, the intercept of the line with the *y*-axis equals $1/V_{max}$ and the intercept of the line with the *x*-axis equals $-1/K_m$.

Summary of key enzyme points:

1. Enzymes are biological catalysts, usually proteins.
2. Enzymes increase the reaction rate by lowering the activation energy.
3. Enzymes do not change the free energy of products or reactants or the ΔG for reactions.
4. Enzymes catalyze reactions but are not themselves consumed.
5. Enzymes do not change reaction equilibria.
6. Enzyme activity requires precise protein folding.
7. Enzymes have active sites that are the site of catalysis.
8. Enzymatic reactions are saturable with increasing substrate concentration.

Enzyme Regulation

Enzymes perform biological catalysis, but each enzyme does not act alone. Enzymes are part of complex interdependent pathways acting in concert to maintain life. Maintenance of life requires that not all enzymes are active at all times, but that the activity of distinct enzymes and pathways is regulated to meet the conditions inside and outside of the body at any time. If all enzymes were active all the time, a great deal of energy would be wasted, with metabolic pathways for synthesis and degradation acting against each other. Such inefficient activity would be rapidly selected against in the environment. Life selects for an efficient use of energy since organisms that use energy efficiently are more likely to survive.

In a pathway that produces a metabolite such as an amino acid, several steps might be required to produce the final product in the pathway from the original starting point. The amino acid will only be needed at certain times, and producing the amino acid when it is not needed would be wasteful and perhaps harmful. To avoid this, the cell regulates the pathway to turn it off when enough of the final product is present. Such negative regulation of a pathway by the final product is called **feedback inhibition** (see figure). The enzyme that catalyzes the first committed step in the pathway is generally the target of inhibition by the final reaction product. (The first committed step is the reaction for which the product has no other function than to form the end-product.) This prevents material from accumulating wastefully at an intermediate step that cannot be used for other purposes.

Feedback Inhibition

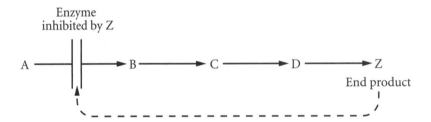

There are several mechanisms by which enzymes are regulated. One mechanism is by small molecules other than the substrate that bind to the enzyme to change its activity. An example of this is observed in feedback inhibition of an enzyme in a pathway. In **competitive inhibition**, a molecule that resembles substrate binds to the active site and blocks substrate from the enzyme (see figure). Adding more substrate will overcome a competitive inhibitor, since the substrate can out-compete the inhibitor if enough of it is added. A competitive inhibitor will change the apparent K_m, since with inhibitor present it takes more substrate to achieve the same level of enzyme activity, but it does not change the V_{max}. A Lineweaver-Burk plot can determine if this is the case in an experiment, if a researcher adds increasing amounts of substrate in the presence or absence of inhibitor and measures the enzyme reaction rate (velocity). If V_{max} is the same with inhibitor present but K_m changes, the inhibitor is competitive.

Enzymes catalyze reactions in part by stabilizing the unstable reaction intermediate. The active site will bind substrate, then induce it to form the intermediate by stabilizing the intermediate form, allowing it to form with less energy. The most effective competitive inhibitors are those that resemble the reaction intermediate, since these will have even higher affinity for the active site than substrate. Molecules that resemble the reaction intermediate may bind the active site so tightly that they remain in the active site and resemble covalent modifiers of the active site.

**Competitive vs. Non-Competitive
Enzyme Inhibition**

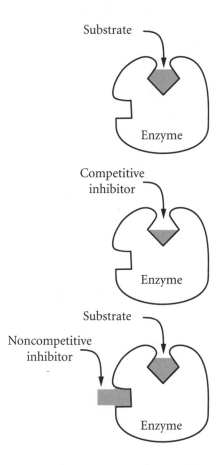

Another type of enzyme regulation is called **non-competitive inhibition**. A non-competitive inhibitor binds to a site on an enzyme other than the active site to change the activity of the enzyme (see figure). The molecule does not compete with substrate and need not resemble substrate chemically since it will not occupy the same site. A non-competitive inhibitor changes the V_{max}, but does not change the K_m. Substrate and inhibitor interact with distinct sites, so adding more substrate will not overcome a non-competitive inhibitor.

Both competitive and non-competitive inhibition are reversible. In **irreversible inhibition**, the enzyme inhibitor covalently modifies the enzyme at the active site to block activity. Nerve gas and insecticides often act as irreversible inhibitors of the enzyme acetylcholinesterase, an important enzyme in nervous impulse transmission. Penicillin is an irreversible inhibitor of an enzyme bacteria use to build their cell wall, and aspirin acts by covalently blocking the active site of the enzyme that forms prostaglandins involved in inflammation. When the irreversible inhibitor binds to the active site, a piece of the inhibitor remains covalently bound to the enzyme, blocking the key residues requires for catalysis. Generally the only way to increase enzyme activity again is to make new enzyme molecules.

Enzyme regulators not only inhibit but also activate enzymes. When a molecule binds to one site on a protein to increase binding at another site, this is termed an **allosteric** interaction. The best known example of this is not an enzyme, but the oxygen transport protein hemoglobin. Hemoglobin contains four polypeptide subunits that form the quaternary structure of the hemoglobin protein. In the absence of oxygen, all four subunits are "tense" and in a conformation with a weak affinity for oxygen. When one subunit binds oxygen, it changes its conformation to a "relaxed" form with higher affinity for oxygen (see figure).

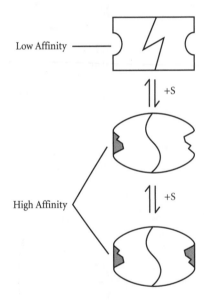

Not only does the affinity of the subunit for oxygen change, but the change in conformation of that subunit is communicated through the protein structure to the other subunits to convert them to the relaxed, high-affinity state as well. The binding of oxygen to subunits 2, 3 and 4 then occurs more rapidly, making the overall process cooperative. When increasing oxygen is plotted against the percent hemoglobin sites that are occupied, the resulting curve has a **sigmoidal** shape (see figure). The shape of the curve also demonstrates that the binding of oxygen to the first hemoglobin subunit is difficult, but occurs more readily as oxygen binds cooperatively to the remaining subunits. Myoglobin is an oxygen transport protein that is very similar to hemoglobin, but has only a single subunit and therefore does not display cooperative binding. If the hemoglobin subunits are dissociated, they bind oxygen without cooperativity like myoglobin, demonstrating the importance of interactions between subunits for allosteric regulation. Many enzymes display similar allosteric cooperativity of substrate binding. Some multisubunit enzymes have subunits that are strictly catalytic and others that are regulatory, acting only to allosterically modulate the activity of the catalytic subunits.

Sigmoidal Curve

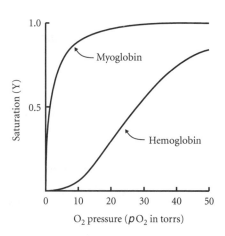

Many enzymes are regulated by covalent modification after they are produced. The most common type of modification is post-translational modification by the addition of a phosphate group to serine, threonine or tyrosine residues in the protein. Phosphorylation of proteins is catalyzed by enzymes called **protein kinases**, each designed to modify the activity of proteins in response to changes in the internal or external environment of the cell. The phosphorylation of proteins can change their activity by changing their structure, either increasing or inhibiting their activity.

Enzymes can also be regulated by cleavage. Some enzymes are secreted in an inactive form called a **zymogen**. Protease enzymes involved in the digestion of proteins are synthesized in the pancreas in an inactive state to prevent the enzymes from digesting the proteins in the tissue of the pancreas. When the proteases' precursor enzymes are secreted, they are cleaved at specific sites and activated in the process. Activation of proteins by proteolytic cleavage also occurs during blood clotting, which involves a cascade of proteases that cleave each other to activate fiber formation by fibrinogen from plasma.

Chapter 1: Review Questions

1. Which of the following constitutes the quaternary level of protein structure?

   (A) bonding between side chains of amino acids
   (B) sequence of amino acids joined by peptide bonds
   (C) beta pleated sheet
   (D) interactions between polypeptide subunits
   (E) alpha helix

2. Which of the following statements is FALSE regarding SDS-polyacrylamide gel electrophoresis?

   (A) proteins are separated by molecular weight
   (B) SDS, a detergent, gives charge to protein
   (C) large proteins move more slowly through the gel
   (D) SDS is used to maintain the three dimensional structure of protein
   (E) small proteins move more quickly through the gel

3. Which of the following nitrogenous bases is NOT found in DNA?

   (A) adenine
   (B) guanine
   (C) uracil
   (D) thymine
   (C) cytosine

4. Many of the unique physical properties of water may be explained by:

   (A) its polar nature
   (B) its low surface tension
   (C) its ability to easily dissolve nonpolar substances
   (D) its low pH
   (E) its ability to contract when it freezes

5. Which of the following statements about enzymes is NOT true?

   (A) enzymes are almost always proteins
   (B) enzyme activity is affected by changes in pH
   (C) enzymes increase the rate of reaction
   (D) enzymes increase the activation energy
   (E) enzymes do not change the free energy of products

6. In one type of enzyme regulation, the presence of the end product of a metabolic pathway inhibits an enzyme which functions in an early step in the pathway. This type of enzyme regulation is called

   (A) feedback inhibition
   (B) competitive inhibition
   (C) non-competitive inhibition
   (D) irreversible inhibition
   (E) none of the above

7. Which of the following involves the binding of a molecule to the active site of the enzyme?

    (A) non-competitive inhibition
    (B) feedback inhibition
    (C) irreversible inhibition
    (D) A, B, and C
    (E) none of the above

8. If the free energy change of a reaction is greater than zero, then the reaction

    (A) is spontaneous
    (B) is nonspontaneous
    (C) is at equilibrium
    (D) is endothermic
    (E) is exothermic

9. Which is NOT a characteristic of proteins:

    (A) can function as enzymes
    (B) peptide bonds
    (C) important in cell signaling
    (D) may be used as an energy source
    (E) contain nitrogenous bases

10. Which of the following are true regarding enzymes?

    (A) altering the three dimensional structure of an enzyme disrupts its activity
    (B) each enzyme is specific for a given reaction
    (C) hydrophobic, ionic, and hydrogen bonds play a role in binding substrate
    (D) A, B, and C
    (E) none of the above

Please see the answers and explanations to these review questions in Section V of this book.

Cells

CHAPTER 2

CELL THEORY

The fundamental unit of life is the cell. Cells are quite small in size, allowing them to exchange material by diffusion efficiently between the interior and exterior. The efficiency of exchange between the interior and exterior is related to the surface-area to volume ratio, and decreases rapidly with increasing cell volume. Prokaryotic cells, the simplest cells, arose first in evolution of the life on Earth, have a minimum of structure inside the cell and are single cell organisms. Eukaryotic cells are larger, more complex, and have a variety of membrane-bound organelles in the cell. Viruses are not truly cells, and will be discussed later.

Until the advent of the microscope in the seventeenth century, scientists were unable to see cells. Matthias Schleiden and Theodor Schwann proposed that all life is composed of cells in 1838, while Rudolph Virchow proposed in 1855 that cells arise only from other cells. The cell theory based on these ideas may be summarized as follows:

1. All living things are composed of cells.
2. All chemical reactions of life occur in cells.
3. Cells arise only from preexisting cells.
4. Cells carry genetic information in the form of DNA. This genetic material is passed from parent cell to daughter cell.

2.1   LIPID BILAYER MEMBRANES

One of the key requirements for cells is to separate the cellular interior from the exterior. Cells must prevent their essential components from escaping into the exterior as well as regulate their content and volume. While maintaining a distinct interior and exterior, the cell must also have the ability to obtain nutrients and information from the environment. The eukaryotic cell not only separates the interior from the exterior of the cell, but it also forms distinct compartments within the cell to carry out specific functions. Lipid bilayer membranes are an essential component of cells that allows them to carry out all of these functions.

The plasma (or cell) membrane encloses the cell and exhibits selective permeability, regulating the passage of materials into and out of the cell. According to the fluid mosaic model, the cell membrane consists of a phospholipid bilayer. Phospholipids have both a hydrophilic phosphoric acid region and a hydrophobic fatty acid region. In a lipid bilayer, the hydrophilic regions are found on the exterior surfaces of the membrane facing water, while the hydrophobic regions are found on the interior of the membrane, facing each other (see figure).

The Lipid Bilayer Model of the Plasma Membrane

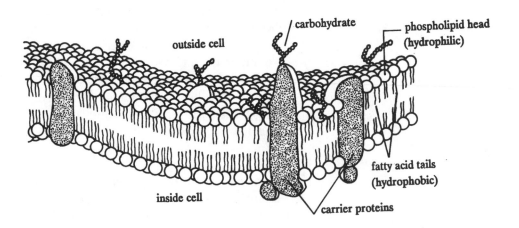

This arrangement of lipids is the most thermodynamically stable arrangement and will form spontaneously when phospholipids are mixed with water. Phospholipids and proteins can diffuse freely sideways within the plane of the membrane, moving to give the membrane a fluidlike quality and thus the term "fluid mosaic." Longer fatty acids and saturated fatty acids give the phospholipid tails more hydrophobic interactions with each other, making the membrane less fluid. Cholesterol molecules embedded in the hydrophobic interior of eukaryotic bilayer membranes modulate the membrane's fluidity.

As a selectively permeable barrier, the membrane allows material in and out of the cell in a selective manner. The hydrophobic interior of the membrane prevents charged or polar molecules from diffusing across the membrane, although some small molecules such as water, oxygen, and carbon dioxide diffuse freely through the membrane. Hydrophobic molecules such as hydrocarbons can also diffuse through the membrane easily, although they have limited solubility in water. If two molecules are equally soluble, then the smaller molecule will cross the plasma membrane faster. Membranes are relatively impermeable to all ions, even small ones such as H^+ and Na^+, in part due to the large sphere of hydrating water molecules around these ions. Charged substances in the cell that are involved in metabolism are retained in the cell since they cannot cross through the hydrophobic interior of the lipid bilayer membrane. Substances that cannot diffuse through the membrane require proteins embedded in the membrane to transport them.

Membrane Proteins

Proteins that cross through lipid bilayer membranes play a key role in many essential cellular activities. The nature of proteins expressed in a membrane to a large extent determines the function of the membrane. The region of proteins that passes through the membrane is generally very hydrophobic, stabilizing its interaction with the membrane interior, and is known as a transmembrane domain. The transmembrane domain forms an alpha-helix that passes from one side of the membrane to the other (see figure). Many proteins have more than one transmembrane domain, and are threaded back and forth through the membrane several times. G-protein coupled receptors, for example, are a class of proteins that carry signals across the plasma membrane into the cell and that have seven transmembrane domains threaded back and forth through the membrane. By having regions on both sides of the membrane, a protein is able to transmit information or material through the membrane.

Integral Membrane Protein

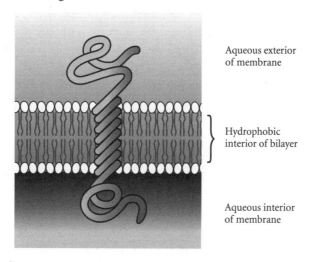

Aqueous exterior of membrane

Hydrophobic interior of bilayer

Aqueous interior of membrane

Transmembrane proteins found in the plasma membrane include proteins involved in carrying information across the cell surface, allowing the cell to respond to its environment, as well as proteins that carry material across the membrane. Cell membrane proteins act as selective pores for ions and receptors that bind signaling molecules outside of the cell and send signals into the cell. They also carry out the functions of cell adhesion and nutrient transport.

Transport Mechanisms

Simple Diffusion

Passive transport is the net movement of dissolved particles down a concentration gradient, from a region of higher concentration to a region of lower concentration. Simple diffusion does not require proteins, but involves the diffusion of material directly through the membrane itself and occurs with a limited number of substances such as water, oxygen, and carbon dioxide.

Osmosis

Osmosis is the simple diffusion of water through a membrane, such as the plasma membrane, from a region of lower solute concentration to a region of higher solute concentration, with water molecules moving to equalize solute concentrations. If a membrane is impermeable to a particular solute, then water will flow across the membrane until the differences in the solute concentration have been equilibrated.

When the cytoplasm of the cell has a lower solute concentration than the extracellular medium, the medium is said to be **hypertonic** to the cell. When a cell such as a red blood cell is placed in a hypertonic solution, the solutes of the extracellular environment cannot diffuse through the plasma membrane so osmosis equalizes the solute concentrations on the interior and exterior of the cell. Since there are more solutes outside of the cell in this case, water will flow out, causing the cell to shrink. On the other hand, when the cytoplasm of a cell has a higher solute concentration than the extracellular medium, the medium is **hypotonic** to the cell. In this case water will diffuse through the membrane into the cell, causing the cell to swell. If too much water flows in, the cell may **lyse**, the cell membrane breaking and releasing the cell's contents. Red blood cells, for example, lyse when put into distilled water. Finally, when solute concentrations are equal inside and outside, the cell and the medium are said to be isotonic. Water flows across the membrane at an equal rate in both directions and there is no net flow of water in either direction.

Facilitated Diffusion/Passive Transport

The net movement of solute down a concentration gradient with the help of proteins in the membrane is known as facilitated diffusion or passive transport. This process does not require external energy. An example of proteins involved in facilitated diffusion is ion channels. Cells must regulate the concentration of ions on the interior of the cell. Since ions cannot diffuse through a membrane, they require a protein for transport across the membrane. In some cases, an ion channel acts simply as a pore, allowing specific ions to flow through the membrane from one side to the other down their concentration gradient. In other cases the ion channel can open and close in response to specific signals. For example, neurons have a variety of ion channels that open and close in response to the binding of neurotransmitter or changes in voltage. When the channel is closed, no transport occurs, but when a specific signal is received, the channel will open and allow diffusion of the ion down its gradient through the ion channel.

In addition to channels, another form of transport protein is carriers. Carriers do not have a channel from one side of the membrane to the other, but have a binding site for solute. When solute binds, the protein changes conformation to release the bound solute on the other side of the membrane. There are three types of carrier proteins: uniport, symport, and antiport. Uniport proteins carry a single solute across the membrane. Symport proteins translocate two different solutes simultaneously in the same direction; transport occurs only if both solutes bind to the proteins. Antiport proteins exchange two solutes by transporting one into the cell and the other out of the cell.

Active Transport

Active transport is the net movement of dissolved particles against their concentration gradient with the help of transport proteins. Movement of material down a concentration gradient is a thermodynamically favorable process, with a $\Delta G < 0$, and occurs spontaneously, but movement against a concentration gradient has a $\Delta G > 0$ and will not occur on its own. In this case, extra energy needs to be put into the process to make it move forward. For example, potassium ions are present at a higher concentration in the cytoplasm than outside the cell. The movement of potassium ions into a cell against a concentration gradient (making them flow "uphill") does not happen spontaneously, but requires energy. If glucose is to be transported into an intestinal cell that has more glucose than the intestine, then energy must be used to transport the glucose into the cell. The most common forms of energy to drive active transport are ATP or a concentration gradient of another molecule.

Endocytosis/Exocytosis

In the processes discussed above, transmembrane proteins transport material through the membrane, while the membrane itself remained relatively unchanged. In other processes carried out by eukaryotic cells, the membrane changes its topology to either internalize or release a spherical section of membrane and its contents. Endocytosis is a process in which the cell membrane invaginates, forming a vesicle that contains material from the extracellular medium (see figure). Endocytosis is mediated by receptors in the plasma membrane and occurs in response to binding of specific material by the receptors in the cell membrane. For example, receptors that bind hormones or other signaling molecules and transmit a signal through the membrane are sometimes internalized by endocytosis when they bind their specific signaling molecule. When internalized, the vesicle can in some cases go to the lysosome where the receptor and the molecule bound to it are degraded. In this case, endocytosis can be a way for the cell to down-regulate a signal, by removing the receptor from the cell surface. In other cases, the receptor can be recycled back to the cell surface.

Endocytosis

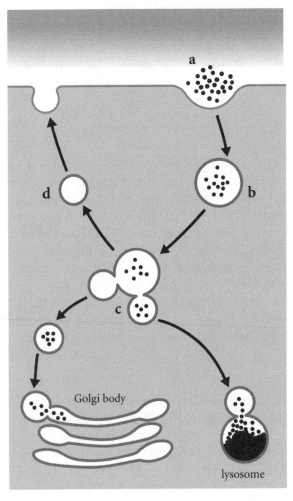

a. Molecules collect within indentations in the plasma membrane.

b. Vesicles form.

c. Molecules are routed to various cell organelles such as the Golgi body or lysosomes.

d. Vesicles return to the plasma membrane.

Golgi body

lysosome

Exocytosis is similar to endocytosis, except that in this case a membrane-bound vesicle inside the cell fuses with the plasma membrane to release its contents to the exterior of the cell. Cells involved in secretion of material, such as the secretion of proteases by the pancreas into the intestine, will use exocytosis to release the secreted material from the interior to the exterior of the cell. In both endocytosis and exocytosis, material is enclosed within the membrane of the vesicle and kept separate from the cytoplasm.

Two additional specialized forms of endocytosis form pockets of membrane to move material into the cell across the plasma membrane: pinocytosis and phagocytosis. Pinocytosis forms small vesicles filled with liquid and solutes from the external environment and brings these into the cell. Phagocytosis involves the engulfing of large particles by cells in sections of membrane. For example, macrophages are immune cells that remove bacteria by engulfing the bacterial cells whole through phagocytosis. The macrophage then goes on to kill the bacteria after they are ingested. Protists sometimes use phagocytosis to consume other microorganisms.

2.2 PROKARYOTIC CELLS

The prokaryotes include two distinct kingdoms of organisms that are recognized today: bacteria and archaebacteria. Only recently, in part through analysis of the genomes of these two groups, have the differences between the bacteria and archaebacteria been realized. In fact, analysis of the ribosomal RNA of these organisms suggests that archaebacteria are more closely related to eukaryotes than they are to bacteria. All prokaryotes are single-celled organisms that lack membrane-bound internal structures.

Although some prokaryotes live in groups of cells as filaments or colonies, each cell is a separate independent organism. Bacteria have several distinct external forms that are often used to distinguish them. Rod-shaped bacteria are known as bacilli and spherical bacteria are known as cocci. Prokaryotes have an outer lipid bilayer cell membrane, but lack membrane-bound organelles and also lack a cytoskeleton. They have no nucleus, but instead, a region of the cell called the **nucleoid** region in which the genome, consisting of a single circular molecule of DNA, is concentrated. The lack of organelles means that the interior of prokaryotes is one continuous compartment, the **cytosol**, in which all of the activities of life occur. Prokaryotes may also contain plasmids, small circular extrachromosomal DNAs containing few genes. Plasmids replicate independently and often incorporate genes that allow the prokaryotes to survive adverse conditions, making them very useful in modern molecular biology to carry genes in and out of bacteria. Prokaryotes have lipid bilayer cell membranes as do eukaryotes, with transmembrane proteins involved in the communication between the interior and the exterior of the cell. The cytoplasm of prokaryotes contains many ribosomes for protein production as does the cytoplasm of eukaryotes, although prokaryotic ribosomes are smaller (see figure).

Prokaryotic Cell

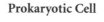
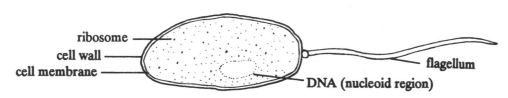

ribosome
cell wall
cell membrane
flagellum
DNA (nucleoid region)

Most prokaryotes have a strong porous cell wall for support of the cell. In bacteria the cell wall is made of peptidoglycan polymers woven into a single large molecule that surrounds the cell. Peptidoglycans are polysaccharides cross-linked by peptides that contain amino acids with the D stereochemistry, while proteins found in organisms normally have L-amino acids. Archaebacteria lack peptidoglycan in their cell wall, one indication of their uniqueness. Some bacteria have an additional outer membrane beyond the cell wall that often contains lipid toxins in pathogenic bacteria. Gram-staining is commonly used to separate bacteria into two groups: gram-negative and gram-positive. Gram-positive bacteria have a thick peptidoglycan cell wall that holds the stain but no outer membrane, while gram-negative bacteria have a thin cell wall and an outer membrane.

A layer of sticky polysaccharides called the capsule is often found around the cell membrane and the cell wall. The polysaccharides may help in the attachment of bacterial cells to outside surfaces and may help pathogenic bacteria to avoid the immune system since the cell wall can be hidden underneath the polysaccharides. Projections from the bacterial surface include flagella involved in movement (although unrelated in structure to the flagella found in some eukaryotic cells) and pili that are involved in attaching cells to other cells, in some cases for the exchange of genetic information. Bacterial flagella have a motor in the base anchored to the cell that causes the flagella to whip around; the flagella itself is composed of a protein called flagellin. Eukaryotic flagella are formed from microtubules with motor proteins to drive movement and are surrounded by the plasma membrane.

Prokaryotes reproduce by a mechanism distinct from eukaryotes. Eukaryotic cells divide by either mitosis or meiosis, and many eukaryotes reproduce sexually, with the union of genetic information from two parents. Prokaryotes reproduce asexually by a mechanism called binary fission, in which a cell replicates its DNA and divides in two. Binary fission does not involve any of the structures or processes of mitosis, which eukaryotes use to perform asexual reproduction. Bacteria do exchange genetic information through conjugation and other methods, and they also can perform recombination, but these occur apart from cell division.

Although prokaryotic cells are simple in structure, they are diverse in their metabolic activities and habitats. One means of distinguishing the life styles of organisms is according to their use of oxygen. **Obligate anaerobes** cannot survive in the presence of oxygen, **facultative aerobes** can survive with or without oxygen, and **obligate aerobes** require oxygen to survive. Another means of classifying organisms is according to their mode of nutrition. **Photoautotrophs** are photosynthetic, using light to generate the energy to produce their own nutrient molecules. Photosynthetic bacteria use the plasma membrane as the site of photosynthesis, although some have infoldings of the membrane called mesosomes. Another type of nutrition is used by **chemoautotrophs.** These organism produce their own nutrient molecules but use energy derived from inorganic molecules to drive nutrient production rather than the power of the sun. Nitrogen fixation, essential to all life, is carried out by chemoautotrophic prokaryotes. Chemoautotrophs are responsible for deep sea thermal vent ecosystems that are independent of light from the surface for energy. A common method of nutrition for all life, prokaryotes as well as eukaryotes, is used by **chemoheterotrophs**, which consume organic molecules for both carbon and as an energy source. If an organism obtains glucose, for example, from its environment and uses it for food and energy, it is a chemoheterotroph.

Although prokaryotes are often called "primitive" due to their simplicity, they are highly evolved in many ways. Bacteria fill every conceivable ecological niche on the planet, including bacteria that live in water that is near boiling and others that live deep in solid rock. The rapid rate of bacterial reproduction, as fast as one cell division per twenty minutes, allows bacteria to evolve extremely rapidly. In fact, some of the features of eukaryotes such as the presence of introns in genes that were once taken as signs of eukaryotic superiority may indicate that eukaryotes have evolved more slowly than prokaryotes and so retained primitive features of earlier precursor cells.

2.3   EUKARYOTIC CELLS AND ORGANELLES

All multicellular organisms (like you, or a tree, or a mushroom) and all protists such as amoebas and paramecia are eukaryotic. The eukaryotes include the protists, fungi, animals and plants. Eukaryotic cells are enclosed within a lipid bilayer cell membrane, as are prokaryotic cells. Unlike prokaryotes, eukaryotic cells contain membrane-bound organelles (see figure). An organelle is a structure within the cell with a specific function that is separated from the rest of the cell by a membrane. The presence of membrane-bound organelles in eukaryotes allows eukaryotic cells to compartmentalize activities in different parts of the cell, making them more efficient. Compartments within a cell can allow the cell to carry out activities such as ATP production and consumption within the same cell and control each independently.

Eukaryotic Cell

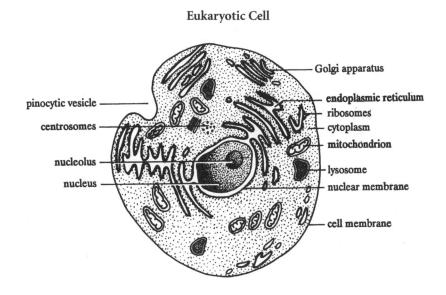

Nucleus

The genetic material, the DNA genome, is found in the largest organelle of the animal cell, the nucleus. The nucleus is separated from the rest of the cell by the **nuclear envelope**, a double-membrane that has a large number of nuclear pores through the envelope for communication of material between the interior and exterior of the nucleus. The pores are large enough to allow proteins to pass through but are also selective in the proteins that are transported into the nucleus or excluded from the nucleus. Special sequences in proteins signal a protein to be imported into the nucleus. While the prokaryotic genome is generally found in a single circular piece of DNA, the eukaryotic genome in each cell is split into chromosomes. Chromosomes contains the DNA genome complexed with structural proteins called histones that help to package the large strands of DNA in each chromosome within the limited space of the nucleus. Genes in the DNA genome are read (transcribed) to make RNA, which is processed in the nucleus before it is exported to the cytoplasm, where the RNA is read in turn (translated) to make proteins. The basic information flow of the cell is DNA to RNA to protein. The DNA genome is replicated in the nucleus when the cell divides. Other metabolic activities such as energy production are excluded from the nucleus. The structure and function of the eukaryotic genome will be presented later in more detail.

A dense structure within the nucleus in which ribosomal RNA (rRNA) synthesis occurs is known as the nucleolus. The nucleolus is not surrounded by a membrane, but is the site of assembly of ribosomal subunits from RNA and protein components. After assembly, the ribosomal subunits are exported from the nucleus to the cytoplasm to carry out protein synthesis.

Ribosomes

Ribosomes are not organelles but are large complex structures in the cytoplasm that are involved in protein production (translation) and are synthesized in the nucleolus. They consist of two subunits, one large and one small. Each ribosomal subunit is composed of ribosomal RNA (rRNA) and many proteins. Free ribosomes are found in the cytoplasm, while bound ribosomes line the outer membrane of the endoplasmic reticulum (ER). Proteins that are destined for the cytoplasm are synthesized by ribosomes free in the cytoplasm, while proteins that are bound for one of several membranes or that are to be secreted from the cell are translated on ribosomes bound to the rough endoplasmic reticulum (rough ER). Prokaryotic ribosomes are similar to those of eukaryotes, composed of rRNA and proteins that form two different size subunits that come together to perform DNA synthesis. Prokaryotic ribosomes are, however, smaller and simpler than eukaryotic ribosomes. Mitochondria and chloroplasts also have their own ribosomes in their interior and carry out synthesis of a few proteins, but the ribosomes of these organelles are distinct from those of the eukaryotic cytoplasm and more closely resemble prokaryotic ribosomes.

Endoplasmic Reticulum

The endoplasmic reticulum (ER) is an extensive network of membrane-enclosed spaces in the cytoplasm. The interior of the ER between membrane layers is called the lumen and at points in the ER the lumen is continuous with the nucleus. If a region of the ER has ribosomes lining its outer surface, it is termed rough endoplasmic reticulum (rough ER); without ribosomes, it is known as smooth endoplasmic reticulum. Smooth ER is involved in lipid synthesis and the detoxification of drugs and poisons and has the appearance of a network of tubes, while rough ER is involved in protein synthesis and is a series of stacked plates. Proteins that are secreted, found in the cell membrane, the ER, or the Golgi are made by ribosomes on the rough ER. Proteins synthesized on the rough ER cross into the lumen of the rough ER during synthesis. A hydrophobic sequence of amino acids at the amino terminus of proteins as they are synthesized determines whether the protein will be sorted into the secretory pathway starting at the rough ER or synthesized in the cytoplasm. Proteins that are secreted will have only one hydrophobic signal sequence, the signal peptide, and will be inserted into the ER lumen when they are synthesized, then released from the cell later. Proteins that are destined to be membrane bound have hydrophobic transmembrane domains that are threaded through the rough ER membrane as the protein is synthesized. When the protein reaches the correct membrane destination along the secretory pathway, additional signals in the protein sequence and structure will cause the protein to stay localized at the correct location.

Small regions of ER membrane bud off to form small round membrane-bound vesicles that contain newly synthesized proteins. These cytoplasmic vesicles are then transported to the Golgi apparatus, the next stop along the secretory pathway.

Golgi Apparatus

The Golgi is a stack of membrane-enclosed sacs, usually located in the cell between the ER and the plasma membrane (see figure on p. 67). The stacks closest to the ER are called the cis Golgi and the stacks farthest from the ER, closer to the plasma membrane, are called the trans Golgi. Vesicles containing newly synthesized proteins bud off of the ER and fuse with the cis Golgi. In the Golgi, these proteins are modified and then repackaged for delivery to other destinations in the cell. For example, the Golgi carries out post-translational modification of proteins through glycosylation, the process of adding sugar groups to proteins to form glycoproteins. Many proteins destined for the plasma membrane have carbohydrate groups added to the surface of the protein facing the exterior of the cell.

After processing in the cis Golgi, proteins are packages in vesicles that move to the next layer in the stack, where they fuse and release their contents. Proteins proceed in this manner from one stack to the next until they reach the trans Golgi. In the trans Golgi, proteins are sorted into vesicles based on signals in different proteins that indicate their final destination. The nature of the signal varies, but includes the protein primary sequence, its structure, and post-translational modifications. Once packaged into vesicles, the vesicles move on to their final destination. The final destination for a protein may include the lysosome, the plasma membrane or the exterior of the cell. Some proteins are retained in the Golgi or the ER. Proteins that are destined for the plasma membrane as transmembrane proteins are inserted in the membrane in the ER as they are synthesized, and maintain their orientation in the membrane as they move from ER to Golgi to vesicle to the plasma membrane. Proteins that are secreted from the cell are inserted in the ER lumen during protein synthesis, and remain in the lumen of the ER to the Golgi, where they form secretory vesicles. The last step in secretion is the fusion of the secretory vesicle with the plasma membrane, releasing the contents of the vesicle to the cellular exterior.

Lysosomes

Lysosomes contain hydrolytic enzymes involved in intracellular digestion that break down proteins, carbohydrates and nucleic acids. For white blood cells, the lysosome may degrade bacteria or damaged cells. For a protist, lysosomes may provide food for the cell. They also aid in renewing a cell's own components by breaking them down and releasing their molecular building blocks into the cytosol for reuse. A cell in injured or dying tissue may rupture the lysosome membrane and release its hydrolytic enzymes to digest its own cellular contents.

The lysosome maintains a slightly acidic pH of 5 in its interior, a pH at which lysosomal enzymes are maximally active. The contents of the lysosome are isolated from the cytoplasm by the lysosomal membrane, keeping the pH distinct from the neutral pH of the cytoplasm. The distinct pH optimum and compartmentalization of lysosomal enzymes prevents them from degrading the rest of the cellular contents.

Peroxisomes

Peroxisomes contain oxidative enzymes that catalyze reactions in which hydrogen peroxide is produced and degraded. Peroxisomes break fats down into small molecules that can be used for fuel; they are also used in the liver to detoxify compounds, such as alcohol, that may be harmful to the body. The peroxides produced in the peroxisome would be hazardous to the cell if present in the cytoplasm, since these molecules are highly reactive and could covalently alter macromolecules such as DNA. Compartmentalization of these activities within the peroxisome reduces this risk.

Mitochondria

Mitochondria are the source of most energy in the eukaryotic cell as the site of aerobic respiration. Mitochondria are bound by an outer and an inner phospholipid bilayer membrane (see figure on page 67). The outer membrane has many pores and acts as a sieve, allowing molecules through on the basis of their size. The area between the inner and outer membranes is known as the intermembrane space. The inner membrane has many convolutions called **cristae**, as well as a high protein content that includes the proteins of the electron transport chain. The area bounded by the inner membrane is known as the mitochondrial **matrix**, and is the site of many of the reactions in cell respiration, including electron transport, the Krebs cycle, and ATP production.

Mitochondria are somewhat unusual in that they are semiautonomous within the cell. They contain their own circular DNA and ribosomes, which enable them to produce some of their own proteins. The genome and ribosomes of mitochondria resemble those of prokaryotes more than eukaryotes. In addition, they are able to self-replicate through binary fission. Mitochondria are believed to have developed from early prokaryotic cells that began a symbiotic relationship with the ancestors of eukaryotes, with the mitochondria providing energy and the host cell providing nutrients and protection from the exterior environment. This theory of the origin of mitochondria, and the modern eukaryotic cell, is called the endosymbiotic hypothesis.

Cytoskeleton

The cytoplasm is the liquid inside the cell surrounding the organelles. The cytoplasm has structure as well, since it contains not only soluble enzymes and ribosomes, but also an intricate and dynamic fibrous network called the **cytoskeleton**. The cell gains mechanical support, maintains its shape, and carries out cell motility functions with the help of the cytoskeleton. The cytoskeleton is composed of microtubules, microfilaments, intermediate fibers, and chains and rods of proteins each with distinct functions and activities.

Microtubules are hollow rods formed of the polymerized protein tubulin that radiate throughout cells and provide it with support. They also provide a framework for organelle movement within the cell. Proteins that act as molecular motors can attach to microtubules and move organelles within the cell. Microtubules have polarity, with distinct ends called + and –, and grow and shrink rapidly, making them dynamic structures. Eukaryotic flagella and cilia contain specialized arrangements of microtubules and molecular motors that aid in the movement of these structures. Centrioles, which direct the separation of chromosomes during cell division, are composed of microtubules, as well as the spindle fibers that move chromosomes during cell division.

Cell movement and support are maintained in part through the action of solid rods termed **microfilaments** composed of actin subunits. Muscle contraction, for example, is based on the interaction of actin with myosin in muscle cells. Microfilaments are active, for instance, in the contraction phase of cell division in which the membrane between dividing cells pinches closed, and they are involved in amoeboid movement in which cytoplasm streams into pseudopods for amoebas to move. Specialized proteins can interact with actin microfilaments to modify their behavior. For example, in some cells such as muscle cells, microfilaments are very stable long term structures, while in other cells they are more dynamic.

The third element of the cytoskeleton is **intermediate filaments**. These cytoskeletal structures are a collection of fibers involved in the maintenance of cytoskeletal integrity. Their diameters fall between those of microtubules and microfilaments. These fibers help a cell to resist mechanical stress and help tissues to maintain structural integrity as well.

Specialized Plant Organelles

Plants lack centrioles, but also have some organelles that are not found in animal cells.

Chloroplasts

Chloroplasts are found only in plant cells and some protists. With the help of one of their primary components, chlorophyll, they function as the site of photosynthesis, using the energy of the sun to produce glucose. The process of photosynthesis will be presented in more detail later. Chloroplasts have two membranes, an inner and an outer membrane. Additional membrane sacs called **thylakoids** inside the chloroplast are derived from the inner membrane and form stacks called **grana**. The fluid inside the chloroplast surrounding the grana is the **stroma**. The thylakoid membranes contain the chlorophyll of the cell.

Like mitochondria, chloroplasts contain their own DNA and ribosomes and exhibit the same semi-autonomy. They are also believed to have evolved via symbiosis of a photosynthetic early prokaryote that invaded the precursor of the eukaryotic cell. In this arrangement, the chloroplast precursor cell provided food and received protection. Photosynthetic prokaryotes today carry out photosynthesis in a manner similar to the chloroplast.

Vacuoles are membrane-enclosed sacs within the cell. Many types of cells have vacuoles, but plant vacuoles are particularly large, taking up 90% of the cell volume in some cases. Plants use the vacuole to store waste products, and the pressure of liquid and solutes in the vacuole helps the plant to maintain stiffness and structure as well.

All plant cells have a cellulose cell wall that distinguishes them from animal cells, which lack a cell wall. The cell wall of plants is also distinct from the peptidoglycan cell wall of bacteria and the chitin cell wall of fungi. The cell wall provides structure and strength to plants.

Comparison of Cell Properties:

Property	Prokaryote	Eukaryote
DNA genome	Small, circular, no histones	Large segmented chromosomes with histones, packaging
Nucleus	None	Yes
Membrane-bound organelles	None	ER, Golgi, Mitochondria, chloroplasts, etc.
ATP production	Plasma membrane	Mitochondria
Cell division/reproduction	Binary fission	Mitosis/meiosis/ sexual reproduction
Cell wall	Yes, in most	Plants and fungi: yes, animals: no
Flagella	Yes: but different structure	Yes: with microtubules
RNA processing	Simple, no splicing	5'-cap, poly-A tail, mRNA splicing
Transcription and translation	Together in cytosol	Separated
Cytoskeleton	None	Yes

2.4   THE CELL CYCLE AND ITS REGULATION

Cells must replicate if an organism is to grow or repair itself, with one cell becoming two cells. The process by which cells divide to form two new daughter cells is known as **mitosis**. Mitotic cell division is a means of reproduction for many eukaryotes, including protists, fungi, and plants. Through mitosis, a cell (or a single-celled organism) is able to clone itself, making identical copies of genome from one generation of cells to the next. For multicellular eukaryotes, mitosis is also the mechanism of growth, development, and replacement of worn-out cells. Meiosis is a form of cell division used to form gamete cells during eukaryotic sexual reproduction. The mechanism of reproduction used by prokaryotes is a simple form of cell division called binary fission.

Mitosis is the division of a cell to produce two identical daughter cells. A key step in mitotic cell division is the replication of the genome so that each of the daughter cells can receive a complete and identical copy of the genome after cell division is complete. Every eukaryotic organism has a characteristic number of chromosomes in each cell. Humans have two copies of 23 chromosomes in each cell, for a total of 46 chromosomes. Cells and organisms with two copies of each chromosome are called **diploid**, while cells with one copy of each chromosome are called **haploid**. A diploid cell with 46 chromosomes that goes through mitosis will produce two diploid daughter cells, each with 46 chromosomes.

Every cell of the body contains a complete copy of the genome in its DNA, but not every cell contributes that DNA to the next generation. Somatic cells are cells of the body that do not contribute to the next generation. Somatic cells reproduce through mitosis. Germ cells are those that do contribute their genome to the next generation during the process of sexual reproduction. In a diploid organism that reproduces through sexual reproduction, germ cells form haploid cells that unite, one from each parent, to form a diploid cell that will develop into a new organism. The haploid cells involved in reproduction are called gametes. A specialized form of cell division called meiosis produces haploid gametes.

The four stages of the mitotic cell cycle are designated G_1, S, G_2, and M. The first three stages of this cell cycle are interphase stages; that is, they occur between cell divisions. The fourth stage, mitosis, incorporates the actual division of the cell.

Cell Cycle

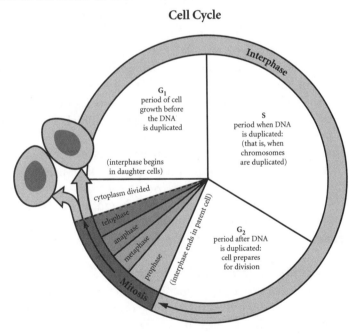

Stage G_1. The G_1 stage in mitosis is characterized by intense biochemical activity, producing the energy required to drive biosynthesis and synthesizing the proteins, carbohydrates and organelles required for the cell to grow.

Stage S. This is the stage during which synthesis of DNA takes place (S is for synthesis). Each chromosome is replicated so that during division, a complete copy of the genome can be distributed to both daughter cells. After replication, the chromosomes consist of two identical sister chromatids held together at a central region called the centromere. The ends of the chromosomes are called telomeres. Cells entering G_2 actually contain twice as much DNA as cells in G_1, since they have completed the replication of their genome, but have not yet divided the DNA into two daughter cells.

Stage G_2. The cell continues to grow in size, as assembly of new organelles and other cell structures continues, in preparation for mitosis.

Stage M (Mitosis and Cytokinesis). Mitosis is the process of nuclear division of the cell followed by the splitting of the cytoplasm and membrane into two completely separate daughter cells through the process of cytokinesis. Mitosis is broken down into four stages: prophase, metaphase, anaphase, and telophase.

The Stages of Mitosis

Prophase. During interphase, the chromosomes are packaged with histones but still are long slender diffuse strands that cannot be seen through a light microscope. In prophase, chromosomes condense into short thick bands that can easily be sorted and moved during cell division. The chromosomes contain both replicated copies of the chromosome formed during S phase and joined together in the middle region called the centromere. Each of the two joined copies is called a **sister chromatid**. During interphase, the centriole has replicated so that (in animal cells) there are two centrioles. During prophase the centrioles separate and move toward opposite poles of the cell. Some of the microtubules of the cytoskeleton break down and new ones form, moving the centrioles apart. The new microtubules begin to form the spindle apparatus, involved in moving the chromosomes during mitosis. During the transition to the next phase, the nuclear membrane begins to break down into small vesicles.

Metaphase. During metaphase, the centrioles arrange themselves at opposite poles of the cell and the nuclear envelope completely dissociates into small vesicles. Kinetochores with attached kinetochore fibers appear at the chromosome centromere. The kinetochore fibers interact with the spindle apparatus, one fiber from each end of the cell attached to each of the sister chromatids in a chromosome and pulling to align the chromosome at the metaphase plate (equatorial plate) halfway between the two ends of the cell.

Anaphase. In anaphase, the spindle pulls the centromeres of each chromatid apart, allowing the sister chromatids to separate. The telomeres are the last part of the chromatids to separate. The sister chromatids are pulled towards the opposite poles of the cell through the shortening of the kinetochore fibers. Once the chromatids are separated from each other, they each are called a chromosome. It is crucial that when the sister chromatids separate, each daughter cell receives one copy of each chromosome. If this does not happen, the cell lacking a chromosome will probably not function properly since it will lack a large segment of genetic information.

Telophase. In telophase, the spindle apparatus disappears. A nuclear membrane reforms around each set of chromosomes, and the nucleoli reappear. The chromosomes uncoil, resuming their interphase form. Each of the two nuclei receives a complete copy of the genome identical to the original genome and to each other.

Cytokinesis. After the cell has divided its DNA in mitosis, it enters cytokinesis, in which the cytoplasm and all the organelles of the cell divide. This division ultimately results in the formation of two daughter cells from one parent cell. Cytokinesis in animal cells results from a division furrow, in which microfilaments with actin contract to pinch the plasma membrane together between the new nuclei to form two new separated daughter cells. In plant cells, cytokinesis is somewhat different. In plant cells, membrane vesicles move between the two new nuclei formed during mitosis. These vesicles fuse to form a new plasma membrane, and help to build a cell plate, the precursor of a new cell wall between the cells.

2.5 REGULATION OF THE CELL CYCLE

While it is important for cells to divide, it is also important that they do not divide continually in an uncontrolled manner. When an organism is growing, cells must proliferate and divide to form tissues at the right time in the right place. Cells must divide mitotically to heal a wound, but they must also stop dividing when healing is complete. Uncontrolled cell division results in cancer, emphasizing the importance of careful control of the cell cycle.

The cell cycle has decision points, or check points, that determine whether a cell will proceed from one stage to the next. There is a decision point between G_1 and S that determines if a cell will enter S to replicate its genome. Some cells such as neurons in the brain appear to be permanently locked into G_1 once they differentiate and will never enter S phase again for the life of the cell or the individual. Another checkpoint at G_2-M determines when the cell exits G_2 and enters M. Generally, once a cell passes through a checkpoint it is committed to completing the next stage in the cell cycle. These checkpoints are determined by the activity of protein kinase enzymes. Protein kinases are enzymes that phosphorylate other proteins and enzymes in the cell to regulate their activity. An important kinase that regulates progression through the cell cycle is the cyclin-dependent kinase (cdk). When active, this kinase phosphorylates proteins that initiate the next step in the cell cycle. Cdk requires other proteins called cyclins to be active. Cyclins derive their name from the cyclical nature of their expression in the cell during the cell cycle, with different cyclin-cdk pairs triggering progression through different cell cycle check points.

The cyclin and cdks are part of a network of signaling systems in the cell. They themselves can be targets of phosphorylation and regulation as well as the regulation they exert on cell cycle factors. External signals such as a growth factor outside the cell can trigger signaling systems that can activate cdk to turn on cell division. Factors that stimulate cell proliferation are called **mitogens**. When a growth factor binds to its receptor in the plasma membrane, the receptor may activate a protein kinase and a cascade that activates cdk, triggering cell division. In normal cells, proliferation is regulated by several factors. One factor that modulates mitogenic signals is the external environment of the cell.

Cancer is a state of uncontrolled cell proliferation. Two principles seem to play a role in the development of cancer—the signals that turn on cell cycle remain on, and the signals that block proliferation remain off. Very often, cancer seems to result from a combination of these. One technique used to study cancer has been to determine the genetic changes that happen in cells that have become cancerous by identifying **oncogenes**, or genes that cause cancer. Oncogenes appear often to be mutated versions of genes involved in the cell cycle. For example, one of the most common human oncogenes found in human tumors involves the **ras gene**, a signaling molecule involved in mitogenic signals. In normal cells, the ras gene will be turned on in response to a growth factor, then turned off again to turn off the signal. In a cancer cell with a mutant ras gene, the ras protein never turns off and as a result the cellular proliferation never stops. Mutant forms of the membrane receptors for growth factors are often found in tumors. The normal receptor for a growth factor may have a protein kinase activity that is only active when the growth factor binds to the receptor, stimulating proliferation. Without growth factor, the kinase will be silent. The oncogenic form of the gene for a growth factor receptor may be a protein kinase that is always active, even if the growth factor is not present. Identifying oncogenes from tumors has allowed scientists to backtrack to learn more about the way the normal cell cycle is regulated.

Tumor suppressors such as p53 and the retinoblastoma gene product are factors that oppose uncontrolled mitogenic signals. These proteins keep the cell cycle off most of the time, and prevent inappropriate cell division. If a cell attempts to divide when it should not, the tumor suppressors should normally block the division and sometimes kill the cell to prevent it from dividing. Mutations in the p53 gene are one of the most common defects found in human cancers. Many cancers appear to result not from one genetic change, but from at least two genetic defects. For example, if a growth factor receptor is mutated and becomes active, but the p53 gene is normal, the cell may be induced to kill itself rather than become cancerous. Only when both the growth factor gene and the p53 gene are mutated can cancer ensue. From there, additional mutations may be required for the cell to become malignant and proliferate into surrounding tissues. It may even require both copies of the gene in a diploid somatic cell to be altered for cancer to occur. For those cells that escape these multiple controls, the immune system appears to remove many early cancerous cells before they proliferate. Fortunately, it is only when the multiple redundant protective systems regulating the cell cycle fail that cancer results.

The changes in cells that bring about these changes can result from inherited defects in genes, or accumulated defects caused by environmental chemicals or radiation that mutate genes over a person's lifetime. For example, an individual who inherits an altered p53 gene may be one step closer to cancer development, increasing their risk by exposure to environmental factors. The environmental factors that influence cancer risk are primarily factors that damage DNA; these factors are called mutagens. Radiation and chemical hazards fall into this category. Luckily, the mutations found in a tumor that are caused by a person's environmental exposure may not be found in the germ line and may not be passed on to future generations. The complex interacting factors involved in cancer formation clearly include genetic and environmental influences, but in many cases a significant proportion of cancers can be avoided through lifestyle choices that individuals make, including exercise, proper diet, and avoidance of cigarettes.

Chapter 2: Review Questions

1. Which of the following is NOT present in an animal cell?

   (A) nucleus
   (B) mitochondria
   (C) cell wall
   (D) DNA
   (E) ribosomes

2. The protein products of tumor suppressor genes

   (A) are present in noncancerous cells
   (B) may signal cell death
   (C) regulate the cell cycle
   (D) cause unregulated cell growth
   (E) A, B, and C

Questions 3–4 refer to the choices below:

   (A) telophase
   (B) anaphase
   (C) prophase
   (D) metaphase
   (E) cytokinesis

3. nuclear membrane disintegrates

4. chromosomes line up along midline of cell

Questions 5–7 refer to the choices below:

   (A) osmosis
   (B) facilitated diffusion
   (C) exocytosis
   (D) endocytosis
   (E) active transport

5. fusion of a vesicle with the cell membrane; contents are released outside cell

6. movement of dissolved particles against the concentration gradient

7. movement of dissolved particles through membrane proteins with the concentration gradient

Questions 8–10 refer to the choices below:

   (A) nucleus
   (B) Golgi apparatus
   (C) lysosome
   (D) plasma membrane
   (E) endoplasmic reticulum

8. semipermeable cell barrier

9. present in a prokaryotic cell

10. modifies and packages proteins

Please see the answers and explanations to these review questions in Section V of this book.

Cellular Energetics

3.1 COUPLED REACTIONS

According to the laws of thermodynamics, chemical reactions and other processes will occur spontaneously if the free energy of the end products is lower than the free energy of the starting materials. Examples of spontaneous processes include ions moving down a concentration gradient and the hydrolysis of ATP. Many cellular processes are not spontaneous on their own however, including biosynthetic reactions, active transport of material against a concentration gradient, movement of material in the cell, and contractile motion. If these processes are not spontaneous, how can they occur?

The cell can drive forward thermodynamically unfavorable reactions by coupling them to reactions that are favorable. In a metabolic pathway containing several reactions, the free energy change (ΔG) for each step can be added together to determine the free energy change for the entire pathway. For example, consider Reaction 1 with a free energy change of +9 kcal/mol, and Reaction 2 with a free energy change of −12 kcal/mol.

Reaction 1:	$A + B \rightarrow C + D$	$\Delta G = +9$ kcal/mol
Reaction 2:	$D \rightarrow E + F$	$\Delta G = -12$ kcal/mol
	$A + B \rightarrow C + E + F$	$\Delta G = -3$ kcal/mol

Reaction 1 will not proceed on its own, but if it is coupled to Reaction 2, the net free energy change for the two coupled reactions is 9 + (12) = −3 kcal/mol. With a negative free energy change ($\Delta G < 0$), the two coupled reactions will proceed spontaneously, with Reaction 2 driving Reaction 1 forward. In this example, the equilibrium for Reaction 1 lies to the left, but the equilibrium for Reaction 2 is to the right, favoring the production of E and F. Reaction 2 will draw D away, converting it to E and F, ensuring that Reaction 1 continues to move to the right, and changing the equilibrium for the coupled reactions.

The free energy for ATP hydrolysis to ADP + P_i is –7.3 kcal/mol under standardized conditions, and greater in the conditions found typically inside the cell. The free energy for hydrolysis of the high-energy phosphate bond to make AMP (adenosine monophosphate) from ADP (adenosine diphosphate) is similar.

$$ATP + H_2O \rightarrow ADP + P_i + H^+ \qquad \Delta G = -7.3 \text{ kcal/mol}$$
$$ADP + H_2O \rightarrow AMP + P_i + H^+ \qquad \Delta G = -7.3 \text{ kcal/mol}$$

With a very negative change in free energy for ATP hydrolysis, it is very favorable, with the equilibrium shifted far to the right side of the equation. A common theme in metabolic pathways is to couple ATP hydrolysis, a very favorable reaction, to other reactions that are not favorable on their own. This is the key function of ATP as the primary energy currency of the cell. Coupling non-spontaneous reactions to ATP hydrolysis can drive reactions forward by giving the pathway as a whole, with ATP hydrolysis included, a net negative free energy change. The biosynthesis of proteins, polysaccharides and nucleic acids requires ATP hydrolysis. Active transport often uses ATP hydrolysis as an energy source.

Although ATP hydrolysis is the most common energy used to drive unfavorable reactions forward, other sources of energy are used as well. Active transport is coupled in some cases not to ATP hydrolysis but the movement of one solute down a concentration gradient. The formation of ATP itself is driven by a concentration gradient of protons in the mitochondria generated by the electron transport chain during oxidative phosphorylation.

The central role of ATP as an energy source makes it an important molecule for all cells. ATP is not stored in the cell, but is constantly generated, consumed, and regenerated again.

3.2 GLYCOLYSIS AND AEROBIC RESPIRATION TO PRODUCE ATP—OVERVIEW

Cells need energy to make proteins or DNA, to move, and to grow. Biosynthesis of macromolecules such as proteins and nucleic acids requires energy, as does active transport or any other activity that is not thermodynamically spontaneous in the cell. The energy used by the cell most commonly for biosynthesis and other activities is chemical energy in the form of ATP, adenosine triphosphate. The triphosphate part of ATP contains three linked phosphate groups that release a great deal of energy when they are hydrolyzed in the cell. The energy released by hydrolysis of these high-energy phosphate bonds in ATP is used to drive reactions or processes that need energy.

Where does ATP come from? All cells, both prokaryotic and eukaryotic, share in common a basic set of biochemical reactions that make ATP using energy captured from the oxidation of glucose. The complete oxidation of glucose is given by the equation:

$$C_6H_{12}O_6 + 6O_2 \rightarrow 6CO_2 + 6H_2O + \text{energy}$$

In this reaction, glucose has much greater free energy than the products, carbon dioxide and water. The free energy change, ΔG, is negative meaning that this reaction will occur spontaneously, with a release of energy. Glucose left sitting in a beaker with oxygen will react to form carbon dioxide and water, but extremely slowly. If a flame is put to the glucose, the reaction will occur very rapidly, with the input of energy in the form of heat. Alternatively, if the reaction is catalyzed by an enzyme, the energy can be captured by the cell in a useful form, the chemical energy found in ATP. The metabolic pathways that oxidize glucose to make ATP in the cell are glycolysis, the Krebs cycle, and electron transport.

3.3   GLYCOLYSIS

Glycolysis is the first step involved in capturing the energy of glucose to make ATP. In eukaryotes, enzymes perform glycolysis in the cytoplasm, and in prokaryotes all reactions occur in the same compartment since there are no organelles. In glycolysis, the sugar glucose with six carbon atoms is broken into two molecules of pyruvate, each with three carbon atoms (see figure). The path from glucose requires ten steps, each catalyzed by a specific enzyme.

In general, metabolism of nutrients for energy oxidizes the nutrient and releases energy. Pyruvate is more oxidized than glucose, so glycolysis releases energy that is captured in ATP. The first reactions of glycolysis use ATP to phosphorylate glucose, a reaction that consumes ATP to drive the reaction forward. The ATP is used to phosphorylate glucose twice to form fructose 1,6,-bisphosphate. One result of phosphorylating glucose is that it is maintained in the cell, since glucose 6-phosphate and other phosphorylated sugars are strongly polar and there is no export mechanism for glucose 6-phosphate from the cell. Fructose 6-phosphate is a six-carbon sugar with a phosphate at each end of the molecule, and is cleaved into two three-carbon sugars in the next step of glycolysis. The two three carbon sugars isomerize to form glyceraldehyde 3-phosphate, which continues in the pathway. Each molecule of glyceraldehyde 3-phosphate is oxidized and phosphorylated to form the next intermediate, with a molecule of reduced NADH carrying some energy. The rest of the pathway involves ATP production from the three carbon phosphorylated intermediates, resulting in pyruvate. For every glucose molecule that enters glycolysis, two ATP molecules are put into the reaction, but four are produced, for a net gain of two ATP molecules per glucose.

The major control point of glycolysis is the third step: the addition of a second ATP to fructose 6-phosphate to form fructose 1,6-bisphosphate. This reaction is catalyzed by the enzyme **phosphofructokinase**. This reaction is first irreversible step in glycolysis, making phosphofructokinase an ideal target for regulation of glycolysis. The term "irreversible" indicates that in the cell the reaction is very favorable in the forward direction, driven to the right by coupling to ATP hydrolysis. When energy in the cell is low, with ATP low and AMP increased, phosphofructokinase will be activated, turning on glycolysis and leading to increased ATP generation.

Glycolysis

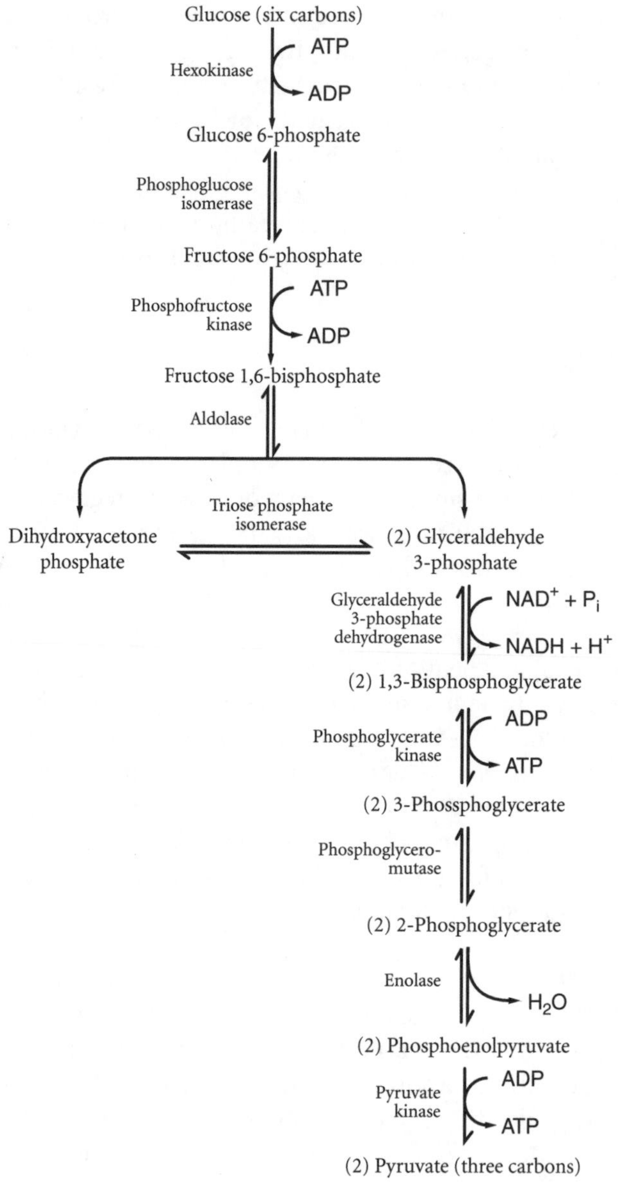

Glucose (six carbons)

Hexokinase — ATP ⟶ ADP

Glucose 6-phosphate

Phosphoglucose
isomerase

Fructose 6-phosphate

Phosphofructose
kinase — ATP ⟶ ADP

Fructose 1,6-bisphosphate

Aldolase

Dihydroxyacetone
phosphate

Triose phosphate
isomerase

(2) Glyceraldehyde
3-phosphate

Glyceraldehyde
3-phosphate
dehydrogenase — $NAD^+ + P_i$ ⟶ $NADH + H^+$

(2) 1,3-Bisphosphoglycerate

Phosphoglycerate
kinase — ADP ⟶ ATP

(2) 3-Phossphoglycerate

Phosphoglycero-
mutase

(2) 2-Phosphoglycerate

Enolase — H_2O

(2) Phosphoenolpyruvate

Pyruvate
kinase — ADP ⟶ ATP

(2) Pyruvate (three carbons)

Another product of glycolysis is the reduction of two molecules of the energy-carrying molecule NAD^+ to NADH. NADH will play a role in ATP generation as well through electron transport. The overall reaction of glycolysis is:

$$\text{Glucose} + 2P_i + 2ADP + 2NAD^+ \rightarrow 2 \text{ pyruvate} + 2ATP + 2NADH + 2H^+ + 2H_2O$$

Since NAD^+ is required for glycolysis to occur and is converted to NADH as part of glycolysis, NAD^+ must be regenerated or glycolysis would run out of NAD^+ and stop, halting ATP production as well (and probably the life of the cell or organism involved). The NAD^+ is regenerated in one of two ways. In the first, NADH goes on to the electron transport chain and is used to produce more ATP, as described in the sections that follow, being converted back to NAD^+ in the process. This process is termed aerobic respiration because it requires oxygen. The second way to regenerate NAD^+ occurs in the absence of oxygen or in anaerobic organisms that do not use oxidative metabolism. This alternate pathway is called fermentation.

3.4   FERMENTATION

In fermentation, NADH produced in glycolysis is regenerated back to NAD^+ in the absence of oxygen to allow glycolysis to continue. Either ethanol or lactic acid is produced as a by-product. Ethanol fermentation is carried out by yeast in the absence of oxygen and is used to make beer and wine. The first step in ethanol fermentation occurs when pyruvate is decarboxylated (loses a CO_2) to become the two-carbon molecule acetaldehyde. NADH from glycolysis reduces acetaldehyde to ethanol and is itself oxidized back to NAD^+. The regenerated NAD^+ allows glycolysis to proceed and to continue ATP production. This process can continue until the alcohol level rises so high that it kills the organisms producing it.

Lactic acid fermentation occurs in some bacteria and fungi and, closer to home, in human muscles during strenuous exercise. Fermentation in muscle allows ATP production to continue when oxygen in the muscle is consumed more rapidly than the blood stream can supply it. In lactic acid fermentation, pyruvate is reduced to lactic acid by NADH, which is itself oxidized to regenerate NAD^+. The NAD^+ then allows glycolysis and ATP production to proceed. During exercise, the lactic acid produced can accumulate in muscle and cause pain as well as a drop in blood pH. When the blood stream is again able to supply sufficient oxygen, the lactic acid will be oxidized back to pyruvate to reenter oxidative metabolism pathways.

3.5   AEROBIC RESPIRATION

Overview

Although glycolysis produces 2 ATP and 2 NADH for every molecule of glucose that enters the pathway, this is not where the eukaryotic cell extracts most of its energy from glucose. Glycolysis is only the beginning. The pyruvate from glycolysis still contains a great deal of energy that is extracted in oxidative metabolism, also called aerobic respiration. Oxidative metabolism of glucose produces a maximum total of 38 ATP in the complete oxidation of one glucose molecule, compared to two ATP from glycolysis (or fermentation) alone (note that the actual total is 36 ATP/glucose in eukaryotes due to ATPs that are used in transporting intermediates across membranes). To accomplish this more efficient energy production, pyruvate from glycolysis is oxidized all the way to carbon dioxide in a pathway called the Krebs Cycle. The Krebs Cycle and the other steps of oxidative metabolism occur in mitochondria. As pyruvate is oxidized in the Krebs cycle, NADH and another high-energy electron carrier called $FADH_2$ are produced. In electron transport, the energy of these high-energy electron carriers creates a pH gradient by pumping protons (H^+ ions) out of the mitochondria. The energy of this pH gradient drives ATP synthesis, and this is the source of most of the ATP produced in the oxidative metabolism of glucose.

Aerobic Respiration

Acetyl-CoA Production from Pyruvate

Under aerobic conditions the pyruvate produced in the cytoplasm in glycolysis is transported through both mitochondrial membranes into the interior of the mitochondria. In the mitochondrial matrix, pyruvate is converted into a two-carbon molecule called acetate by an enzyme called **pyruvate dehydrogenase**, with the release of CO_2. In the reaction catalyzed by pyruvate dehydrogenase, the acetate group is linked to a large carrier molecule called **coenzyme A**, to make acetyl-coenzyme A. The oxidation of pyruvate to acetyl-CoA also produces more NADH that will go on to make ATP via electron transport and the generation of the proton gradient.

$$Pyruvate + CoA + NAD^+ \rightarrow acetyl\text{-}CoA + CO_2 + NADH$$

Acetyl-CoA has one of two fates in the cell: it may go into the Krebs cycle to produce energy or it can be used in lipid biosynthesis. Although acetyl-CoA comes from glucose through glycolysis, there is no pathway for glucose to be regenerated from acetyl-CoA, making this an irreversible one-way reaction and a key regulatory point. If energy in the cell is abundant, as measured by the amount of ATP in the cell, then acetyl-CoA production and the Krebs cycle can slow down.

The Krebs cycle is a series of reactions that further oxidize the pyruvate from glycolysis and extract energy to make the high-energy electron carriers called NADH and $FADH_2$ (see figure). This pathway is also sometimes called the citric acid cycle or tricarboxylic acid cycle. Acetyl-CoA enters the Krebs cycle by combining with a four-carbon intermediate (oxaloacetate) to make the six-carbon citrate (or citric acid). In every complete loop of the Krebs cycle, two CO_2 molecules are produced and leave the cycle for every atom of acetyl-CoA that enters the cycle (see figure). This leaves the net level of intermediates constant and regenerates the four-carbon intermediate that combines with another acetyl-CoA in the next round of the cycle. Every cycle also produces three NADH, one $FADH_2$ and one ATP. The NADH and $FADH_2$ are both carriers of high-energy and play a role in the creation of ATP through the electron transport chain. Since two acetyl-CoAs are produced from every glucose, the net result of the Krebs Cycle is (for each glucose that enters glycolysis):

2 Acetyl-CoA + $6NAD^+$ + 2FAD + 2ADP + $2P_i$ + $4H20$ → $4CO_2$ + 6NADH + $2FADH_2$ + 2ATP + $4H^+$ + 2CoA

Aerobic Respiration

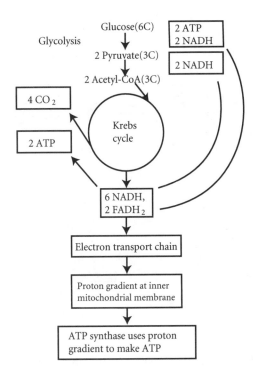

Electron Transport

Glycolysis directly produces two ATP per glucose, and the Krebs cycle directly produces two more ATP per glucose. However, most of the energy of glucose is not extracted directly into ATP. The high-energy electron carriers NADH and $FADH_2$ contain much more energy that is extracted by the oxidation of glucose in glycolysis and the Krebs cycle. Electron transport is the mechanism used to convert the energy held by these carriers into a more useful form.

The electron transport chain is a series of proteins and electron carriers located in the inner mitochondrial membrane. NADH and $FADH_2$ transfer the high-energy electrons they carry to the chain and are oxidized back to NAD^+ and $FADH^+$, respectively. The high-energy electrons that enter the electron transport chain are transferred from one carrier to another, moving through the system and transferring energy to the carriers along the way in a series of oxidation-reduction reactions. The carriers alternate between oxidized and reduced forms, and the ultimate electron acceptor at the end of the chain is oxygen, which is reduced to water. This oxygen is the oxygen needed for aerobic respiration. As the electrons flow through the electron transport chain, some of their energy is used at three steps in the electron transport pathway to pump H^+ ions out of the mitochondria. The pumping of these H^+ ions out of mitochondria creates a pH gradient across the inner mitochondrial membrane.

Oxidative Phosphorylation

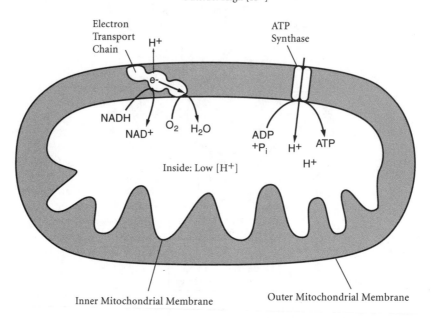

Outside: High [H^+]

Electron Transport Chain

ATP Synthase

H^+

e^-

NADH

NAD$^+$ O$_2$ H$_2$O

Inside: Low [H^+]

ADP +P$_i$

H^+ ATP

H^+

Inner Mitochondrial Membrane

Outer Mitochondrial Membrane

In glycolysis and the Krebs cycle, the energy stored in the chemical bonds of glucose is converted to form a few high-energy ATP bonds directly, and much more energy is converted to reduce the high-energy electron carriers NADH and FADH$_2$. In electron transport, the energy of NADH and FADH$_2$ is converted again, this time to the energy in the proton gradient between the inside and the outside of the mitochondria. Since these protons cannot diffuse through the membrane, this energy is stored as a concentration gradient and the cell can harness it to make ATP.

A protein called ATP synthase is found in the inner mitochondrial membrane and harvests the energy of the pH gradient to produce the bulk of the ATP from aerobic respiration (see figure). ATP synthase allows the protons on the outside of the mitochondria to flow down their concentration gradient back into the mitochondria. It does not allow them to freely diffuse through the membrane, however. It harnesses the energy of proton movement down the gradient to drive ATP production from ADP, making 3 ATP for every NADH that enters electron transport and 2 ATP for every FADH$_2$ that enters electron transport. The use of the proton gradient to generate ATP in mitochondria is called the **chemiosmotic principle**. In this way the thermodynamically favorable process of allowing protons to flow down a concentration gradient is used to drive the thermodynamically unfavorable reaction producing ATP.

One test of the chemiosmotic principle was to see if a pH gradient alone is sufficient for mitochondria to produce ATP. In an experiment to answer this question, mitochondria were isolated from the cell and deprived of pyruvate so that the Krebs cycle could generate reduced electron carriers (NADH and FADH$_2$). If the mitochondria had ADP and phosphate, but the pH was basic outside the mitochondria, no ATP would be produced. If acid was added to the solution containing mitochondria, the mitochondria immediately began to produce ATP from the ADP and phosphate. Also, mitochondria with broken membranes did not make ATP. This experiment and others demonstrated that the pH gradient in the mitochondria is an essential component of ATP production.

Uncoupling agents can make the inner mitochondrial membrane permeable to protons, so that protons can move down their gradient back into mitochondria without passing through the ATPase. When this happens, the proton gradient is destroyed and ATP production ceases. The lack of ATP stimulates the Krebs cycle and electron transport in these circumstances however. Earlier in this century, an uncoupling agent, 2,4 dinitrophenol, was used as a diet aid. By uncoupling the pH gradient, glucose is consumed but ATP is not produced, increasing the rate of glucose metabolism. Uncoupling the pH gradient from ATP production creates a futile cycle in which the energy of the pH gradient is released as heat rather than captured as ATP.

The efficiency of energy production by aerobic respiration can be calculated. Comparing the 2 ATP per glucose to the 38 generated in aerobic respiration, it is easy to see the much greater efficiency of energy extraction in aerobic respiration. A significant fraction of the energy of glucose oxidation is also lost to the environment as heat. The greater efficiency of energy extraction in aerobic respiration compared to fermentation means that less glucose (or other energy sources) must be acquired from the environment. An organism that utilizes fermentation only will need to consume 19 times more glucose than an organism that uses aerobic respiration.

The Krebs cycle, electron transport, production of the proton gradient, and ATP production are regulated together for coordinated energy production. If oxygen is removed, all of these processes stop, since they are all dependent on each other and regulated by each other. The concentration of ATP regulates these pathways, as a form of feedback inhibition. If the metabolic needs of the cell increase (low ATP), then all of these processes increase in rate. In this way the cell regulates ATP production and the associated pathways to meet its needs.

3.6 PHOTOSYNTHESIS

To survive, all organisms need energy. ATP is an energy intermediary used to drive biosynthesis and other processes. ATP is generated in mitochondria using the chemical energy of glucose and other nutrients. Where does the chemical energy of glucose come from? The energy foundation of almost all ecosystems is photosynthesis. Plants are autotrophs, or self-feeders, that generate their own chemical energy from the energy of the sun through photosynthesis. The chemical energy that plants get from the sun is used to produce glucose. This glucose can then be burned in plant mitochondria to make ATP, which is used to drive all of the energy-requiring processes in the plant, including the production of proteins, lipids, carbohydrates, and nucleic acids. Similarly, animals can eat plants to extract the energy for their own metabolic needs. In this way, photosynthesis is the energy foundation of most living systems.

Photosynthesis occurs in plants in the chloroplast, an organelle that is specific to plants. In prokaryotes that perform photosynthesis, there are no chloroplasts; instead, photosynthesis occurs in association with the plasma membrane or infoldings of the membrane. Chloroplasts are found mainly in the cells of the mesophyll, the green tissue in the interior of the leaf. The leaf contains pores on its surface called stomata that allow carbon dioxide in and oxygen out to facilitate photosynthesis in the leaf. The chloroplast has an inner and outer membrane; within the inner membrane there is a fluid called the **stroma**. In addition, the interior of the chloroplast contains a series of membranes called the **thylakoid** membranes that form stacks called **grana**. The structure of the chloroplast is similar in many ways to that of the mitochondrion, with an inner and outer membrane in which the inner membrane is impermeable to most ions.

Photosynthesis can be summarized by this equation:

$$6CO_2 + 12H_2O + uv \text{ light} \rightarrow C_6H_{12}O_6 + 6O_2 + 6H_2O$$

Photosynthesis involves the reduction of CO_2 to a carbohydrate. It can be characterized as the reverse of respiration, in that reduction occurs instead of oxidation. Photosythesis has two main parts, the **light reactions** and the **Calvin-Benson cycle** (also called the Calvin cycle). The light reactions occur in the interior of the thylakoid while the Calvin-Benson cycle occurs in the stroma.

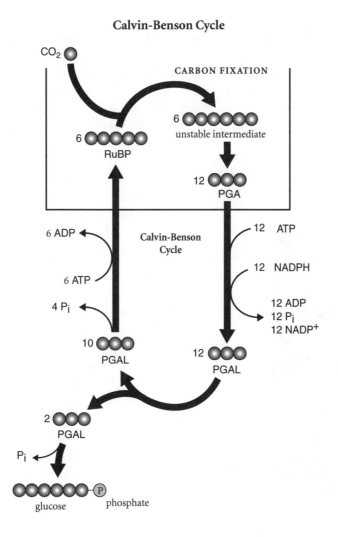

Calvin-Benson Cycle

Light Reactions

The first part of photosynthesis is made up of light reactions, in which light energy is used to generate ATP, oxygen, and the reducing molecule NADPH. The molecule that captures light energy to start photosynthesis is a pigment called chlorophyll in the thylakoid membranes of the chloroplast. Chlorophylls absorb most wavelengths of visible light, with the exception of green. Since chlorophyll does not absorb green, it reflects green light, making plants appear green. Chlorophyll is used by two complex systems called photosystems I and II in the thylakoid membrane. Each photosystem is a complex assembly of protein and pigments in the membrane. When photons strike chlorophyll, electrons are excited and transferred through the photosystems to a **reaction center**. When electrons reach the reaction center, the reaction center gives up excited electrons that enter an electron transport chain where they are used to generate chemical energy as either reduced NADPH or ATP.

Two different processes occur in the photosystems, **cyclic photophosphorylation** and **non-cyclic photophosphorylation**. Both are used to generate ATP, but in different ways. The ATP in turn is used to generate glucose in the dark reactions. Cyclic photophosphorylation occurs in photosystem I to produce ATP. In the cyclic method, electrons move from the reaction center, through an electron transport chain, then back to the same reaction center again (see figure). The reaction center in photosystem I includes a chlorophyll called P700 because its maximal light absorbance occurs at 700 nm. This process does not produce oxygen and does not produce NADPH.

Cyclic Photophosphorylation

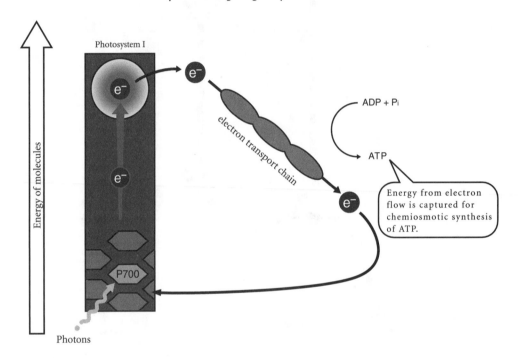

Non-cyclic photophosphorylation starts in photosystem II (see figure). In non-cyclic photophosphorylation, chlorophyll pigment absorbs light and passes excited electrons to a reaction center, a process equivalent to cyclic photophosphorylation. The photosystem II reaction center contains a P680 chlorophyll, distinct from photosystem I. From the photosystem II reaction center, the electrons are passed to an electron transport chain. In this case however the electrons are not returned to the reaction center at the end of the electron transport chain but are passed to photosystem I. Photosystem II replaces the electrons it lost by getting them from water, producing oxygen in the process. The electrons that enter photosystem I in this case are used to produce NADPH.

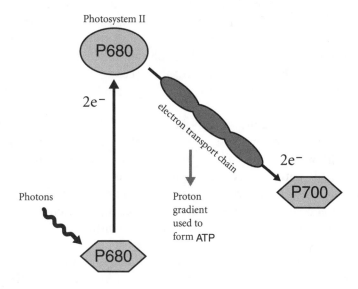

Noncyclic Photophosphorylation:
Photosynthesis Light Reactions

So far, we have not addressed the mechanism used to produce ATP during photosynthesis. As the electrons work their way through the electron transport chains, protons are pumped out of the stroma and into the interior of the thylakoid membranes, creating a proton gradient. This proton gradient generated by an electron transport chain is similar to the pH gradient created in mitochondria during aerobic respiration and is used in the same way to produce ATP. Protons flow down this gradient back out into the stroma through an ATP synthase to produce ATP, similar once again to mitochondria. The NADPH and ATP produced during the light reactions are used to complete photosynthesis in the Calvin cycle, using carbon from carbon dioxide to make sugars.

The oxygen produced in the light reactions is released from the plant as a byproduct of photosynthesis. Starting about 1.5 billion years ago, photosynthesis helped to create the oxygen rich atmosphere found on earth today, which allowed the evolution of animals requiring the efficient energy metabolism provided by aerobic respiration. The oxygen produced through photosynthesis today maintains Earth's oxygen and is a key to the continued functioning of the biosphere.

Calvin-Benson Cycle

The first portion of photosynthesis, in which ATP, oxygen and NADPH are produced is sometimes called the light reactions since light energy is required to energize electrons and drive the reactions forward. In the remaining part of photosynthesis, the energy captured in the light reactions as ATP and NADPH is used to drive carbohydrate synthesis. This cycle, also known as the **Calvin-Benson cycle**, fixes CO_2 into carbohydrates, reducing the fixed carbon to carbohydrates through the addition of electrons. The NADPH provides the reducing power for the reduction of CO_2 to carbohydrate, and air provides the carbon dioxide. CO_2 first combines with, or is fixed to, ribulose bisphosphate (RuBP), a five-carbon sugar with two phosphate groups. The enzyme that catalyzes this reaction is called **rubisco** and is the most abundant enzyme on earth. The resulting six-carbon compound is promptly split, resulting in the formation of two molecules of 3-phosphoglycerate, a three-carbon compound. The 3-phosphoglycerate is then phosphorylated by ATP and reduced by NADPH, which leads to the formation of phosphoglyceraldehyde (PGAL). This can then be utilized as a starting point for the synthesis of glucose, starch, proteins, and fats.

Rubisco is not completely specific for CO_2 however, and will also catalyze a reaction with oxygen and ribulose bisphosphate in a process called **photorespiration**. This reaction produces a two-carbon molecule called glycolate, that passes to peroxisomes where it is oxidized to CO_2. This pathway reverses some of the energy captured in photosynthesis, undoing carbon fixation without producing energy. It is not clear if there are any advantages to this, but the consequences can be predicted. By reversing a percentage of carbon fixation, photorespiration reduces the overall efficiency of photosynthesis.

Some plants have evolved changes in photosynthesis that bypass photorespiration to increase the overall efficiency of photosynthesis. This altered process is called C_4 **photosynthesis**. C_4 plants have an enzyme called PEP carboxylase in their leaves, which catalyzes carbon dioxide fixation with phosphoenolpyruvate. The product of this reaction is oxaloacetate, a 4-carbon molecule (thus the name C_4 photosynthesis). Since a different enzyme is used for carbon fixation and PEP carboxylase does not react with oxygen, there is minimal photorespiration in C_4 plants, making photosynthesis in C_4 plants more efficient.

C_4 plants perform the Calvin-Benson cycle as do C_3 plants. In C_4 plants, however, carbon fixation and the Calvin-Benson cycle occur in different cells. C_4 plants have different leaf structure than C_3 plants to accommodate the different biochemistry they use (see figure). **Mesophyll** cells in C_4 plants line the surface of the leaves and carry out carbon fixation to produce oxaloacetate. The oxaloacetate diffuses from the mesophyll cells into neighboring cells in the interior of the leaf called **bundle sheath cells**. In bundle sheath cells, CO_2 is removed from oxaloacetate, producing pyruvate as well as CO_2. The pyruvate is transported back to the mesophyll cells, and the CO_2 is used in the bundle sheath cells to enter the Calvin-Benson cycle. The carbon dioxide enters the Calvin-Benson cycle with catalysis by rubisco, as in C_3 plants, so that C_4 plants actually perform carbon fixation twice. The movement of oxaloacetate from mesophyll to bundle sheath cells tends to pump CO_2 into bundle sheath cells, maintaining the high CO_2 concentration required to avoid photorespiration.

C$_4$ Photosynthesis

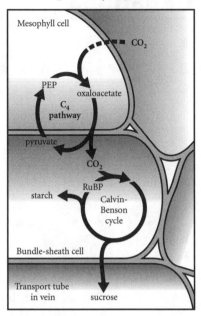

C$_4$ plants tend to live in dry, hot environments; examples of C$_4$ plants are corn and crab grass. The carbon fixation step catalyzed by rubisco does not occur in hot conditions with low CO$_2$. In hot environments, plants, close the stomata on their leaves and limit the supply of CO$_2$. In C$_3$ plants these conditions favor photorespiration, but C$_4$ plants carry on carbon fixation under these conditions with PEP carboxylase. C$_3$ plants are more efficient at cooler temperatures, however. In ecosystems in which C$_3$ and C$_4$ plants compete, the plant which dominates may depend on the temperature of the environment.

Another class of plants called **CAM** plants has also adapted to perform photosynthesis in hot, dry conditions. CAM plants keep their stomata closed during the day to conserve water, and like C$_4$ plants, keep up carbon fixation using PEP carboxylase. CAM plants, however, perform carbon fixation during the night when the stomata are open and store the CO$_2$ for use in the Calvin-Benson cycle. The products of carbon fixation are stored for use during the day.

Chapter 3: Review Questions

1. The product(s) of the light reactions of photosynthesis is/are

   (A) pyruvate
   (B) glucose
   (C) ATP and NADPH
   (D) CO_2 and H_2O
   (E) ribulose biphosphate (RuBP)

2. How many ATP are produced via the chemiosmotic principle for every molecule of NADH that transfers high-energy electrons to the electron transport chain?

   (A) 1
   (B) 2
   (C) 3
   (D) 4
   (E) 5

3. How many ATP are produced via the chemiosmotic principle for every molecule of $FADH_2$ that transfers high-energy electrons to the electron transport chain?

   (A) 1
   (B) 2
   (C) 3
   (D) 4
   (E) 5

4. Which of the following generates the most ATP?

   (A) fermentation
   (B) glycolysis
   (C) active transport
   (D) substrate-level phosphorylation
   (E) oxidative phosphorylation

5. Pyruvate is a product of

   (A) the electron transport chain
   (B) glycolysis
   (C) Krebs cycle
   (D) the Calvin cycle
   (E) oxidative phosphorylation

6. Oxygen is reduced to water in what stage of aerobic respiration?

   (A) glycolysis
   (B) Krebs cycle
   (C) electron transport chain
   (D) all of the above
   (E) none of the above

Questions 7–9 refer to the following:

   (A) stomata
   (B) cytoplasm
   (C) chlorophyll
   (D) thylakoid
   (E) stroma

7. site of the light reactions

8. regulates entrance of CO_2

9. site of the Calvin cycle

10. In which of the following is water split to form hydrogen ions, oxygen, and electrons?

   (A) photosystem II
   (B) photosystem I
   (C) cyclic photophosphorylation
   (D) the Calvin cycle
   (E) fermentation

Please see the answers and explanations to these review questions in Section V of this book.

HEREDITY AND EVOLUTION

Section III

One of the characteristics shared by all life is that living organisms reproduce themselves to produce offspring that resemble their parents. The fact that offspring inherit physical traits from parents has been intuitively obvious to humans for a long time, but the mechanism by which traits were inherited by offspring from parents has only become evident in recent history. Genes transmit traits and are encoded by the molecule called DNA, making understanding the structure and function of DNA essential to understanding life. The traits encoded in the DNA genome are passed from one generation to the next down through time, providing for the constancy of life. The species that inhabit the earth have changed dramatically over time and change is still evident today. Evolution is the driving force for the diversity of species in the past and today, and the ever-changing nature of life forms on the planet.

In this section, we will explore heredity, molecular genetics, and current theories of evolutionary development.

Heredity

How are new organisms created, and traits transferred from parents to offspring? In asexual reproduction, a new organism is produced through mitosis from cells of the parent organism, and all offspring are exact copies of the parent. This is the form of reproduction used by all prokaryotes and in some instances by eukaryotes. Sexual reproduction is more complicated and requires more energy, using genetic material from two parents to create offspring. The advantage of sexual reproduction, however, is the genetic variation it creates in a species, allowing for more adaptability to a changing and variable environment. Classical genetics describes the patterns of inheritance from one generation to another, and the study of meiotic cell division allows understanding of the underlying mechanisms behind the patterns of inheritance. Modes of reproduction and patterns of inheritance will be explored in this chapter.

4.1 EUKARYOTIC CHROMOSOMES

The genome consists of the DNA an organism possesses and encodes genes responsible for the characteristics of a species and variation of individuals within a species. In eukaryotes, the genome is split into several linear segments called chromosomes. Each species has a characteristic number of chromosomes. Humans have 23 different chromosomes, fruit flies have 4, and dogs have 39. Chromosomes that are not involved in sex determination are called **autosomes**, while chromosomes that are involved in sex determination are called **sex chromosomes**. Humans have 22 autosomes and one pair of sex chromosomes, the X and the Y chromosomes. Most cells of the body in humans (and other animals) are diploid, meaning that every chromosome is present in a pair, for a total of 46 chromosomes in humans. The presence of two copies of the genome in diploid organisms has important consequences for the way genes are expressed. Since there are two copies of the autosomal chromosomes in a diploid organism, there are also two copies of most genes. If there are two copies of each gene, then both copies together determine how the gene will be expressed in an organism. For example, a defective gene in a haploid cell or organism will be reflected as a defect in the organism. The defective gene must be expressed since there is no alternative. In a diploid organism, however, the second copy of the gene provides a backup, so that one copy can be defective and not affect the organism. The interaction of two copies of genes in diploid organisms is the basis of recessive and dominant traits in Mendelian genetics.

Sex determination in humans and other mammals involves the X and Y chromosomes. Human females have two X chromosomes (XX) and human males have one X and one Y chromosome (XY). Other species sometimes determine sex differently. In grasshoppers, for example, females have two X chromosomes (XX), while males have only one (XO). Humans are diploid for all 22 autosomal chromosomes, and females are diploid for the sex chromosome as well (two homologous X chromosomes), but males are not diploid for the sex chromosomes since they have only one X chromosome. The Y chromosome does not provide a second copy of the genetic information on the X chromosome. The X and Y chromosomes play important roles in inheritance patterns in humans. In cases where a gene is found on the X chromosome, females will be diploid for the gene, but males will not. Males with a defective gene on the X chromosome will always express the defect since they are haploid for genes on the X chromosome.

An individual's chromosomes can be observed during mitosis when they are highly condensed. To look for chromosomal abnormalities, it is possible to stain mitotic cells and take a picture of the chromosomes. Each individual chromosome can be separated in the photo and matched with its homologous chromosome based on the size and shape of each chromosome in the photograph. This visual examination of chromosomes is called a **karyotype**. The presence of a chromosome without a homolog may indicate that monosomy has occurred. In **monosomy**, only one copy of a chromosome is present.

Each eukaryotic chromosome consists of a single molecule of linear double-stranded DNA. Along its length each chromosome has thousands of genes, each of which codes for a gene product, usually a protein, although RNAs such as ribosomal RNA or transfer RNA are also encoded by genes. The genes and their products are responsible for the heritable traits observed in genetics. Different chromosomes contain different genes.

Chromosomes do not consist of DNA alone, but also of proteins. DNA has a large number of negative charges in its backbone from phosphate groups that tend to repel each other, giving DNA an extended conformation. A single DNA molecule is millions of base-pairs long, but must still fit within the narrow confines of the eukaryotic nucleus. Histones are proteins with a large number of positive charges provided by lysine and arginine amino acid residues. The positive charges in the histones neutralize the negative charges in the DNA backbone, allowing the DNA to coil tightly around the histone proteins. Histones form a sphere with about 200 base pairs of DNA wrapped twice around it in a structure called a nucleosome (see figure). Nucleosomes in turn are packaged in higher order structures in chromosomes, with strings of nucleosomes condensed together, and sections of DNA forming loop-shaped domains. During mitosis, additional folding results in the highly condensed state of chromosomes that can be observed under the microscope.

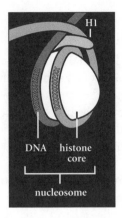

DNA is wrapped around histone proteins, forming structures called nucleosomes.

The central region of each chromosome contains a **centromere**. The centromere is involved in separation of chromosomes during meiosis and mitosis. Spindle fibers attach to a structure called the **kinetochore** to draw chromosomes apart using motor proteins. The ends of the chromosomes are called **telomeres**, and are largely repetitive sequences that cap the ends of the DNA molecules to prevent damage to the chromosomes.

The bulk of the genetic information in eukaryotes is carried in the chromosomes as described above, but exceptions to this rule include the genetic information carried in mitochondria and chloroplasts. These organelles contain their own unique independent genome, distinct from the nuclear genome. The genome of these organelles is small, circular, not packaged with histones, and resembles the genome of prokaryotes more than the nuclear genome of eukaryotes. Mitochondria replicate within the eukaryotic cell by a mechanism resembling binary fission, and replicate independently of replication of the nuclear genome in meiosis and mitosis. Changes in the genome of these organelles can affect the traits of the organism.

4.2   ASEXUAL REPRODUCTION

Asexual reproduction is any method of producing new organisms in which fusion of nuclei from two individuals (fertilization) does not take place. In asexual reproduction, only one parent organism is involved in reproduction, and offspring are genetically identical to their parents. Asexual reproduction serves primarily as a mechanism for perpetuating less complex organisms and plants, especially in times of low population density. Asexual reproduction can allow more rapid population growth than sexual reproduction, but does not create genetic diversity as does sexual reproduction. The benefits of asexual reproduction can include avoiding the energy expenditure associated with finding a mate. The only genetic variation found in a population that reproduces asexually is the variation introduced through mutation or through exchange of genetic information unrelated to reproduction, such as changes in the genome introduced by a virus.

Asexual Reproduction in Primitive Organisms

Binary fission is the mechanism of asexual reproduction in prokaryotes in which a single prokaryotic cell doubles its size, replicates its genome, then splits in two, with each daughter cell receiving one copy of the genome. This type of reproduction can occur very rapidly, since prokaryotic cells and genomes are small and the process is much less complicated than mitosis. Under optimal growth conditions, bacteria can replicate as quickly as one generation every twenty minutes through binary fission.

In some cases, eukaryotes perform asexual reproduction. **Budding** is a form of asexual mitotic reproduction that occurs in yeasts, a single-cell eukaryotic fungus that involves an unequal division of cytoplasm (cytokinesis) and equal division of the nucleus (karyokinesis). The parent cell forms a smaller daughter cell with less cytoplasm than the parent. Eventually, the daughter organism becomes independent and is released. Although budding is common in unicellular organisms like yeast, it also occurs in some multicellular organisms such as hydra, forming small identical copies of the whole parent organism. If a sponge is broken into small pieces, each of the small pieces may develop into a small complete sponge.

Asexual Reproduction in Animals

Asexual reproduction is not common in animals, although it does bring with it certain benefits. It is suitable for animal populations that are widely dispersed, as animals who practice asexual reproduction do not need to find a partner. The two major types of asexual reproduction found in animals are **parthenogenesis** and **regeneration**. In parthenogenesis, an egg develops in the absence of fertilization by sperm. This form of reproduction occurs in bees and ants: unfertilized eggs develop into haploid females, and fertilized eggs develop into diploid males. Regeneration is the ability of certain animals to regrow a missing body part. Sometimes part of an animal grows into a complete animal. For example, planaria (flatworms), earthworms, lobsters, and sea stars can all regenerate limbs or entire organisms. This process is similar to vegetative propagation in plants.

Asexual Reproduction in Plants

Many plants utilize asexual reproduction to increase their numbers. Undifferentiated tissue (meristem) in plants provides a source of cells from which new plants can develop. Vegetative propagation offers a number of advantages to plants, including speed of reproduction. The advantages to the farmer are lack of genetic variation and the ability to produce seedless fruit. This process can occur either naturally or artificially.

Natural forms of vegetative propagation include:

Bulbs. Bulbs are short vertical stems under the ground with thick modified leaves that store food. Examples include tulips and onions.

Tubers. These modified underground stems have buds, such as the eye of a potato, which develop into new plants through mitotic cell division.

Stolons. Stolons, or runners, are plant stems that run above and along the ground, extending from the main stem. Near the main plant, new plants develop and produce new roots, as well as upright stems at intervals (as in lawn grasses).

Rhizomes. Rhizomes are underground stems that move horizontally through the earth to give rise to new plants. They reproduce through new upright stems that appear at intervals, eventually growing into independent plants. The iris is a rhizome.

Artificial forms of vegetative propagation include:

Cutting. When cut, a piece of stem of some plants will develop new roots in water or moist ground. Examples include the geranium and the willow. Plant growth hormones like auxins accelerate root formation in cuttings.

Layering. The stems of certain plants, when bent into the ground and covered by soil, will take root. The connection between the main plant and this offshoot can then be cut, resulting in the establishment of a new plant. Blackberry and raspberry bushes reproduce in this manner.

Grafting. Desirable types of plants can be developed and propagated using this method, in which the stem of one plant (scion) is attached to the rooted stem of another closely related plant (stock). One condition for successful grafting is that the cambium (the tissue in stem that is not differentiated and allows stems to grow thicker) of the scion must be in contact with the cambium of the stock, since these two masses of undifferentiated cells must grow together to make one. Grafting does not allow for any mixing of hereditary characteristics, since the two parts of the grafted plant remain genetically the same.

4.3   SEXUAL REPRODUCTION

Sexual reproduction involves the union of two haploid cells, one from each parent, to produce diploid offspring with two different copies of each chromosome. Sexual reproduction is the method of reproduction used by most complex, multicellular organisms. Utilizing this method enables such organisms to ensure genetic diversity and variability in their offspring through the introduction of new combinations of alleles. Variability can allow a species to adapt and survive in a changing environment.

The haploid cells that are involved directly in producing new organisms in sexual reproduction are called gametes. When the male gamete (the sperm) and the female gamete (the egg) join, a zygote is formed that develops into a new organism that is genetically distinct from both its parents. The diversity created by sexual reproduction occurs in part during gamete production and in part through the random matching of gametes to make unique individuals. Meiosis is the form of cell division that produces the haploid gametes from diploid cells. The mechanisms that generate diversity during sexual reproduction become clear with study of the key steps in meioisis.

4.4 MEIOSIS

In sexual reproduction, two parents contribute to the genome of the offspring and the end result is genetically unique offspring. To do this requires that each parent contribute a cell with one copy of each chromosome to create a new cell with two copies of each chromosome that will develop into a new organism. **Meiosis** (see figure on pages 104–105) is the process whereby these sex cells, gametes, are produced. To produce haploid gametes, the number of chromosomes must be reduced in half from the diploid number. For this reason meiosis is sometimes called **reductive cell division**. The goals that must be accomplished in meiosis are to produce haploid gametes from diploid cells, to produce diversity in the gametes and the organisms they produce, and to ensure a complete copy of the genome in all of the gametes produced.

Many of the same structures and processes occur in meiosis and mitosis. In meiosis, however, DNA replication is followed by two rounds of cell division rather than one. The first round of division (meiosis I) produces two intermediate daughter cells. The second round of division (meiosis II) results in four genetically distinct haploid gametes. In this way, a diploid cell produces haploid daughter cells.

Each meiotic division has the same four stages as mitosis, although it goes through each of them twice (except for DNA replication): in Meiosis I, the first meiotic cell division, and Meiosis II, the second meiotic cell division. The stages of meiosis are detailed in the following paragraphs.

Interphase I

In the first meiotic interphase, the genome of the diploid gamete precursor cell is replicated in a manner similar to S phase in the mitotic cell cycle. In addition, the cell growth required for cell division also occurs during this phase of the meiotic cell cycle.

Prophase I

During this stage, chromatin condenses into chromosomes, the spindle apparatus forms, and the nucleoli and nuclear membrane disappear. Homologous chromosomes (matching chromosomes that code for the same traits, one inherited from each parent), come together and intertwine in a process called synapsis. Since at this stage each chromosome consists of two sister chromatids, each synaptic pair of homologous chromosomes contains four copies of the genome, and is therefore often called a tetrad.

Meiotic Recombination

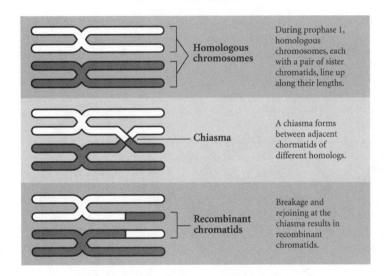

Homologous chromosomes — During prophase 1, homologous chromosomes, each with a pair of sister chromatids, line up along their lengths.

Chiasma — A chiasma forms between adjacent chromatids of different homologs.

Recombinant chromatids — Breakage and rejoining at the chiasma results in recombinant chromatids.

During prophase I homologous chromosomes break at corresponding points and exchange equivalent pieces of DNA in a process called crossing over or **meiotic recombination** (see figure above). Note that meiotic recombination occurs between homologous chromosomes and not between the identical sister chromatids of the same chromosomes that are produced during DNA replication prior to meiosis I. The result of meiotic recombination is that genes can be exchanged between homologous chromosomes. Meiotic recombination would not make much difference if the two copies of the gene that were exchanged were identical. However, individuals often have two slightly different copies a gene called alleles, having received different copies from each parent. Since there are many genes laid out along the length of a chromosome, recombination can change the combinations of alleles found on a given chromosome, creating new combinations of alleles that were not found in either parent organism. The chromosomes are joined at points called **chiasmata** where the crossing over occurs. Recombination among chromosomes is one mechanism in meiosis that results in increased genetic diversity within a species. The great expenditure of energy required to create diversity through meiosis provides evidence of the strong evolutionary pressure favoring diversity in a species.

MEIOSIS I

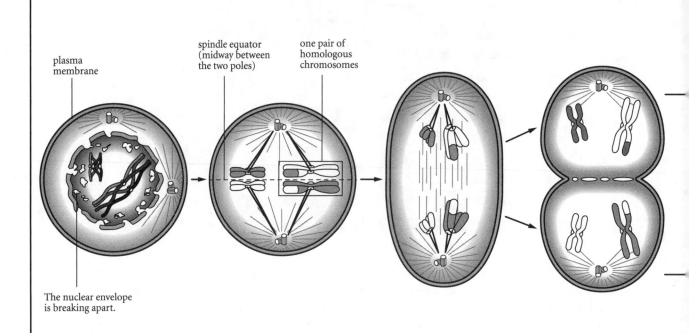

plasma membrane

spindle equator (midway between the two poles)

one pair of homologous chromosomes

The nuclear envelope is breaking apart.

| PROPHASE I | METAPHASE I | ANAPHASE I | TELOPHASE I |

MEIOSIS II

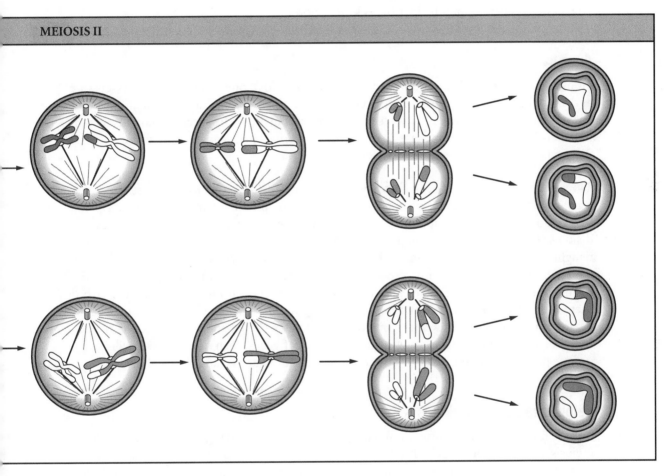

PROPHASE II   METAPHASE II   ANAPHASE II   TELOPHASE II

Prophase I can be a very lengthy process due in part to the complexity of recombination. In addition, during the formation of female gametes (ova) during human reproduction, the cells are paused in prophase I for years, until an ovum matures later in life.

Metaphase I

Homologous pairs (tetrads) align at the equatorial plane of the dividing cells, and each pair attaches to a separate spindle fiber by its kinetochore.

Anaphase I

Homologous pairs of chromosomes (not sister chromatids), one from each parent, separate and are pulled to opposite poles of the cell by spindle fibers. Each daughter cell at this point has one chromosome that each contains two sister chromatids. The sister chromatids remain together during meiotic anaphase I, and separate during the second round of meiotic cell division.

The separation of homologous chromosomes during meiosis is called disjunction, and it accounts for a Mendelian law of genetics, the **law of independent assortment**. During disjunction in meiotic anaphase I, each chromosome of paternal origin separates (or disjoins) from its homolog of maternal origin, and either chromosome can end up in either daughter cell. The distribution of homologous chromosomes to the two intermediate daughter cells is random with respect to parental origin. The daughter cells, and at the completion of meiosis, the gametes, will have a complete set of chromosomes, but in each cell there will be different combinations of homologous chromosomes due to independent assortment of chromosomes during meiosis. Since the maternal and paternal copies of each homologous chromosome can have different alleles of genes, each daughter cell will have a unique pool of alleles provided by a random mixture of maternal and paternal origin.

Telophase I and Cytokinesis

A nuclear membrane may form around each new nucleus of the daughter cells resulting from meiosis I. At this point, each chromosome still consists of sister chromatids joined at the centromere. Next, the cell divides through cytokinesis into two daughter cells, each of which receives a nucleus containing the haploid number of chromosomes. Between cell divisions there may be a short rest period, or interkinesis, during which the chromosomes partially uncoil.

In general, meiosis II is similar to mitosis, except that it is not proceeded by a round of DNA replication and results in haploid cells rather than diploid cells as happens in mitosis in a diploid organism.

Prophase II

The centrioles migrate to opposite poles and the spindle apparatus forms.

Metaphase II

The chromosomes line up along the equatorial plane once again. The centromeres divide, separating the chromosomes into pairs of sister chromatids.

Anaphase II

The sister chromatids are pulled to opposite poles by the spindle fibers.

Telophase II

A nuclear membrane forms around each new haploid nucleus. Cytokinesis follows, and two daughter cells are formed. Thus, by the time meiosis is completed, four haploid daughter cells are produced from one diploid cell.

The random distribution of homologous chromosomes in meiosis, coupled with crossing over in prophase I, enables individuals to produce gametes with almost limitless genetic combinations of alleles. Every gamete gets one copy of each chromosome, but the copy of each chromosome found in a gamete is random. For example, each gamete has a chromosome #9, but this chromosome can be either of the two copies of this chromosome from an individual (the maternal and paternal copies). With 23 chromosome pairs, there are 2^{23} possible combinations, not including the additional genetic diversity created by recombination. If there are 1000 genes at which the maternal and paternal homologous chromosomes differ, then there are 2^{1000} possible combinations of alleles after recombination. This is why sexual reproduction produces genetic variability in offspring, as opposed to asexual reproduction, which produces identical offspring. The possibility of so many different genetic combinations increases the capability of a species to evolve and adapt to a changing environment.

4.5   LIFE CYCLES IN SEXUAL REPRODUCTION

Animals generally begin life as a diploid single-celled zygote formed as the result of fusion between two haploid gamete cells. The zygote develops through mitotic cell division into the multicellular adult organism, which then produces haploid gametes through meiosis. Fusion of gametes from two parents produces the next generation of zygotes and adult organisms. In the animal life cycle, the diploid adult form occupies most of the life cycle of the organism and the haploid cell, the gamete, is merely a transient stage from one diploid generation to the next (see figure).

General plant life cycle

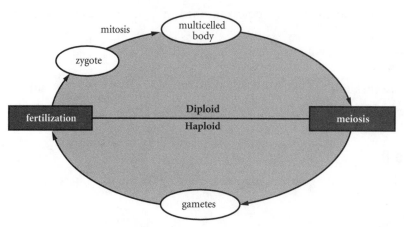

General animal life cycle

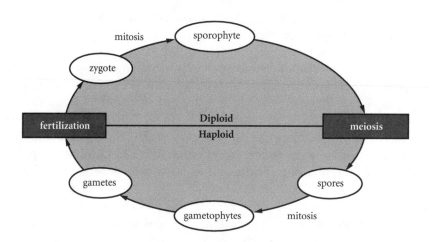

Plants have a very different life cycle, although it still includes haploid and diploid forms. In plants the haploid form called the **gametophyte** forms, and is responsible for the production of haploid gametes (see figure). The fertilization of gametes in plants creates the other component of the plant life cycle, the zygote that develops into a **sporophyte**. Sporophytes are diploid and through meiosis form spores, haploid cells that are resistant to environmental extremes. When conditions are correct, spores will grow through mitosis into gametophytes to complete the cycle and create the next generation of gametes.

4.6   HUMAN GAMETE FORMATION

In humans, and other animals, specialized organs called gonads produce gametes, functional sex cells. Male gonads, called testes, produce sperm in the seminiferous tubules through the process called spermatogenesis, while female gonads, called ovaries, produce eggs through oogenesis. The details of these processes will be presented later, but they will be briefly described here as a prelude to the patterns of genetic inheritance.

Spermatogenesis is the development of sperm. A cell that is committed to gamete formation is called a **germ cell**, while a cell that is part of the body but does not contribute its genome to gamete formation is called a **somatic cell**. Germ cells are diploid and reproduce through mitosis. In spermatogenesis, germ cells divide mitotically to produce **spermatogonia**, still diploid precursors (see figure). Spermatogonia divide mitotically to produce **primary spermatocytes**, the cells that enter meiosis. After DNA replication, primary spermatocytes enter meiosis I and go through recombination, then separate homologous chromosomes to produce **secondary spermatocytes** with two copies of the genome in sister chromatids. Secondary spermatocytes enter meiosis II to separate sister chromatids and produce haploid spermatids with a single copy of the genome. Spermatids must go through a maturation process after they complete meiosis before they are fully developed. Sperm production occurs throughout the reproductive life of human males, and in spermatogenesis meiosis of a cell proceeds without a lengthy pause, as occurs in oogenesis.

Spermatogenesis

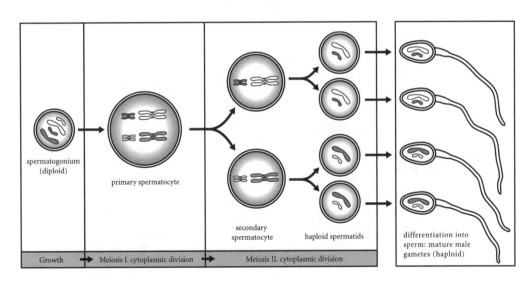

The testes of humans are located outside the abdominal cavity because they must remain cooler than the rest of the body to ensure proper development of sperm. The sperm itself has a head that contains DNA and a tail for motility; each sperm contains a single haploid copy of the genome. In humans, sperm determine the sex of the offspring by contributing either an X or Y chromosome. A specialized sac at the tip of the sperm called the acrosome contains enzymes that allow the sperm to break through the protective layers around the egg.

The egg develops in a discontinuous process called **oogenesis**. Oogenesis begins during embryonic development, with the production of the female germ cells, oogonia. In humans, the proliferation of oogonia cells stops before birth, in contrast to the production of spermatogonia throughout male adult life. When the oogonia stop proliferating, they form **primary oocytes** that enter meiosis. Primary oocytes in humans enter meiotic prophase I during development and then are arrested in this state for very long periods of time. The discontinuous nature of oogenesis with a lengthy pause in prophase I is also in contrast to spermatogenesis, in which cells progress steadily through meiosis. In adult women, one primary oocyte each month completes meiosis I in response to the hormones of the menstrual cycle. When the oocyte completes meiosis I, the cytoplasm of the cell is not divided equally. One daughter cell, now called the **secondary oocyte**, receives virtually all of the cytoplasm and will proceed to produce the viable ova. The other daughter cell, the first **polar body**, is considerably smaller and degenerates. In humans, the secondary oocyte progresses into meiosis II, but is paused again. In humans, the secondary oocyte does not complete meiosis until after fertilization. At fertilization, the secondary oocyte completes meiosis, goes through another unequal division of cytoplasm, forms the second polar body, and the haploid **ootid** that develops into the ova. In humans, the male nuclei from the sperm and the female nuclei from the ova fuse at this time, producing a diploid zygote.

Oogenesis

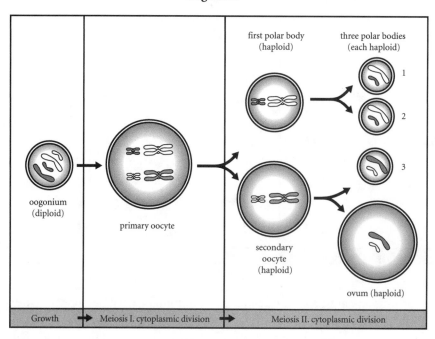

Ova contain only X chromosomes since the mother has only X chromosomes to contribute to gametes during meiosis. One salient difference between oogenesis and spermatogenesis is that females are born with all the eggs they will ever have, while males produce fresh sperm daily. This is the reason that genetic anomalies are more common in the eggs of older women; these anomalies have had years to accumulate while sperm have a short life span.

4.7 MEIOTIC ERRORS

Meiosis is an intricate process. For normal gametes to be produced, homologous chromosomes must progress through meiosis I and separate correctly. Similarly, sister chromatids must separate correctly during meiosis II for normal gametes to be formed. Failure of either homologous chromosomes or sister chromatids to separate correctly is called **nondisjunction**. The result of nondisjunction is a gamete that has either too many copies of a chromosome or that lacks a copy of one chromosome. If a gamete resulting from non-disjunction is involved in fertilization to produce a diploid zygote, the zygote will also have either one too many or one too few chromosomes.

Having either an extra copy or one too few copies of the chromosome is a major change in the genome and has profound affects on the individual. **Trisomy** (three copies of a chromosome) or **monosomy** (one copy of a chromosome) usually result in embryos that do not complete development. In humans, however, some do survive, although with significant abnormalities in development. Trisomy of chromosome 21 does create a viable embryo but causes Down's syndrome, with underdeveloped intelligence, changes in the heart, and other health problems. Changes in the sex chromosomes are often tolerated as well, resulting in Turner syndrome, for example. Turner syndrome occurs in individuals who have only one X chromosome, what is termed XO. These individuals are fairly normal in appearance, phenotypically females, although they are sterile. XXY individuals, with an extra X chromosome, have Klinefelter syndrome, and appear as males, although they are sterile.

4.8 INHERITANCE PATTERNS: MENDELIAN GENETICS

Around 1865, based on his observations of seven characteristics of the garden pea, Gregor Mendel developed the basic principles of genetics. Mendel first described traits in pea plants in pairs such as wrinkled seed and smooth seed. He then mated combinations of these plants and examined the offspring to determine the quantitative nature of inheritance of these traits and arrived at the principles of genetic inheritance that still form the foundation of genetics: dominance, segregation, and independent assortment. Although Mendel formulated these principles, he was unable to propose any mechanism for hereditary patterns, since he knew nothing about chromosomes, genes or meiosis. His work was largely ignored until the early 1900s, when new researchers investigating inheritance came to the same conclusion as Mendel: that traits must be carried by two copies of the genetic information that separate during gamete formation, confirming the power of the method Mendel used to examine the nature of inheritance.

Some of the basic rules of gene transmission and expression are:

1. Genes are elements of DNA that are responsible for observed traits.

2. In eukaryotes, genes are found in large linear chromosomes, and each chromosome is a very long continuous DNA double helix. Humans have twenty-three different chromosomes, with two copies of each chromosome in most cells.

3. Each chromosome contains a specific sequence of genes arranged along its length.

4. Each gene has a specific location on a chromosome (locus).

5. Diploid organisms have two copies of each chromosome and therefore two copies of each gene (except for the X and Y chromosomes in males).

6. The two copies of each gene can have a different sequence in an organism and a gene can have several different sequences in a population. These different versions of a gene are called **alleles.**

7. The type of alleles of an organism has, its genetic composition, is called the **genotype.**

8. The appearance and physical expression of genes in an organism is called the **phenotype.**

9. Types of alleles include **dominant** and **recessive** alleles. A dominant allele expresses its phenotypic effect in an organism regardless of the second allele present. A recessive allele will not express its phenotypic effect in a diploid organism if the other allele for the gene is a dominant one.

10. A **homozygous** individual has two copies (two alleles) of a gene that are identical and a **heterozygous** individual has two different alleles for a gene.

11. The phenotype of an individual is determined by the genotype.

Dominance of Phenotypic Traits

A **pure-breeding** organism is one in which related animals that are bred to each other always produce offspring with the same traits as the parents. For example, pure-breeding pea plants that produce wrinkled peas always produce offspring with wrinkled peas if they are crossed (mated) with each other. Pure-breeding organisms are homozygous for alleles at the trait being observed, as well as other alleles after many generations of selective breeding. If two members of a pure-breeding strain are mated, their offspring will always have the same phenotype as the parents since they are all homozygous for the same allele.

What happens if two different pure-breeding strains that are homozygous for two different alleles are crossed? In an example such as two different alleles for flower color, what often occurs is that all of the offspring of the cross match the phenotype of one parent and not the other. For example, if a pure-breeding red flower strain is crossed with a pure-breeding white flower, the result will be offspring that are all red. Where did the allele coding for the white trait go? Did the allele for the white trait disappear from the offspring?

If it is true that both parents contribute one copy of a gene to each of their offspring, then the allele cannot disappear. Offspring receive one copy of each gene from each parent, so they must all be heterozygous, with a white allele from one parent and a red allele from the other parent. Despite having both alleles, however, they only express one: the red allele. Red in this example is a **dominant** allele since one copy of this allele is always expressed phenotypically in the heterozygotes, regardless of the identity of the second copy of the gene. White in this example is a **recessive** allele of the flower color gene, since it is not expressed in the heterozygous offspring in this cross.

Dominant and recessive alleles behave in the same manner in humans. Humans are diploid organisms with two copies of each of their 23 chromosomes (with the exception of the X and Y chromosome in men). Each gene is present in two copies that can either be the same, or different. For example, a gene for eye color could have two alleles: B or b. B is a dominant allele for brown eye color and b is a recessive allele for blue eye color. There are three potential genotypes: BB, Bb, or bb. BB individuals and Bb individuals have brown eyes, and only bb individuals have blue eyes. Bb individuals have brown eyes since the B allele is dominant and the recessive b allele is not expressed phenotypically in the heterozygote.

The description of dominant and recessive traits was one of Mendel's contributions to genetics, and led him to one of his main conclusions about the nature of the genes responsible for traits: that they are present in two independent copies in diploid organisms. If the two copies are independent, then the recessive copy is still present in the heterozygote even if the allele is not expressed phenotypically. The recessive trait can be expressed again in a later generation. An alternative explanation for the behavior of traits would be that the mixing of red and white flowers is like mixing paints and results in "blending" of traits that cannot be separated again once mixed. To distinguish between these, Mendel performed countless test crosses between pea plants to examine the interactions of recessive and dominant alleles.

Test Crosses

Often, a geneticist will study the transmission of a trait in a species such as flies or pea plants by performing crosses (matings) between organisms with defined traits. For example, an investigator may identify two possible phenotypes for flower color in pea plants: pink and white. Pink plants bred together always produce pink offspring and white plants bred together always produce offspring with white flowers. It is likely that the differences in flower color are caused by different alleles in a gene that controls flower color. Which of these traits is determined by a recessive or dominant allele, however? One cannot tell based on the color alone which trait will be dominant or recessive. Either pink or white could be dominant, or neither.

The way to determine if an allele is dominant or recessive is by performing a **test cross**. A test cross is a controlled mating procedure between two organisms to characterize the behavior and identity of the alleles that an organism carries. In the above example, a test cross could be used to determine whether the pink allele or the white allele is dominant. Since the pink plants always produce pink offspring when they are bred together, the pink strain is termed "pure-breeding" and is homozygous for the P allele (PP genotype has a pink phenotype). Similarly, white plants are pure-breeding and homozygous for the p allele (pp genotype has a white phenotype). Is P or p dominant? What will be the phenotype of a plant with the Pp genotype? To answer these questions, a geneticist can perform a series of crosses.

When performing a test cross, a useful tool to analyze the results is a **Punnett Square**. To construct a Punnett Square, first determine the possible gametes each parent in the cross can produce. In the example above, a PP parent can make gametes with either of the two P alleles and the pp parent can only make gametes with the p allele:

PP parent: Gametes have either one P allele or the other P allele.

Punnett Square

	Pp x pp Parent 1	
	P	p
Parent 2 p	Pp pink	Pp pink
p	pp white	pp white

50% white offspring,
so Parent 1 is heterozygous

pp parent: Gametes have either one p allele or the other p allele.

The next step is to examine all of the ways that these gametes could combine if these two parents were mated together in a test cross. This is where the Punnett Square comes in. On one side of the square, align the gametes from one parent, and on the other side of the square align the gametes from the other parent. At the intersection of each potential gamete pairing, fill in the square with the diploid zygote produced by matching the alleles. In this example, all of the offspring of this cross are going to be heterozygous.

In a cross between these two pure-breeding strains, pink and white, if all of the offspring are pink, what does this reveal about the nature of these alleles? The offspring of this cross can be called the F_1 generation. Using a Punnett Square to predict the genotypes of the F_1 offspring:

F_1 Punnett Square

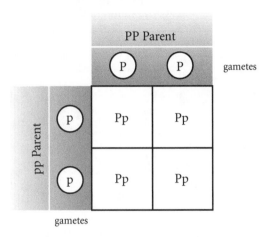

The F_1 offspring all have the Pp genotype and the pink phenotype. If the heterozygous Pp plant has the same phenotype as the homozygous PP plant, the p allele is not expressed and the P allele is dominant over the p allele. What will occur if two of these F_1 plants are crossed? A Punnett Square can be used again to predict the genotypes in the F_2 generation.

Parent 1: P and p gametes are produced.

Parent 2: P and p gametes are produced.

F_2 Generation Punnett Square:

F_2 Punnett Square

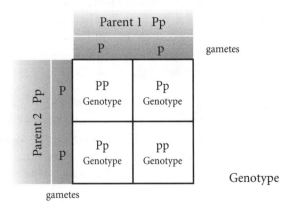

Since we know that the P allele for pink is dominant, we can use the genotypes to predict phenotypes of the F_2 generation. PP homozygotes will be pink, and Pp heterozygotes will also be pink since P is dominant. pp plants will be white like the original pure-breeding white plants. Filling in the square above with these phenotypes:

**F_2 Punnett Square
with Phenotypes**

	P	p
P	PP pink	Pp pink
P	Pp pink	pp white

The ratios of the different genotypes and phenotypes in the Punnett Square should match the statistical probability of producing these in real life by a cross of this type. For example, if two heterozygous Pp plants are crossed, 75% of the offspring will be pink and 25% white. This is predicted from the Punnett Square based on the ratio of 3:1 for phenotypes that will produce pink (3) to white (1). In real life, the actual numbers observed in a test cross are almost never exactly the predicted ratios due to statistical uncertainty, but the larger the number of offspring analyzed, the closer the results observed will approximate the predicted values. The large number of offspring produced by organisms like pea plants or fruit flies are one reason why they are good to use in genetic studies.

Another form of test cross can be used to determine the genotype of an organism. For example, a pink pea plant can have either the homozygous PP genotype or the heterozygous Pp genotype. Since P is dominant over p, it is not possible to tell from the phenotype of the plant if it is PP or Pp. A test cross can distinguish between these plants however. In this test cross the organism in question would be crossed against a pure-breeding recessive plant like a white plant with a pp genotype. If the plant in question is PP and is crossed with the pp plant, 100% of the offspring from the test cross will have the Pp genotype and the pink phenotype. If the plant is heterozygous Pp, however, then from a Punnett Square, the offspring can be predicted to be 50% pink and 50% white.

The fact that traits in pure breeding strains could disappear in the F_1 generation and then reappear in the next generation in test crosses indicated to Mendel with his pea plants that traits are not blended, but merely hidden. The ratios Mendel observed in test crosses indicated to him that traits are determined by two units that separate independently between generations. He referred to this phenomena as the law of **independent segregation**.

Law of Segregation

By examining the ratios of phenotypes in test cross offspring, Mendel deduced that there must be two units that determine inherited traits. For each of the seven traits in pea plants he observed, the same 3:1 ratios of dominant and recessive phenotypes were observed in F_2 test cross offspring, leading him to this conclusion. Mendel was unaware of meiosis or chromosomes, but his prediction based solely on analysis of trait behavior in crosses exactly matches the behavior predicted based on the mechanisms of meiosis. The two independent "units" of inheritance that Mendel hypothesized were responsible for each trait are the two alleles present for a gene in a diploid organism. During meiosis the homologous chromosomes that contain the two alleles for a gene are separated into different cells, with one copy of a given gene in each gamete. For example, when a heterozygous Pp plant is forming gametes, the P and the p alleles can separate into different gametes and act independently during a cross.

Law of Independent Assortment

Mendel performed not only crosses involving one trait, but more complicated test crosses involving more than one trait. For example, one of the traits that Mendel studied was for seed shape with a dominant round allele and a recessive wrinkly phenotype. A second trait that Mendel studied was plant height, with a dominant tall allele and a recessive allele that led to short plants. Either of these traits produced the expected ratios of phenotypes in offspring according to the law of independent segregation. A heterozygous round seed plant crossed with another heterozygous round seed plant would always produce a 3:1 ratio of round:wrinkled phenotypes and similarly a cross between heterozygous tall plants would produce a 3:1 ratio of tall:short plants. What happens however if animals that differ at both of these genes are crossed?

Such a cross between animals that vary at two different genes is called a **dihybrid** cross. The **Law of Independent Assortment** describes the relation between different genes in a dihybrid cross. If the gene that determines plant height is on a different chromosome than the gene for seed shape, then the alleles for these traits will separate into gametes independently of each other during meiosis. This independent behavior of alleles for two different genes during meiosis is reflected in the ratios of offspring observed in dihybrid test crosses. For example, if two heterozygous tall plants with round seeds are crossed, the offspring will be 3 tall: 1 short and 3 round: 1 wrinkled. The exact distribution of both traits in the offspring can be predicted from a Punnett Square:

Dihybrid Cross

		Parent 1, Tall-Round seed Tt Rr			
		TR	Tr	tR	tr	Gametes
	tr	Tt Rr	Tt rr	tt Rr	tt rr
Parent 2, Tall-Round seed Tt Rr	tR	Tt RR	tT Rr	tt RR	tt Rr
	Tr	TT Rr	TT rr	Tt Rr	Tt rr
	TR	TT RR	TT Rr	Tt RR	Tt Rr

Gametes

The simplest approach to an independent assortment problem is to consider each of the genes separately, determine the predicted Mendelian ratios for each of the traits alone, and then use the laws of probability to find the probability of each trait combined. The probability that two events will both happen can be calculated by multiplying the probability of the first event times the probability of the second event. For example, in the cross above, the predicted Mendelian phenotype ratios are 1/4 of plants have wrinkled seeds and 1/4 of plants are short. The probability that a plant will be both wrinkled and short is $1/4 \times 1/4 = 1/16$, which is the ratio predicted in the Punnett Square above, with plants that have the homozygous ttrr genotype. Similarly, if the odds of a plant having round seeds is 3/4 and the odds of a plant being tall in the cross are also 3/4, then the odds of a plant being both tall and having round seeds can be obtained by multiplying $3/4 \times 3/4 = 9/16$, which again matches the phenotypes predicted from the genotypes in the Punnett Square above. In this type of dihybrid cross, the predicted and observed phenotypes are in the ratios 9 tall-round: 3 tall-wrinkled: 3 short-round: 1 short-wrinkled. If a significant deviation from these ratios is observed in a dihybrid test cross, this indicates that independent assortment has not been observed as a result of genes being located near each other on the same chromosome.

Linkage

There are about 100,000 genes in the human genome spread across 23 pairs of chromosomes. Each chromosome therefore contains many thousands of genes. One of the assumptions for independent assortment to occur between two genes is that two genes are found on different chromosomes. If two genes are located near each other on the same chromosome, then the alleles for these genes will stay together during meiosis. This phenomenon in which alleles fail to assort independently because they are on the same chromosome is called **linkage**.

The traits that Mendel studied in pea plants are all located on different chromosomes, so they exhibited independent assortment. If, however, the genes for plant height and seed shape were located near each other on the same chromosome, they would display linkage and would not assort independently. In this case, the following would be observed in a cross between a homozygous tall-round plant and a homozygous short-wrinkled plant. In the F_1 generation, all of the plants are tall-round heterozygous. This does not distinguish the F_1 generation from the previous example without linkage, since all of the offspring still display the dominant phenotype for both traits:

Linkage in F_1 Generation

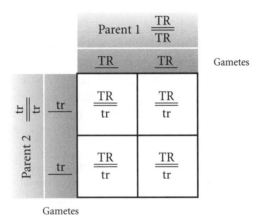

However, if two of the heterozygous F_1 offspring are crossed with each other, the following is observed if the two genes are linked on the same chromosome:

F_2 Dihybrid Cross with Linkage

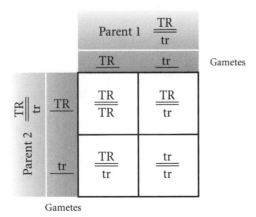

Rather than the 9:3:3:1 ratio expected in a dihybrid cross without linkage, the ratio of phenotypes observed is 3:1 tall-round to short-wrinkled, with no offspring that are tall-wrinkled or short-round. The failure of independent assortment caused by linkage between genes does not allow the alleles for these genes to separate during meiosis, so the same allele combinations on chromosomes remain together through the F_1 generation into the F_2.

Another factor that affects linkage is the recombination between homologous chromosomes that occurs during meiosis. Even if two genes are located on the same chromosome, they will not necessarily be linked 100% through meiosis. If recombination occurs in the DNA between the two genes in the chromosome, this reduce the linkage between genes. If the genes are further apart, more recombination will occur between them, and less linkage will be observed. If the genes are far enough apart on the chromosome, then recombination between the genes may be so frequent that they will display no linkage and will assort independently despite being located on the same chromosome.

The fact that the linkage between genes is related to their distance from each other can be used to map the position of genes relative to each other on a chromosome. By performing a test cross and counting the number of offspring, a geneticist can determine how many recombination events occurred between the genes and from this can estimate the distance of two genes from each other. In the above cross, in which the height and seed shape genes were linked on a chromosome together, recombination would be indicated by the appearance of tall-wrinkled or short-round offspring in the F_2 generation and these would be called the recombinant phenotypes. If none of these phenotypes are observed, no recombination is detected between the genes and they are probably located very near each other on the chromosome. If more of these phenotypes are observed, more recombination has occurred between the genes and they are further apart on the chromosome. If the genes are far enough apart, the 9:3:3:1 phenotype ratio will be observed, with the genes behaving as if they have no linkage.

Inheritance Patterns

Genetic traits can be classified according to whether they are transmitted on autosomal chromosomes or sex chromosomes, and whether they are dominant or recessive. The following patterns of inheritance may be observed:

Autosomal dominant: An autosomal dominant gene will always be expressed in offspring, regardless of the cross. The trait will be expressed in every generation and there are no carriers (individuals that carry an allele for a trait without expressing the trait). Autosomal dominant traits that are lethal or deter reproduction are rapidly selected against in the population. They occur either as a result of infrequent spontaneous mutation of a gene, or because the gene causes a condition that does not affect reproduction. For example, the gene responsible for Huntington's disorder has been identified as an autosomal dominant trait. The onset of the disease related to the faulty gene does not occur until after the reproductive years, allowing the responsible allele to be passed from one generation to another.

Autosomal recessive: An autosomal recessive trait often encodes a defective gene product, such as an enzyme that has no activity due to a mutation in the gene. In a heterozygote, the normal copy of the gene may provide enough activity that the faulty copy has no effect on the phenotype. Heterozygotes for autosomal recessive traits often have a normal phenotype and are called carriers for this reason. Only the homozygote for an autosomal recessive trait will express the condition. Since heterozygote carriers do not express autosomal recessive conditions, the trait can skip one generation but reappear in the next. Even if the trait is lethal in homozygotes, recessive traits can persist for a long time in a population through the heterozygotes. One reason that inbreeding is a problem is that it greatly increases the probability that harmful recessive alleles shared by relatives will become homozygous and be expressed.

X-linked recessive: Another common pattern of inheritance is for X-linked recessive genes. Females are diploid for the X chromosome but males have only one copy of the X chromosome. If males receive an X chromosome with a gene for a recessive disorder or trait, they will express the trait always since they do not have a second copy of the gene. A female that receives one copy of a recessive X-linked gene will not express the trait if the other gene copy is normal (dominant, wild type). For this reason males display X-linked traits far more commonly than women. X-linked recessive traits have very distinctive patterns of inheritance. Males are never carriers of these traits, but females can be. Children of a female carrier have a 50% chance of receiving the recessive allele. Males can never inherit a recessive X-linked trait from their father, since their X-chromosome is always from their mother. A common example of an X-linked trait is color blindness, caused by changes in a gene located on the X chromosome for a pigment protein involved in color vision. Another X-linked recessive trait is hemophilia.

Ethical restraints do not allow geneticists to perform test crosses on humans. Instead, they must rely on examining matings that have already occurred, using tools such as pedigrees. A pedigree is a family tree depicting the inheritance of a particular genetic trait over several generations. By convention, males are indicated by squares, and females by circles. In a pedigree, matings are indicated by horizontal lines, and descendants are listed below matings, connected by a vertical line. Individuals affected by the trait are generally shaded, while unaffected individuals are unshaded.

The following pedigrees illustrate two types of heritable traits: recessive disorders and sex-linked disorders. When analyzing a pedigree, look for individuals with the recessive phenotype. Such individuals have only one possible genotype: homozygous recessive. Matings between them and the dominant phenotype function as test crosses; the ratio of phenotypes of the offspring allows deduction of the dominant genotype. In any case in which mostly males are affected, sex-linkage should be suspected.

Recessive Disorder

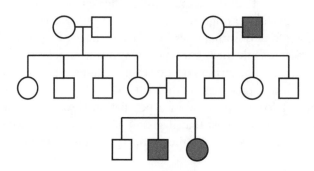

Sex-linked Disorder

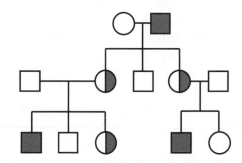

Non-Mendelian Inheritance Patterns

While Mendel's laws hold true in many cases, these laws cannot explain the results of certain crosses. Sometimes an allele is only incompletely dominant or, perhaps, codominant. **Incomplete dominance** is a blending of the effects of contrasting alleles. Both alleles are expressed partially, neither dominating the other. An example of incomplete dominance is found in the snapdragon flower. When a red flower (RR) is crossed with a white flower (WW), a pink blend (RW) is created. When two pink flowers are crossed, the yield is 25 percent red, 50 percent pink, and 25 percent white (phenotypic and genotypic ratio 1:2:1). In **codominance**, both alleles are fully expressed without one allele dominant over the other. An example is blood types. Blood type is determined by the expression of antigen proteins on the surface of red blood cells. The A allele and the B allele are codominant if both are present and combine to produce AB blood. The allele for blood type A, I^A, and the allele for blood type B, I^B, are both dominant to the third allele, i. I^A and I^B may appear together to form blood type AB; however, when both are absent, blood type O results.

To summarize:

- I^A = gene for producing antigen A on the red blood cell
- I^B = gene for producing antigen B on the red blood cell
- i = recessive gene; does not produce either antigen

And these genes combine in various ways to form the following possible genotypes and blood types (phenotypes):

- $I^A I^A$ or $I^A i$ = Type A blood
- $I^B I^B$ or $I^B i$ = Type B blood
- $I^A I^B$ = Type AB blood, with A and B alleles codominant
- ii = Type O blood

Traits encoded in the mitochondrial genome display another unique pattern of inheritance termed **maternal** or **cytoplasmic inheritance**. When a sperm and ovum fuse during fertilization, all of the cytoplasm of the zygote is contributed by the ovum and none by the sperm. All of the mitochondria of the zygote and the resulting individual are derived from the mother, regardless of the sex of the offspring. Traits encoded by mitochondrial genes do not follow the Mendelian rules of inheritance but are always passed from a mother to all of her children and are never transmitted by the father.

There are many circumstances in which the patterns of inheritance do not follow Mendelian rules. One complicating factor is interactions between genes, in which a second gene determines whether a first gene is expressed or not. These interactions are called **epistasis**. Many traits are polygenic, or affected by many different genes. Height, for example, is not determined in most humans by one gene but probably is the result of many genes acting together. The phenotype is therefore not always clear-cut, but a bell-shaped curve of phenotypes reflecting a statistical blending of many different contributing factors. Disorders caused by single gene are far easier to identify in humans than those caused by many genes acting together. Traits such as height are also not purely genetic but can be strongly influenced by the environment. Studies of identical twins are often used to determine if a trait is genetic in nature or is the result of a similar environment or life style. Two twins, for example, will tend to grow to the same height but if the twins are raised in different environments and one individual suffers malnutrition as a child, this environmental difference will cause them to grow to different heights. The degree to which a trait is influenced by the environment can be described by the penetrance of an allele. The more penetrant an allele is, the less it is influenced by the environment. An allele that is 100% penetrant is not influenced by the environment at all in its expression.

Chapter 4: Review Questions

1. Which of the following is not a form of asexual reproduction?

   (A) binary fission
   (B) budding
   (C) parthenogenesis
   (D) regeneration
   (E) meiosis

2. If an individual who is homozygous dominant for a trait mates with an individual who is homozygous recessive for that trait, their offspring will be

   (A) all homozygous dominant
   (B) all homozygous recessive
   (C) 1/2 homozygous dominant and 1/2 homozygous recessive
   (D) all heterozygous
   (E) 1/2 homozygous dominant and 1/2 heterozygous

3. The AB blood type in humans is best described as

   (A) incomplete dominance
   (B) codominance
   (C) sex-linkage
   (D) polygenic inheritance
   (E) cytoplasmic inheritance

4. All of an individual's mitochondria are descendants of his or her

   (A) father's mitochondria
   (B) mother's mitochondria
   (C) father and mother's mitochondria
   (D) mitochondria can be inherited from either mother or father
   (E) all mitochondria are alike

5. An allele that has 0% penetrance

   (A) is heavily influenced by the environment
   (B) is often influenced by the environment
   (C) is sometimes influenced by the environment
   (D) is rarely influenced by the environment
   (E) is never influenced by the environment

Questions 6–9 refer to the following:

   (A) genotype
   (B) phenotype
   (C) allelle
   (D) epistasis
   (E) heterozygote

6. Interaction between two or more genes

7. Observable appearance reflecting the expression of genes

8. Genetic makeup of an individual

9. An organism which has two different forms of a gene

10. Nondisjunction can occur in _____ and may result in _____ .

   (A) anaphase; extra chromosomes
   (B) prophase; extra chromosomes
   (C) anaphase; extra chromosomes or missing chromosomes
   (D) prophase; extra chromosomes or missing chromosomes
   (E) metaphase; extra chromosomes or missing chromosomes

**Please see the answers and explanations to these review questions
in Section V of this book.**

Molecular Genetics

Plants, animals, and bacteria may differ in their form, biochemistry, and lifestyle, but they all share the same basic biology that underlies the inheritance and expression of traits. Genes encode traits and transmit traits, but what are genes and how do they perform these actions? The following section will present the structure of DNA and the mechanisms by which the information in the genome is decoded. Viruses and the science of DNA manipulation are also presented as topics in which modern molecular biology has advanced practical and basic scientific knowledge.

5.1   THE GENOME AND GENE EXPRESSION

Mendel characterized the inheritance of traits through generations and Darwin described the effects of natural selection on organisms before the nature of genetic material responsible for the traits had been identified. Many scientists once believed that proteins had to be the genetic material, since DNA had only four simple components and they did not believe it was complex enough to transmit genetic information. Through many elegant experiments, however, it was proven that DNA is the genetic material in all but certain viruses, and with the elucidation of the structure of DNA by Watson and Crick in 1953, it became clear how DNA could play this role.

One experiment to identify DNA as the genetic material was performed by Oswald Avery in the 1940s while studying strains of bacteria involved in pneumonia. The bacteria are found in either a virulent form or a nonvirulent form that does not cause disease because it lacks a polysaccharide capsule. The virulent bacteria were killed, their DNA extracted and added to nonvirulent bacteria; the result was that some of the nonvirulent bacteria would acquire the virulence trait (see figure). If the DNA was digested in this experiment, the genetic information was not transferred. Also, proteins or RNA would not transfer the trait, supporting the hypothesis that DNA is the genetic material.

Avery—Evidence That DNA is the Genetic Material

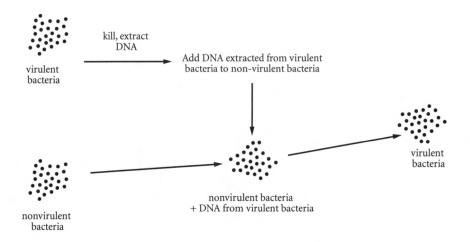

Another key experiment to identify DNA as the genetic material was performed by Hershey and Chase in 1952 while working with a virus that infects bacteria (a bacteriophage). It was known that viruses contained genetic information that they could transfer to bacteria, instructing the bacteria to produce more copies of the virus. However, the nature of the genetic information was not clear. The viruses had two main components, a protein coat and nucleic acids in their interior. The protein coat of the virus could be radiolabeled with a radioactive sulfur isotope, and the nucleic acids could be radiolabeled with phosphorous (see figure). When the bacteria were incubated with radiolabeled virus, only the radiolabeled phosphorous was found inside bacteria, while the sulfur-labeled protein remained outside the bacteria. Virus progeny only contained the radiolabeled phosphorous from the parent virus and not the sulfur from the parent, indicating once again that DNA and not protein is the genetic material.

**Hershey and Chase—More Evidence That
DNA is the Genetic Material**

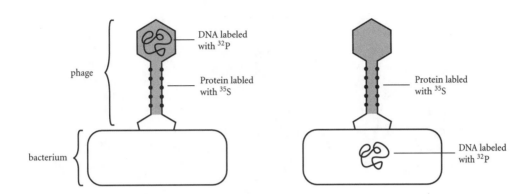

The basic outline of information flow in all living organisms is sometimes called the **Central Dogma** (see figure). The Central Dogma includes the following concepts that are the foundation of modern molecular biology:

1. DNA is the genetic material, containing the genes that are responsible for the physical traits (phenotype) observed in all living organisms.

2. DNA is replicated from existing DNA to produce new copies of the genome.

3. RNA is produced from DNA in a process called transcription.

4. RNA serves as the message used to decode the genetic information and synthesize proteins according to the encoded information. This process of protein synthesis is called translation.

The Central Dogma—Information Flow

5.2 DNA STRUCTURE

DNA is a polymer built from simple building blocks called nucleotides, of which there are four types: adenine (A), guanine (G), thymine (T), and cytosine (C) (see figure). Each nucleotide contains three parts, a five-carbon sugar (deoxyribose), a phosphate group, and a nitrogenous base that distinguishes each of the four nucleotides (A, G, T, or C). There are two types of bases: purines and pyrimidines. The purines are larger, with two rings in each base, and include adenine and guanine. The pyrimidines have one ring and in DNA are thymine and cytosine. The size of the bases is important, since this affects the way the bases fit together to make DNA.

Nucleotides

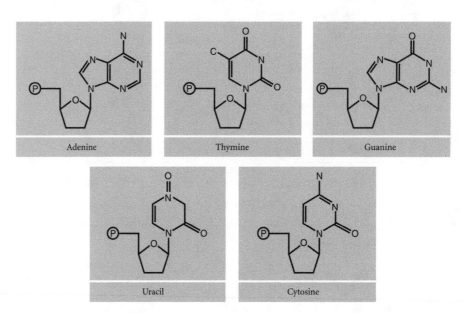

Adenine

Thymine

Guanine

Uracil

Cytosine

To make DNA, nucleotides are polymerized, or joined together in long regular strands of nucleotide building blocks. The phosphate group on a nucleotide forms a covalent bond to the sugar group on the next nucleotide to make a phosphate-sugar backbone in the polymer with the base groups projecting to one side, exposed. One end of a polynucleotide is termed 5' (pronounced "5 prime"), referring to an exposed OH group on the deoxyribose sugar at the #5 position, and the other end is called the 3' (pronounced "3 prime") end, referring to the exposed OH at the 3 position on deoxyribose. The bonds between nucleotides in the backbone are called phosphodiester bonds formed between the 3' hydroxyl of one nucleotide and the 5' hydroxyl of the next nucleotide in the chain. When nucleotides are polymerized by enzymes in the cell, polymerization always proceeds from the 5' toward the 3' end with new nucleotides added onto the 3' end. One polymer strand alone forms half of a DNA double helix—the other half is another strand oriented in the opposite direction (antiparallel). The two strands bind together to form the familiar double-stranded DNA helix.

Single-Stranded DNA

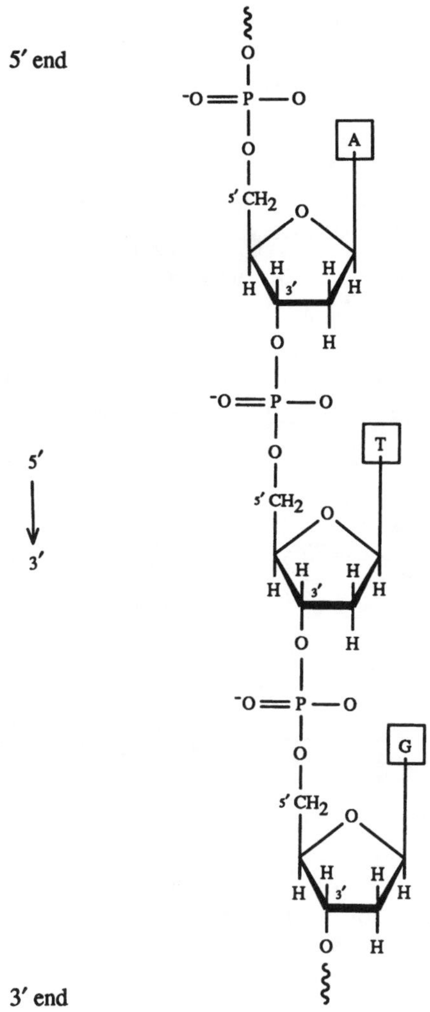

5′ end

5′

↓

3′

3′ end

The formation of the double helix containing two DNA polynucleotide strands requires that the base pairs in each strand hydrogen bond, or base pair, to the other strand in a very specific and restricted manner. The base pairing between the two strands in the double helix involves formation of hydrogen bonds between specific pairs of nitrogenous bases in each strand. An adenine (A) in one strand can only hydrogen bond to a thymine (T) in the other strand to form an A-T base pair, and guanine (G) in one strand only hydrogen bonds to cytosine (C) in the other strand to form a G-C base pair (see figure). This specificity is determined by the way that the base pairs hydrogen bond with each other, with A and T forming two hydrogen bonds and G-C forming three. Also, each base pair must include one purine (big—two rings) and one pyrimidine (small—one ring) to fit into the interior of the double helix. When the bases in two strands match correctly, the bases stack like plates one on top of the other on the inside of the double helix, with each strand wrapped around the other, and the phosphate-sugar backbones facing outward. The two complementary strands of DNA are always oriented in opposite directions, with one strand oriented 5' to 3' in the double helix and the complementary strand oriented in the 3' to 5', or antiparallel, orientation.

Structure of DNA

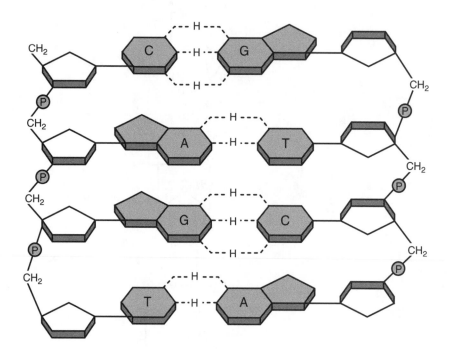

One of the consequences of the structure of DNA is that the amount of adenine in double-stranded DNA must equal the amount of thymine since these two bases are always paired together; the same must be true of cytosine and guanine. Another useful consequence of the double helical structure of DNA is that the two strands in the helix can be split by increasing temperature. The hydrogen bonds that hold the strands together are not covalent and can be disrupted at higher temperatures. The two separated polynucleotides remain intact, however, and if the temperature is allowed to cool, the polynucleotides will often spontaneously reform the original double helix again, with the base pairs forming the correct hydrogen bonds along the length of the molecule. The process of allowing separated polynucleotides to come together in a double helix is called **annealing**. If stands of RNA and DNA are complementary, then RNA can also anneal with DNA. Since G-C base pairs have three hydrogen bonds while A-T base pairs only have two, G-C base pairs require more energy to disrupt and DNA with lots of G-C base pairs must be melted at a higher temperature than A-T rich DNA.

The annealing of DNA strands is a very specific and cooperative process. If base pairs in a strand do not form the correct hydrogen bonds, they disrupt annealing, and prevent the double helix from forming. The specificity of annealing is often used by biologists to look for specific pieces of DNA in a mixture. If RNA or DNA fragments are separated on a gel according to their size, and then transferred to a membrane, the membrane can be incubated with a radiolabeled fragment of DNA to allow annealing between the DNA in solution and the nucleic acid mixture on the blot (see figure). If annealing occurs, the radiolabeled DNA can be detected bound to the membrane. If DNA is on the membrane, this procedure is called a Southern Blot, while if RNA is on the membrane, this procedure is called a Northern Blot.

Southern Blot

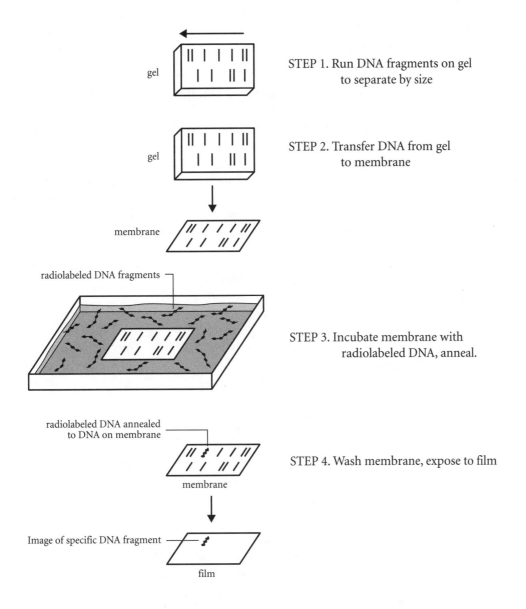

STEP 1. Run DNA fragments on gel to separate by size

STEP 2. Transfer DNA from gel to membrane

STEP 3. Incubate membrane with radiolabeled DNA, anneal.

STEP 4. Wash membrane, expose to film

5.3   DNA REPLICATION

When Watson and Crick deduced the structure of DNA, they immediately saw the implications for the mechanisms involved in replication of the genome. The precise base pairing of the DNA double helix means that each of the two strands contains a complete and complementary copy of the encoded information found on the other strand. An A on one strand means that the opposite strand must contain T in the equivalent position. If the two strands are separated, breaking apart the hydrogen bonds between bases that hold the strands together, then a single strand has all of the information needed to form a new matching strand. As the new strand of DNA is made, all that must be done is to match up the bases on the new strand to have the correct base pairing with the old strand. The strand that is read to make a new complementary copy is called the template.

During cell division, this is exactly what happens (see figure). Each daughter cell must get a complete and accurate copy of the genome after mitosis. To replicate the genome, the two strands of DNA are separated one region at a time, and new DNA is synthesized using each of the separated old strands as a template. When the process is complete, two complete double-stranded copies of the genome have been generated, one from each strand of the original copy. In this mechanism of DNA replication, each new copy of the genome contains one old strand of DNA from the original double helix and one new strand of DNA synthesized using the old DNA as a template. This type of DNA replication is called semi-conservative since one old strand is maintained (conserved) while a new strand is made from the template of the old strand.

DNA Replication

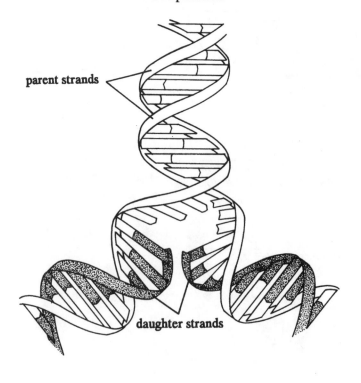

parent strands

daughter strands

There are many enzymes involved in DNA replication, including different forms of **DNA polymerase** that catalyze the synthesis of new DNA based on an existing DNA template. During replication, the double helix is unwound by **helicase** enzymes and opened up one region at a time to form a structure called the **replication fork** where exposed single-stranded DNA serves as a template. DNA synthesis always occurs in the same direction, from 5' to 3', and the two strands in the double helix are oriented in opposite directions. As a result, DNA synthesis on one strand (the **leading strand**) at the replication fork proceeds continuously as the DNA unwinds, but the DNA synthesis on the other strand (the **lagging strand**) must occur in short stretches of DNA called **Okazaki fragments**. DNA replication also cannot start except from a piece of nucleic acid that is already annealed to the template called a primer. Therefore each Okazaki fragment starts at the 5' end with an RNA primer that is later removed. The last step in replication is for an enzyme called **DNA ligase** to covalently join the newly synthesized polynucleotide fragments into a single, continuous strand.

DNA Replication

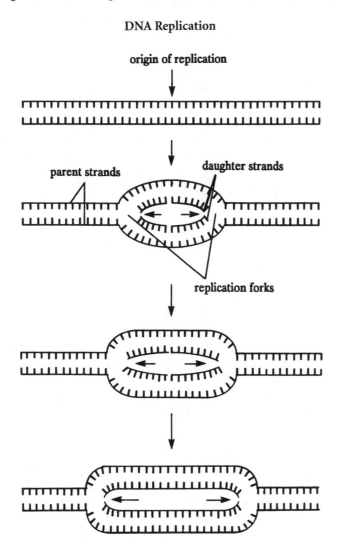

5.4 DNA REPAIR

The DNA in the genome encodes all of the information the cell needs to function. Every enzyme in the cell is encoded in the genome by a specific gene. If there are mistakes in the genome, then the correct enzymes will not be made and the cell or organism may not be normal. DNA replication and the synthesis of new DNA could cause mistakes to occur in the genome if the process were not 100 percent accurate. Exposure to certain chemicals, light, and radiation can all alter DNA and potentially introduce mistakes into the genome. During the growth and life of an organism, cells go through many rounds of cell division, making it important that mistakes are not formed in the genome during each round of DNA replication.

The structure of DNA provides a way to keep the genome free of mistakes during DNA replication. If the genome is altered, or a mistake is made during DNA replication, then the base pairs will not fit properly into the normal double helical structure of DNA. For example, if A is paired with C in a double helix, the two will not hydrogen bond and a bubble will form in the DNA structure, indicating a mistake.

There are enzymes that can detect and fix these mistakes, proofreading and correcting the genome to form the correct base pairing once again. DNA polymerase proofreads each base pair it incorporates as it is added to the growing polynucleotide chain. If the base does not fit into the double helix, DNA polymerase will back up, remove the mistake, and put in the correct nucleotide. If mistakes get past DNA polymerase, then another repair system checks for mismatches in newly synthesized DNA and removes them, putting in the correct nucleotide. Occasional mistakes do still occur, resulting in changes in the genome called mutations that can lead to cancer, but these are much rarer than they would be in the absence of DNA repair and proofreading.

5.5 THE GENETIC CODE

Part of the Central Dogma is that DNA contains genes that are transcribed into messenger RNA, which is in turn translated into proteins. How do the four base pairs in DNA encode the twenty amino acids found in a protein polypeptide chain? The order of the four bases in DNA is the basis of this encoded information, and is called the **genetic code**. Within a gene, every set of three base pairs encodes one amino acid in a polypeptide chain. This three base pair unit of the genetic code is called a **codon**. Since there are four possible base pairs at each of the three positions in a codon, there are a total of $4 \times 4 \times 4$ possible codons, or 64. There are in fact 64 different codons in the genetic code used in genes, more than enough to encode the twenty amino acids (see table). More than one codon encodes some amino acids. For example, the codons GAA and GAG both encode the amino acid glutamate in a protein. This use of multiple codons to encode a single amino acid is called **degeneracy**. There is also one codon that indicates the start of a protein (AUG, which also codes for methionine) and three codons that indicate the end of a protein chain (stop codons). If the DNA sequence of a gene is known, the sequence of the protein it codes for can be predicted based on the genetic code. The code is almost universal for all life, although mitochondria and a few organisms have slight changes in the code.

The Genetic Code

Second Base

		U	C	A	G	
	U	UUU⎱ Phe UUC⎰ UUA⎱ Leu UUG⎰	UCU⎱ UCC⎱ Ser UCA⎰ UCG⎰	UAU⎱ Tyr UAC⎰ UAA⎱ *Stop* UAG⎰ *Stop*	UGU⎱ Cys UGC⎰ UGA⎰ *Stop* UGG⎰ Trp	U C A G
First Base (5′)	C	CUU⎱ CUC⎱ Leu CUA⎰ CUG⎰	CCU⎱ CCC⎱ Pro CCA⎰ CCG⎰	CAU⎱ His CAC⎰ CAA⎱ Gln CAG⎰	CGU⎱ CGC⎱ Arg CGA⎰ CGG⎰	U C A G	Third Base (3′)
	A	AUU⎱ AUC⎱ Ile AUA⎰ AUG⎰ *Start* or Met	ACU⎱ ACC⎱ Thr ACA⎰ ACG⎰	AAU⎱ Asn AAC⎰ AAA⎱ Lys AAG⎰	AGU⎱ Ser AGC⎰ AGA⎱ Arg AGG⎰	U C A G
	G	GUU⎱ GUC⎱ Val GUA⎰ GUG⎰	GCU⎱ GCC⎱ Ala GCA⎰ GCG⎰	GAU⎱ Asp GAC⎰ GAA⎱ Glu GAG⎰	GGU⎱ GGC⎱ Gly GGA⎰ GGG⎰	U C A G

5.6   MUTATIONS

DNA replication is a highly accurate process, but errors do occasionally occur and these errors can cause changes in the DNA genome. These changes in the genome are called **mutations**. Mutations include gene mutations that change one or a few base pairs in one gene and changes that alter sections of chromosomes. Gene mutations can change a single base pair, insert new base pairs (insertion) or delete base pairs (deletion).

Point Mutations

A point mutation occurs when a single nucleotide base is substituted by another nucleotide base. If the substitution occurs in a noncoding region, or if the substitution changes into a codon that codes for the same amino acid due to the genetic code degeneracy, there will be no change in the resulting protein's amino acid sequence (a **silent mutation**, see figure on page 137). A **missense mutation** results in a change in a protein's amino acid sequence with effects that range from insignificant to lethal, depending on the effect the substitution has on the protein. A conservative mutation such as changing a leucine to an isoleucine, may have little if any effect on a protein's function. A **nonsense mutation** is one that changes a codon in a gene to a premature stop codon, shortening the protein produced from the gene. A nonsense mutation usually has a fairly severe effect on protein function since a large piece of the protein may be absent.

Frameshift Mutations

Base pair insertions involve the addition of nucleotides, while base pair deletions involve a loss of nucleotides. Such mutations usually have serious effects on the encoded protein, since nucleotides are read as a series of triplets. The addition or loss of nucleotide(s) (except in multiples of three) will change the reading frame of the mRNA, and is known as a **frameshift mutation** (see figure). All of the codons after the frame shift will be read incorrectly, usually having a severe effect on the activity of the gene product. If a deletion or insertion occurs in a multiple of three base pairs, the effect will be much smaller since the reading frame will be maintained.

Chromosomal mutations can affect many genes. These mutations include **deletions**, **inversions**, **translocations**, and **duplications**. Errors during recombination can lead to deletion or duplication of genes or sections of chromosomes. Duplication of genes is probably the source of families of closely related genes in evolution. In an inversion, a section of a chromosome is removed and reinserted into the chromosome in the opposite direction. Translocation involves the removal of a section of the genome from one chromosome and its reattachment to another chromosome. In any of these cases, genes at or near the location of the chromosomal rearrangements are most likely to be affected.

Mutations are commonly caused by errors in DNA replication, by radiation, and by chemicals. Chemicals that cause mutations in the genome are called **mutagens.** Mutagens include chemicals that resembles the bases in DNA, and are incorporated into DNA, chemicals that bind DNA between bases, thereby bending and distorting the DNA structure so that errors occur during replication, and reactive chemicals that covalently damage DNA. Free radical oxygen derivatives like superoxide produced during cellular metabolism can covalently damage DNA, causing strand breaks, for example. Some mutagens are also carcinogens, or substances that cause cancer, although the terms mutagen and carcinogen do not have the same meaning.

Although mutations can be harmful to the individual or have no effect, mutations are an essential part of evolution. Mutations are the only source of new genetic alleles that did not exist previously. A small number of these mutations may in some context provide a selective advantage to an individual. The responsible allele may, over time, spread through a population if natural selection favors it. Without mutation, evolution would not occur.

Mutation

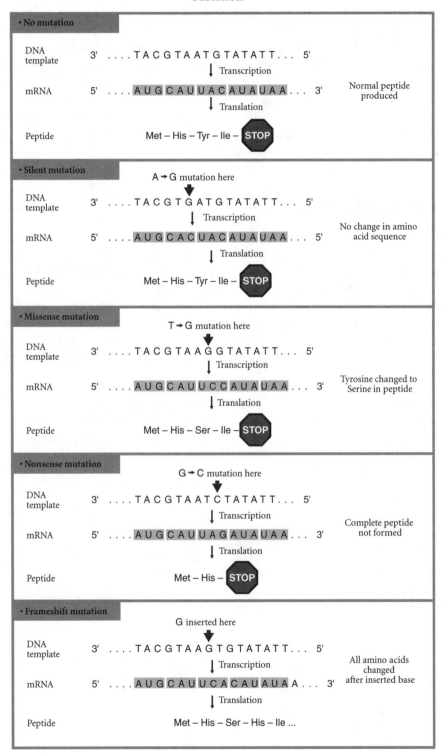

Examples of human genetic disorders include phenylketonuria (PKU), a molecular disease that involves the inability to produce the proper enzyme for the metabolism of phenylalanine, resulting in the accumulation of a toxic degradation product (phenylpyruvic acid). Sickle-cell anemia is another example of a genetic disorder. Widespread in Africa, this disease cripples red blood cells. Patients are unable to synthesize normal hemoglobin, and their red blood cells become sickle (crescent) shaped. Their hemoglobin is incapable of carrying as much oxygen as normal hemoglobin. The disorder can be traced to the presence of valine (GUA or GUG) instead of glutamic acid (GAA or GAG) in the hemoglobin of individuals with sickle-cell anemia, which in its turn is a result of the substitution of just one nitrogen base, A for T, in the DNA molecule.

5.7   RNA COMPARED TO DNA

In the central dogma, RNA is produced by reading DNA. Like DNA, RNA is a polymer of nucleotides. Both DNA and RNA are nucleic acids, and the structure of RNA is very similar to that of single-stranded DNA. During RNA synthesis, the nucleotides in RNA are matched to base pair with the DNA template similar to the base pairing of DNA with DNA during DNA replication. There are, however, a number of important differences. These differences include the use of the sugar ribose in the RNA backbone rather than deoxyribose, the presence of the base uracil in RNA rather than thymine, and the fact that RNA is usually single-stranded, while DNA is double-stranded. Also, RNA is not proofread when it is made, unlike DNA.

DNA-Unique Features:	RNA-Unique Features:
• Double-stranded except when replicating	• Nearly always single-stranded
• Deoxyribose sugar in the nucleotides	• Ribose sugar in nucleotides
• Thymine base forms a thymine-adenine base pair (T-A)	• Uracil base instead of thymine. The base pair is uracil-adenine (U-A)
• Replicates DNA → DNA	• Does not normally replicate (except in the case of some viruses)
• Only one type of DNA per organism. This DNA acts as the original source of information, acting as a master record. Its information is copied onto RNA molecules.	• Three types of RNA (mRNA, tRNA, rRNA)

There are three types of RNA: **messenger RNA** (mRNA), **ribosomal RNA** (rRNA) and **transfer RNA** (tRNA). mRNA encodes protein messages that are to be decoded in protein synthesis. rRNA is involved in translation (protein synthesis). Each tRNA plays a role in protein synthesis, with an anticodon that recognizes one of the three base pair codons in mRNA and brings the amino acid that matches that codon to the translation process. tRNAs are relatively short, do not encode any proteins, have a compact, complex 3D structure, including base pairing within the molecule, and have one end specialized to be bound to amino acids.

5.8   TRANSCRIPTION

For the genotype to produce a phenotype, the coded information in the genome must be converted into proteins. This process does not occur directly between DNA and proteins however, but has mRNA as a messenger between DNA and proteins. **Transcription** is the process in which genes in the DNA genome are used as a template to produce an mRNA message that is subsequently decoded to produce a protein. The enzyme that synthesizes RNA in transcription, **RNA polymerase**, uses single-stranded DNA as a template to read the gene, matching base pairs as it synthesizes new RNA from the DNA template (G matching with C and U matching with A). The production of RNA from the DNA template is similar to the production of DNA from a DNA template by DNA polymerase. RNA is synthesized in transcription in one direction only, from the 5' end of the polymer to the 3' end. Prokaryotes have one type of RNA polymerase while eukaryotes have three, a different RNA polymerase for mRNA, tRNA and rRNA production.

The messenger RNA is not proofread as it is produced. As a result, some mRNA molecules probably contain errors, but the effect of errors in mRNA molecules is much less severe than errors in DNA replication. An mRNA molecule is transient in the cell, so that errors in an mRNA will only affect a small number of protein molecules, while errors in DNA replication would affect all proteins later produced based on that gene.

There are probably 100,000 genes in the human genome, but not all of these are expressed in every tissue. Genes can be turned on or off by regulating gene transcription according to the needs of the cell and the organism. Transcription is turned on or off by regions of DNA near the start of the gene called **promoters** (see figure). Promoters are short sequences of DNA that bind proteins that increase transcription of the gene by RNA polymerase. In eukaryotic cells, mRNA is produced in the nucleus and is translated into proteins in the cytoplasm. Before the mRNA is translated, however, it is usually modified. The modifications include the addition of a special cap to the 5' end of the mRNA, the addition of poly-A tail to the 3' end, and the removal of RNA sections that do not encode a protein, a process called **splicing**. The part of the RNA molecule that encodes the protein message is called an **exon** and the part between coding blocks, the part that is removed, is called the **intron**. Once splicing and processing are complete, the mRNA can be exported from the nucleus through the nuclear pores and is ready for translation.

Eukaryotic Transcription

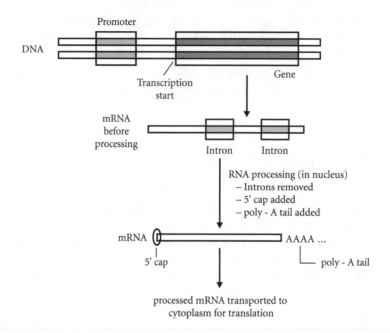

Prokaryotic genes do not have introns and do not go through splicing. Since there is no nucleus in prokaryotes, transcription and translation occur in the same place and ribosomes can start to translate an RNA before transcription is even complete. Prokaryotic genes are often found with several related genes next to each other in the genome as an **operon**. They are regulated together and may be transcribed together. Prokaryotic genes are polycistronic while eukaryotic genes are monocistronic, with only one gene per RNA message.

5.9   TRANSLATION

Protein translation is the process in which the genetic code in mRNA is used to assemble amino acids in the correct sequence to make a protein. Translation occurs in the cytoplasm. After mRNA is processed and spliced in the nucleus and is ready to be translated, it is exported through a nuclear pore to the cytoplasm. To initiate translation, the mRNA is bound by a ribosome at the site on the mRNA where protein synthesis will begin. The start site of translation is the start codon AUG in the mRNA module, which codes for the amino acid methionine. Since all proteins have AUG as the start codon, all proteins have methionine as the first (or N-terminal) amino acid.

In the processed mRNA, each three base pair codon codes for a specific amino acid that will be included in the protein amino acid chain. How does the ribosome match amino acids up to the correct codon? There are intermediary molecules, molecular 'middle-men,' that match each amino acid up to its codons in mRNA. These middle-men are tRNAs (see figure). Each tRNA has a specific amino acid bound covalently at one end. At the other end, the tRNA has a three base pair region called the **anticodon** that will match up and hybridize to the correct codon in mRNA.

Enzymes called **aminoacyl-tRNA synthetases** attach amino acids to the correct tRNAs in a very accurate manner. If a tRNA has the wrong amino acid attached, it will be built into a protein and the wrong protein sequence be made, possibly making the rest of molecule unable to perform its function. There is proofreading in the production of activated tRNAs, but not in protein synthesis once an amino acid is built into a protein chain.

Protein Synthesis

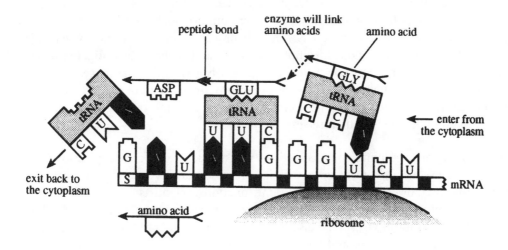

After the ribosome recognizes the first codon, the start codon, it matches up the next codon in the mRNA to the tRNA with the correct anticodon (see figure). If the mRNA has the three base pair code GAU for aspartate, then the tRNA for aspartate, with CUA in its anticodon and aspartate bound to the other end, will match up to the codon on the ribosome. The ribosome will then join aspartate to the end of the growing protein chain, move down the mRNA one codon, and start again to match up another tRNA to the next codon in the mRNA. Once the next tRNA is bound to the ribosome, with its anticodon matching the mRNA codon, the next amino acid will be transferred from the bound tRNA to the end of the protein chain.

With each step in protein synthesis, the ribosome matches another tRNA to the correct mRNA codon, adds the next amino acid to the end of the protein chain, forms a peptide bond in the growing polymer, and releases the tRNA. The tRNA will be recycled by the addition of the correct amino acid once again. When the ribosome reaches a stop codon in the mRNA, it stops translation and releases the mRNA, which can then be transcribed again or degraded by enzymes.

mRNA that produces proteins bound for the cytoplasm is translated by ribosomes in the cytoplasm. The mature proteins can then be released directly into the cytoplasm. Other proteins that are destined for the endoplasmic reticulum (ER), Golgi, or the plasma membrane or that are to be secreted are synthesized by ribosomes bound to the rough ER. When the protein is synthesized, it is inserted into the ER. The protein's sequence tells the ER where to send the protein. From the ER, the newly synthesized proteins are packaged into small spheres of membrane called vesicles. These vesicles then move to the Golgi, where the proteins are further modified, then on to the plasma membrane where they are either secreted or remain as transmembrane proteins.

5.10 VIRUS STRUCTURE AND FUNCTION

Viruses are small packages of nucleic acid in a protein coat that replicate themselves in cells. Viruses are not cells, have no cytoplasm and carry out no biochemical activity of their own. They do not produce or consume DNA outside of a cell and are completely dependent on living within a cell to carry out metabolic processes and replicate as **obligate intracellular parasites**. Although they are able to replicate themselves with the assistance of cells, viruses are not generally considered to be living organisms since they are not cells and lack any metabolic activity of their own. Viruses invade foreign cells and take over the machinery of the cell to produce copies of themselves.

Although they are not living, the mechanisms viruses use to alter gene expression in cells, to alter the cell cycle, and control other cellular processes has revealed a great deal about the mechanisms used by cells to perform the same functions. Scientists have often found that the simplicity of viruses provides insight into mechanisms used by the more complex host cells they infect. The study of viruses has also revealed a great deal about the mechanisms involved in diseases caused by viruses and has allowed viruses to be used as vectors in recombinant technology to introduce engineered genes into cells. There are a great variety of viruses that infect every organism ranging from bacteria to humans. They all have a similar basic structure, however, with a nucleic acid genome in the viral interior, surrounded by a protein coat called the **capsid** (see figure). Outside of the cell a complete virus capable of infection is called a **virion**. The protein coat protects the genome from the external environment and helps the viral genome to attach to cells and get into the cells. The capsid is usually formed from units of one or a few proteins that repeat over and over again in the structure. Viruses have a large variety of structures, but follow two basic plans, forming either a long tubular helix with the genome in the middle or a geometric icosahedral structure. Many animal viruses have a lipid bilayer coat called the envelope that surrounds the capsid and is derived when the virus leaves a cell by budding through the plasma membrane. Viruses without the envelope are called **naked**.

Virus Structure

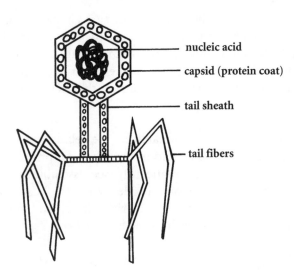

One of the central features common to viruses is that their protein package is generally very small and can only contain a limited amount of genetic material. Viruses have evolved many ingenious mechanisms to maximize the space in their genome. Viruses have very few base pairs in their genome that do not code for a gene product, and some that code for more than one gene product. Unlike eukaryotes, some viruses have multiple overlapping reading frames and other mechanisms that conserve space.

To infect a cell, a virus must first gain entry to the cell. Animal viruses bind to receptors in the plasma membrane to gain entry to the cell. Specific envelope proteins on the surface of the virus bind to the cellular proteins. The cellular proteins that the virus recognizes are not there for the convenience of the virus, but are common cellular proteins such as plasma membrane receptors that the virus has adapted to recognize. The expression of cellular proteins determines the types of cells and tissues that a virus can infect. The infection of the HIV virus in immune cells reflects the expression in these affected immune cells of the receptor the HIV virus recognizes to gain entry into cells. Once a virus binds to the cellular receptors, it enters the cell either by fusion of its envelope with the plasma membrane to release the viral capsid and genome into the cytoplasm by being internalized through endocytosis. Once internalized the virus releases its genome.

All viruses have a nucleic acid genome, but the nature of the genome varies widely. The genome can be either DNA or RNA, double or single-stranded, linear or circular. The type of genome used by virus is one of the distinguishing features of each virus. RNA viruses have smaller genomes than DNA viruses. The lack of proofreading of RNA leads to a very rapid rate of evolution of RNA viruses. One of the common traits of all viruses is that they must use the cellular machinery for translation to produce viral proteins for the viral capsid and other viral proteins. Viral mRNA is found in the life cycle of all viruses.

Bacteriophage Life Cycle

Some of the most extensively studied viruses are viruses that infect bacteria, bacteriophages. These viruses do not have an envelope, since they are not produced by budding, and do not enter the cell by fusion or endocytosis. Bacteriophages typically have complex tail assemblies composed of virus proteins that bind to the bacterial cell wall and inject the phage genome through the cell wall into the bacteria. Once in the bacteria, the viral genome can enter either a **lytic cycle** or a **lysogenic cycle**.

In the lytic cycle, the phage commandeers the cellular biosynthetic machinery to produce new copies of the viral genome and to produce viral proteins to make more virus. As viral proteins accumulate, they spontaneously self-assemble into the capsid, tail and other components with the viral genome inside. When the host cell is full of capsid viruses, an enzyme is produced to degrade the bacterial cell wall and burst the cell, releasing the newly synthesized viruses to infect neighboring bacterial cells (see figure).

Bacteriophage Life Cycle

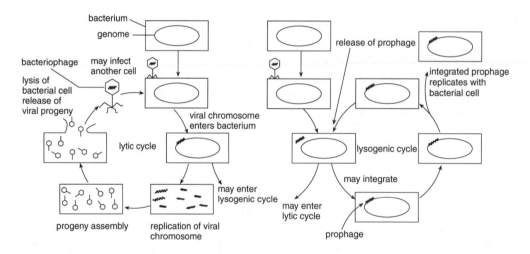

In the lysogenic cycle, however, the phage does not produce viral gene products, but is integrated into the bacterial genome where it remains hidden. The bacteria may reproduce for many generations, carrying the phage genome through DNA replication, with phage remaining integrated in the bacterial genome. In response to changing conditions in the cell, the phage can later excise itself and enter the lytic cycle to produce more copies of the virus.

Double-Stranded DNA Viruses

Double-stranded DNA viruses have a genome that is similar to the genome of the host cells they infect. They use host cell systems for transcription to produce mRNA of virus-encoded genes, including capsid proteins and specialized enzymes, for replication of the viral genome. These viruses often replicate in the nucleus and use host machinery for mRNA processing as well. The apparatus for transcription is abundant and active in most cells, so these systems are readily available for the DNA virus to exploit towards its own end. The virus wants to replicate itself rapidly and continue the infection; most animal cells, however, do not rapidly replicate and do not provide the apparatus required for replication. The regulation of the cell cycle is one feature that DNA viruses must contend with. To overcome these obstacles, DNA viruses will either replicate in cells that are already dividing or will express proteins that release the block on the cell cycle, inducing cells to enter S phase and proliferate. DNA viruses will also express their own proteins to replicate their genome. DNA viruses that are more complex and make more of their own gene products to carry out replication have larger genomes but are also less reliant on the cell.

The DNA virus has a life cycle with different viral proteins expressed at different time periods after infection begins, such as early genes that are expressed soon after infection, and late genes expressed later in the infection. Early genes in a DNA virus would include proteins that release cell cycle blocks to stimulate cellular proliferation, allowing viral replication. The production of proteins to overcome the blocks on the cell cycle can lead to some cancers. Examples of double-stranded DNA viruses include adenovirus associated with respiratory infections, and the herpes virus.

RNA Viruses

While cells do not always produce DNA, they always make proteins. RNA viruses can either be +strand, meaning that the genome can act as mRNA when it enters the cell, or −strand, which is in the opposite orientation as mRNA. +strand RNA viruses are read as mRNA when they enter the cell, and one of the gene products they produce is an RNA-dependent RNA polymerase that is used to synthesize a −strand RNA from the +strand template. Animals do not produce enzymes which are capable of making RNA from an RNA template, so the virus must encode this enzyme. Once −strand RNA is produced, it can in turn be used to make new +strand RNA genomes for new viruses.

−strand RNA cannot be transcribed when it enters the cell since it is in the wrong orientation to be translated. It requires a viral enzyme to produce a +strand RNA using the −strand as a template. Since the viral RNA is not translated when it enters the cell, the virus must carry this enzyme into the cell when it enters. The −strand replication enzyme is packaged with the genome inside the capsid so that when the virus infects a cell, the enzyme can produce +strand copies of the genome and the +strand can then serve in the cell as mRNA.

Retroviruses

Retroviruses have an RNA genome that is a +strand, but when they enter the cell, the RNA is first copied into double-stranded DNA that integrates into the genome of the host cell. The viral genome is transcribed to make mRNA for viral proteins and new copies of the virus RNA genome. Retroviruses need an enzyme called **reverse transcriptase** to copy DNA from RNA, an activity not found in animal cells, and they must carry this enzyme with them when they enter the cell. Since the virus integrates into the host genome, it can escape immune detection. HIV is the best known example of a retrovirus that causes disease in humans.

5.11 RECOMBINANT DNA TECHNOLOGY

One of the most powerful results of the advances in molecular genetics has been the creation of new technologies that allow the manipulation of DNA and the possibility to alter the genetic composition of organisms. Mendel characterized inheritance, Watson and Crick identified the structure of DNA, and recombinant technology has allowed the manipulation of DNA Recombinant DNA technology has had practical applications in medicine and has also fostered rapid advances in the understanding of human biology. This technology has enabled scientists to produce new medicines, explore the mutations associated with cancer and cystic fibrosis, and has even led to the cloning of sheep.

Many of the essential tools in recombinant technology are enzymes that have been isolated from nature. One of the first of these is a class of enzymes called **restriction enzymes**. Restriction enzymes cleave DNA at specific sequences, a very useful property in the manipulation of DNA. For example, the bacteria *E. coli* produces a restriction enzyme called EcoR1 that cleaves double stranded DNA at the sequence: 5'-GAATTC-3'. Bacteria produce restriction enzymes as a protective measure to digest the DNA of viruses that attempt to infect them. The recognition sites for restriction enzymes are typically palindromic (they read the same on both strands) and often leave single-stranded overhangs that are also very useful (see figure).

Restriction Enzymes

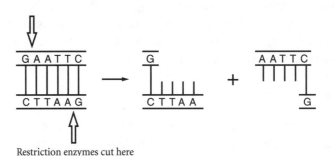

Restriction enzymes cut here

If a scientist is interested in a human gene that has an EcoR1 digestion site in the genome on each side of the gene, he can digest the genome with EcoR1, and collect a series of fragments with EcoR1 sites on each end. If another piece of DNA like a bacterial plasmid has been prepared by digestion with EcoR1, and the gene of interest is mixed with the plasmid, the single-stranded overhanging sticky ends of both DNAs digested with EcoR1 can anneal with each other. DNA ligase can be used to seal the phosphodiester bonds in the DNA at the annealed sticky ends, forming a complete DNA double helix with the gene inserted. The bacterial plasmid can then be replicated in bacteria, replicating the human gene (see figure on following page).

There are a large number of restriction enzymes that have now been identified with a broad range of recognition sequences. If a scientist has DNA of interest, he can examine its sequence for one of hundreds of restriction enzyme sites, select the proper enzyme, digest, isolate the desired fragments, then anneal them into new piece of DNA, such as a plasmid. The restriction enzymes act like molecular tweezers that allow a scientist to pull out desired pieces of DNA. The plasmid in this case is used as a vector, a vehicle to carry a piece of DNA of interest. Viruses are often used as vectors due to their ability to carry genetic material into cells.

Recombinant DNA

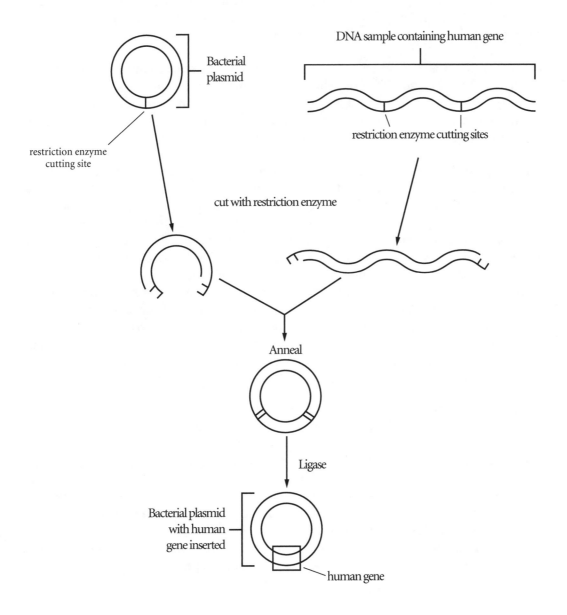

To create the desired DNA strand, a scientist will often need to separate a desired DNA fragment from many other fragments. For example, if the genome is digested with a restriction enzyme, it will produce a large number of fragments, not just the one gene of interest. Gel electrophoresis is used to separate DNA fragments on the basis of size. Electrophoresis for nucleic acids is similar to SDS-polyacrylamide gel electrophoresis used for proteins (see chapter 1). For DNA, however, the gel is formed from agarose, rather than polyacrylamide, and no SDS is used. The DNA remains double-stranded in the gel. When an electric field is applied, the negative charge of DNA causes DNA to migrate toward the positive end of the electrode; smaller DNA fragments move through the gel more quickly than larger fragments. The gel can be stained with ethidium bromide to visualize DNA, and the fragments of interest removed based on their size.

The ability to cut and move DNA into vectors has allowed scientists to clone genes, then characterize their function. For example, a scientist may identify a piece of DNA containing a gene for a receptor in the plasma membrane. The scientist can then cut out the gene with restriction enzymes, ligate the gene into a vector and insert the gene into cells to see if the product of the gene produces a protein that alters the behavior of the cells.

Other ways to study and manipulate DNA include the ability to determine the sequence of base pairs in a piece of DNA, called **DNA sequencing**. The Human Genome Project, now underway, has the goal of sequencing the entire human genome, estimated to contain over 100,000 genes. To replicate a specific section of DNA, a method called the **polymerase chain reaction** (PCR) is often used. In PCR, polynucleotides synthesized by a chemist are annealed to a section of DNA, and a DNA polymerase then synthesizes DNA in both directions. Through repeated rounds of annealing, synthesis and melting, the DNA region is amplified exponentially. PCR can be used to produce a significant quantity of DNA starting from only a few molecules of DNA. PCR and other techniques can be used to mutate genes of interest. For example, a gene can be isolated with restriction enzymes, ligated into a vector, sequenced, and expressed in cells to see that the gene encodes a growth factor. The scientist can then go back to the gene to introduce specific mutations, put the changed genes back into cells and determine if the mutations change the function of the growth factor.

The power of recombinant DNA technology has been applied to medicine to identify the genes responsible for many diseases like cystic fibrosis and may offer in the future the ability to cure genetic diseases by altering the genome through gene therapy. Recombinant technology has already resulted in the production of new medicines such as erythropoietin (Epo). Epo is a protein that stimulates the production of red blood cells. Cancer patients and other often suffer from anemia caused by a lack of red blood cells. The Epo that is given to these patients is the result of the identification of the gene for the Epo protein, followed by introduction of the Epo gene into a vector and the introduction of the vector into cells to produce large quantities of the protein. Recombinant technology is also being introduced into agriculture to tailor crops for desired properties such as pest resistance.

The power of recombinant DNA technology in areas such as medicine is also accompanied by many challenges. Many experiments that biologists will soon be able to perform may be ethically undesirable. The first cloning of a large mammal, the sheep Dolly, was recently reported, raising the prospect of human cloning as a future possibility. Cloning of humans may soon be technologically feasibly but the ethics are still being debated. Scientists may in the future have the ability to alter the human genome to cure genetic diseases, but if this ability is used to "improve" the genome, to try to create individuals with more "desirable" traits, this would also provide an ethical dilemma. Recombinant technology is likely to continue to bring both advances and new challenges in the future.

Chapter 5: Review Questions

1. DNA is composed of nucleotides. Which of the following best describes the components of a nucleotide?

   (A) ribose and a phosphate group
   (B) ribose and a nitrogenous base
   (C) deoxyribose and a phosphate group
   (D) ribose, deoxyribose and a nitrogenous base
   (E) deoxyribose, a phosphate group, and a nitrogenous base

2. What type of bonds joins one strand of DNA to another strand of DNA, forming a double helix?

   (A) ionic bonds
   (B) hydrogen bonds
   (C) polar covalent bonds
   (D) hydrophobic interactions
   (E) nonpolar covalent bonds

3. Which scientist(s) identified DNA as the genetic material of inheritance?

   (A) Mendel
   (B) Watson and Crick
   (C) Darwin
   (D) Avery
   (E) Okazaki

4. Which of the following statements is false?

   (A) Mutations are important in evolution.
   (B) Errors in DNA replication can cause mutations.
   (C) Mutagens are substances which cause cancer.
   (D) Products of cellular metabolism can cause mutations.
   (E) DNA polymerase proofreads newly synthesized DNA.

Questions 5–8 refer to the following:

   (A) uracil nucleotide
   (B) guanine nucleotide
   (C) translation
   (D) transcription
   (E) splicing

5. takes place only in eukaryotes

6. forms three hydrogen bonds when linked with a cytosine nucleotide

7. DNA → RNA

8. found only in RNA

9. Viruses

   (A) have a clearly defined nucleus
   (B) have DNA or RNA and mitochondria
   (C) have DNA or RNA and ribosomes
   (D) have DNA or RNA, ribosomes, and a protein coat
   (E) have DNA or RNA and a protein coat

10. In gel electrophoresis, DNA fragments migrate toward the _____ electrode; the _____ the fragment, the faster it moves through the gel.

   (A) negative; smaller
   (B) negative; larger
   (C) positive; smaller
   (D) positive; larger
   (E) none of the above

Please see the answers and explanations to these review questions in Section V of this book.

Evolutionary Biology

From the origin of life and even before, evolution has been the force responsible for the ongoing development of the species that inhabit the earth. The origin of life, evidence for evolution, and the mechanisms involved in evolution are described in this chapter.

6.1 EARLY EVOLUTION—THE ORIGIN OF LIFE

Prebiotic Development of Replicating Molecules

Fossils reveal much about the various forms of life that have inhabited earth through geologic history, but do not indicate the mechanisms by which life first arose. The cell theory indicates that all life comes from other life, and the belief that life today could arise from non-living matter by **spontaneous generation** was refuted by Louis Pasteur in the 19th century. When the earth was first formed with the rest of the solar system five billion years ago, it consisted only of rock and other substances. There is no evidence of life in rocks formed from the first billion years of the earth's history. Where did the first life come from? The first life on the early earth must have come from non-living material.

The earth coalesced in the early solar system and was heated by radioactive decay and compression to form a molten metal interior that remains today. Above this molten interior floats the crust, including the continents and the ocean basins. When the earth cooled sufficiently, the crust solidified and the oceans formed. The atmosphere of the early earth was released from the interior of the planet as it formed and was captured by gravitation; this early atmosphere included gases such as methane, nitrogen, carbon dioxide, ammonia and water. There was no free molecular oxygen in the early earth's atmosphere, an important factor that resulted in a reducing atmosphere rather than the oxidizing atmosphere that is found on earth today.

Before life was able to form on earth, it is likely that organic precursors of life were present that then spontaneously assembled themselves into more complex structures that were able to replicate and evolve. To see if organic precursors could form in the conditions found on the primitive earth of 4 billion years ago, the gases found in this atmosphere were mixed in a chamber by Stanley Miller, and the mixture was then treated with electrical discharges to simulate lightning. After a period of time, liquid from the vessel was examined and found to contain amino acids like those found in proteins, and nitrogenous bases like those found in DNA, the building blocks of life. Simple sugars could also be produced in similar conditions. In today's environment, these molecules would not persist due to changes in the environment and the presence of living organisms that would consume these materials. In the prebiotic (before life) earth, however, such molecules could accumulate in the oceans and ponds and reach higher concentrations over very long periods of time.

The next step toward living systems would be for the building blocks to polymerize. Molecules that polymerized more efficiently and that were stable once formed would accumulate more rapidly and predominate in the prebiotic environment. If a polymer could catalyze the production of copies of itself, it would have an advantage and would replicate itself to predominate in the abiotic soup of the oceans, using the precursors it found there as raw material for its own propagation. Such a molecule would need to be both a template to produce a copy of itself and a catalyst to allow the reaction to occur.

Proteins form most enzymes found in life today, but it is unlikely that proteins could use themselves as templates to reproduce themselves. Nucleic acids function as information templates today, but were not known to function as enzymes. How could the first genomes have replicated without protein catalysts and how could the catalysts be produced without templates to encode them? The most likely answer would be that the same molecule served as both template and catalyst for self-replication.

It was long thought that only proteins could act as catalysts. However, is now known that RNAs can and do act as enzymes as well. RNAs have been identified that can cleave other RNAs at specific sites, and that can catalyze elongation of RNA polymers as well. Self-cleaving introns have been identified in several organisms. Scientists have produced RNA enzymes starting from scratch. RNA can act in a catalytic manner by folding into a complex 3-D structure to form a catalytic site in the same manner that a protein enzyme does. RNAs with catalytic activity and the ability to self-replicate may have been able to catalyze other reactions as well. Natural selection would have acted even in the prebiotic world since RNAs that replicated, were stable and catalyzed favorable reactions would become more abundant, and continue to evolve. Self-replicating RNAs may have represented the first genomes.

RNA still plays a central role in the expression of genes today, perhaps as molecular relic of RNA-based life forms. Although organisms today use DNA as their genome, RNA is essential for the expression of information in the DNA genome. mRNA, rRNA and tRNA may represent the remains of a time when life originated and was RNA-based. During the evolution of the first life forms, RNA began to encode the synthesis of proteins and protein enzymes replaced RNA enzymes, since the chemical diversity of the amino acid side chains in proteins allows more diverse and efficient enzymes to evolve. The abundant use of nucleotide cofactors by modern enzymes may represent a vestige of a time when all enzymes were RNA-based. Similarly, DNA later replaced RNA as the genome, since DNA is more chemically stable and able to pass on accurate genetic information from one generation to the next.

Self-replicating RNA would be one step toward a living organism, the origin of the genome. Another major step would the formation of cells. To carry out metabolic reactions of any sort, a cell must have a membrane to separate itself from the outside environment. A self-replicating RNA that was able to concentrate material for replication inside a membrane would replicate much more efficiently. Once again, observations suggest that this process can occur spontaneously in the absence of life. If proteins, polysaccharides, nucleic acids and lipids are mixed together, they will form drops of concentrated polymers called **coacervates**. Coacervates have a distinct interior and exterior and with lipid present form a lipid membrane that resembles the cellular plasma membrane in its ability to concentrate material from the exterior and retain molecules in the interior. Natural selection would have selected for coacervates that were most efficient at acquiring material and catalyzing reactions. When coacervates were efficient and grew large they would split and divide their contents into smaller coacervate droplets that passed on the properties of the larger coacervate to favor continued growth, division and evolution.

The First Cells and Early Evolution

Comparison of all life today to see what all life shares in common can suggest the nature of the earliest cell that gave rise to all future life, called the **progenote** (see figure). It is accepted now that the three major divisions of life are the eukaryotes, the eubacteria and the archaebacteria, and it is likely that they were all descended from the progenote. The ribosomal RNAs have been remarkably conserved among all life and comparison of the ribosomal RNA sequence indicates the relationship between the three major groups compared to the progenote. The progenote was likely to be prokaryotic; have no organelles; have a DNA genome; perform translation and transcription using RNA as an intermediary, the modern genetic code; have ribosomes; and possess lipid bilayer membranes. All life uses the same stereochemistry in its building blocks, with L-amino acids and D-sugars, indicating that the basic chemistry of living organisms was similar. Eukaryotes and archaebacteria have introns today, suggesting that the progenote did as well. It appears likely that eubacteria lost introns under selective pressure to eliminate unnecessary DNA.

Origin of Life

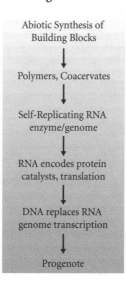

Abiotic Synthesis of
Building Blocks

↓

Polymers, Coacervates

↓

Self-Replicating RNA
enzyme/genome

↓

RNA encodes protein
catalysts, translation

↓

DNA replaces RNA
genome transcription

↓

Progenote

Viruses may in some cases provide clues about the molecular basis of life as well. Viruses may have used an RNA genome as well. At some point in the history of life, the RNA genome of these organisms was copied into a DNA genome. Enzymes that perform this activity are not found in eukaryotes but are found in the retroviruses.

The first life was heterotrophic, using the material from the environment for nutrition and energy. The primitive heterotrophs slowly evolved complex biochemical pathways that enabled them to use a wider variety of nutrients to synthesize their own building blocks, and to be less dependent on the environment. They also evolved anaerobic respiratory processes to convert nutrients into energy. However, these organisms required more nutrients than they could synthesize. Life would have ceased to exist if autotrophic nutrition had not developed. The pioneer autotrophs developed primitive photosynthetic pathways in which solar energy was captured and used in the synthesis of carbohydrates from carbon dioxide and water.

The autotrophs fixed carbon dioxide during the synthesis of carbohydrates and released molecular oxygen as a waste product, converting the atmosphere from reducing to oxidizing over billions of years. Some molecular oxygen was converted to ozone, which screens out high-energy radiation. Thus, living organisms destroyed the conditions that made possible the origin of life. Once molecular oxygen became a major component of the earth's atmosphere, both heterotrophs and autotrophs evolved the biochemical pathways of aerobic respiration. Aerobic respiration is much more efficient at energy production than fermentation, making possible the development of more complex multicellular life.

With competition between cells, precursors of the eukaryotes began to consume smaller prokaryotes, and at one point the consumed cell was able to persist in a symbiotic relationship with the larger cell. The resulting organism was the precursor of the modern eukaryote, and the internalized prokaryotic cells evolved into the mitochondria and the chloroplast. The mitochondria and chloroplast still retain a genome today, as well as ribosomes and transcription/translation systems, molecular relics of their previous independent existence. Over time, most of the genes of the mitochondrial genome were transferred to the nucleus, leaving only a few remaining genes that are transcribed and translated in the mitochondria. The eukaryotes have since developed into complex life forms, including all multicellular life found on earth. The earliest eukaryotic fossils are about 2 billion years old, while the oldest prokaryotic fossils appear about 3.5 billion years ago, indicating that 1.5 billion years passed between the development of prokaryotes and eukaryotes. Another 900 million years passed before multicellular life developed about 1 billion years ago. Life appears to have evolved rapidly on earth, soon after conditions allowed it, but another 3 billion years elapsed before multicellular organisms appeared. The slow evolution of more complex organisms may have been limited by the atmospheric oxygen content required to support the metabolism of larger organisms.

Physical Factors in the Evolution of Life

As life has evolved on earth, it appears likely that physical factors have had a strong influence on the development of life. These have included the movement of the continents, changes in earth's orbit around the sun, changes in solar brightness, major volcanic events, and impacts with comets or meteors. The history of life indicates that the earth passed through several periods of major extinction in which a large percentage of species disappeared within a relatively short period of time. These mass extinctions appear to correlate with climate changes induced by one or more factors.

The continents are thin plates of solid rock floating on top of the thick molten mantle beneath. Movement of the mantle, perhaps caused by convection currents in the earth's interior, drive the continents and the ocean basins to move over very large periods of time in a process known as continental drift. The map of the globe has changed dramatically over geologic time and the process continues today, as evidenced by occasional earthquakes driven by the movement of continental plates. As the continents have moved, they have at times moved together and then apart again, changing ocean currents, climate, and sea level.

Major changes in the earth's climate have occurred through the history of life and changed life significantly. The earth has been through periods of glaciation in which ice sheets covered the poles and much of the continents. Glaciation at times has locked up so much water as ice that the ocean sank by 50 meters. Cold climate periods have alternated with much warmer climates than today. The current climate of earth is relatively cold in comparison to most of the earth's past history. These changes in climate may be traced in part to regular changes in earth's path around the sun. Continental drift has also changed the weather. About 250 million years ago, the continents came together to form one supercontinent called Pangea. With changes in climate caused by changing ocean currents and the distance of the continental interior from the moderating influence of the oceans, the collision of the continents caused immense volcanic eruptions that cooled the climate, caused glaciation, and induced a mass extinction.

Another mass extinction 65 million years ago appears to have been caused by a meteor impact. The Cretaceous period ended abruptly at this time. A layer of iridium metal in sediments, an element that is common in meteors, coincides with this time, and a potential crater for such an impact has been identified near the Yucatan region of Mexico. Other impacts may have contributed to mass extinctions as well.

The Geologic Periods of Life's History

The first prokaryotic cells evolved about 4 billion years ago, followed by eukaryotes about 2.5 billion years later (see Timeline). Multicellular organisms first originated several hundred million years later. By the **Precambrian** era 600 million years ago, multicellular organisms were flourishing in the seas, with plankton and protists feeding on algae and prokaryotes. Most Precambrian organisms had soft bodies and left few fossils although a few valuable examples of Precambrian invertebrate fossils have been discovered

Origin of Life - Timeline

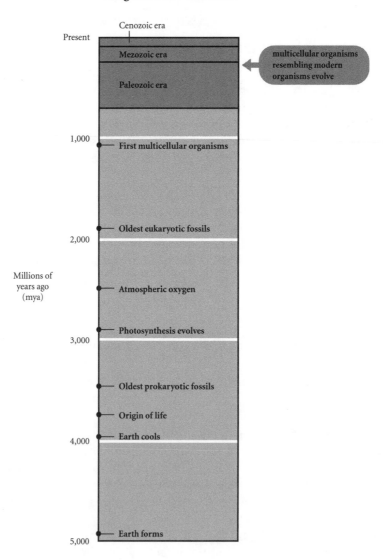

The **Cambrian** era followed at about 540 million years ago, in which the climate and increased oxygen levels led to an explosion in the variety of multicellular life. A rich fossil deposit in the Burgess Shale in Canada has revealed the presence of the first representatives of all major animal phyla found today, including the chordates. Many other fossil organisms appear to represent distinct body plans and evolutionary lines that have not survived. The rich diversity of the Cambrian era suggests that in the rapid explosion of life at this time, diverse body plans evolved, some of which were selected against through random influences or natural selection, leaving behind the ancestors of modern phyla.

In the **Ordovician** period 500 million years ago, plants colonized land for the first time. Glaciation at the end of the period 440 million years ago triggered one of the first mass extinctions. In the **Silurian** period 440 million years ago, ocean animals recovered from the extinction and the first terrestrial animals, arthropods appeared. In the **Devonian** period 400 million years ago, fish diversified and life on land increased, both plant and animal, with the first gymnosperms and the first amphibians. Another extinction marked the end of the Devonian followed by the **Carboniferous** period 345 million years ago, in which insects evolved wings and reptiles evolved from amphibians. Reptiles evolved through the **Permian** period up to 245 million years ago, at which time a mass extinction was caused by the formation of a single large continent and associated volcanic activity.

Reptiles continued to evolve through the **Triassic** period (245–195 million years), as gymnosperms became the dominant plant life on land, the **Jurassic** period(195–138 million years), and the **Cretaceous** period (138–66 million years). During the Jurassic, mammals evolved and bony fish dominated the oceans. In the Cretaceous period, flowering angiosperms evolved. The end of this period was marked by a mass extinction caused by a meteor impact in which all large land vertebrates became extinct, including the dinosaurs.

The mammals of the Cretaceous were small, but after mass extinction ended the period, mammals developed into a large variety of new forms that dominated the land. The continents started to resemble the modern continents, and through this period birds and reptiles evolved into more modern forms. The modern period, the **Quaternary** era, began about 2 million years ago with major glaciation that continued up to about 10,000 years ago with the last glacial retreat.

6.2   EVIDENCE OF EVOLUTION

Scientists have uncovered various types of evidence that evolutionary change has occurred over time, some of which are outlined below.

The Fossil Record

Fossils are preserved remains or impressions in rocks from life forms in the past. Fossils provide the most direct and abundant evidence of evolutionary change. These ancient remains are generally found in sedimentary rocks and were formed when animals settled in sediment that was later covered by many additional layers of sediment. This sediment turned to rock with time and pressure. Fossils generally contain an imprint of the hard, bony parts of organisms. In some exceptional cases organisms may die in sediments that are very low in oxygen such as swamps, and the soft parts of the body may leave an imprint as well, providing information about external coatings and internal anatomy. Sedimentary deposits are laid down in bodies of water over time, with older deposits found in the bottom layers and progressively more recent deposits closer to the surface. The age of a fossil can to some extent be estimated based on the layer of sediment it is found in. Additional clues to the age of a fossil can be determined based on radioactive dating of the fossil or the surrounding rock. Carbon dating is of use for organic material that is no more than a few-thousand years old. ^{14}C is a radioactive isotope of carbon with a half-life of 5,700 years. ^{14}C is created in the atmosphere by solar radiation and is in equilibrium with nonradioactive ^{12}C. Living organisms have the same ratio of $^{14}C/^{12}C$ as the atmosphere, but as soon as organisms die, they stop taking in ^{14}C from the atmosphere. Their age can be determined by the ratio of $^{14}C/^{12}C$ in their body as long as there is ^{14}C remaining, making this isotope useful only for a few thousand years, but not the vast scope of time required for early life. Charcoal is often dated using ^{14}C, placing the date of human encampments. Other isotopes such as potassium/argon are used to determine the age of rocks over billions of years old.

A great deal has been learned about the history of life through fossils, but the fossil record is incomplete. Only a small fraction of past species that have existed have been observed in fossils. The conditions required for optimal fossil preservation limit the number of fossils that form. Most fossils are of hard-bodied animals, and fewer fossils of soft-bodied organisms are found. A vertebrate is more likely to leave a fossil than a worm or a jellyfish. Despite the limitations of the fossil record, a great deal has been revealed about patterns of evolution. Further back in the fossil record, a greater variety of organisms appear compared to today. Intermediate forms in evolution have been observed in several cases and document the gradual evolution of species in several cases, including whales, horses, and man.

Comparative Anatomy

Animals with similar anatomy may be related in their evolutionary descent or may be related by function. When we compare the anatomies of two or more living organisms, we can not only form hypotheses about their common ancestors, but we can also find clues that shed light upon the selective pressures that led to the development of certain adaptations, such as the ability to fly. Comparative anatomists study homologous and analogous structures in organisms.

Homologous structures have the same basic anatomical features and evolutionary origins. They demonstrate similar evolutionary patterns with late divergence of form due to differences in exposure to evolutionary forces. Examples of homologous structures include the wings of a bat, the flippers of a whale, the forelegs of a horse, and the arms of a human, all of which were derived from a common mammalian ancestor but evolved into different forms for different functions. The evolution of different forms from a shared ancestor is called **divergent evolution**.

Analogous structures have similar functions but may have different evolutionary origins and entirely different patterns of development. The wings of a fly (membranous) and the wings of a bird (bony and covered in feathers) are analogous structures. Analogous structures resemble each other but are not be used as a basis for phylogenetic classification. **Convergent evolution** is the mechanism by which analogous structures arise in different organisms to perform similar functions.

Comparative Embryology

Close study of different organisms' embryonic development can tell us much about their evolutionary ancestry. For example, during embryonic development, human resembles the embryonic development of other vertebrates, suggesting a common ancestry and developmental history between humans and other organisms. It does not suggest that human early development is identical to that of these organisms. The human embryo develops gill slits, suggesting a relation to the fishes and other vertebrates, but in humans the gill slits evolve later in development and perform other functions. Sea squirt (tunicate) larvae and amphibian embryos both have a notochord that the adults lack, indicating they are both members of the chordate phylum derived from a common ancestor. The earlier the stage at which embryonic development begins to diverge, the more dissimilar the mature organisms are. For example, it is difficult to differentiate between the embryo of a human and that of an ape until relatively late in the development of each embryo. Other related animals diverge soon after the early cell cleavages following fertilization. Other embryonic evidence of evolution includes such characteristics as teeth in an avian embryo (recalling the reptile stage); the resemblance of the larvae of some mollusks (shellfish) to annelids (segmented worms); and the tail of the human embryo (indicating relationships to other mammals).

Molecular Evolution

Most organisms demonstrate the same basic needs and share the same metabolic processes. They require the same basic nutrients and contain similar cellular organelles and energy storage forms (ATP). The energy-producing pathway glycolysis, for example, evolved early in the history of life and is basically the same in most organisms. The closer the organisms are in an evolutionary sense, the greater the similarity of their genetic information. The large ribosomal RNA has been conserved enough that it can be used to compare eubacteria, archaebacteria and eukaryotes to determine their evolutionary relationship. Protein-coding genes from different organisms can be compared for the same purpose. The more similar the genetic sequences are, the more closely related the animals are. Proteins with rigid structural requirements evolve slowly over time. Some changes, however, have no effect on the function of a protein and will accumulate over time in a gene. The rate of change in a gene can be calculated to derive a "molecular clock" based on the rate of genetic change of a gene between species to derive the time since they shared a common ancestor.

Vestigial Structures

Vestigial structures are structures that appear to be useless in the context of a particular modern-day organism's behavior and environment. It is apparent, however, that these structures had some function in an earlier stage of a particular organism's evolution. They serve as evidence of an organism's evolution over time, and can help scientists trace its evolutionary path. There are many examples of vestigial structures in humans, other animals, and plants. The appendix, small and useless in humans, assists digestion of cellulose in herbivores, while the animal-like tail is reduced to a few useless bones (coccyx) at the base of the spine of humans.

6.3   MECHANISMS OF EVOLUTION

Populations in Evolution—Hardy Weinberg Equilibrium

Evolution is a change not in an individual, but in a population or a species over many generations. A well known incorrect model of evolution called Lamarckian evolution postulated that evolution was the result of change in an individual in response to the environment, such as a giraffe evolving a long neck because it stretches its neck to eat, but this is not the case. Animals do not evolve based on the desirability of a trait to an individual, its use, or its disuse. To examine evolution, it is necessary to think not just of single organisms, but of all of the individuals of a given species that interact, called a population. Genetics states that individuals have a genotype that is responsible for their phenotype, their physical expression of heritable traits. The population is composed of many genotypes and phenotypes. The sum total of all of the alleles in the population is called the **gene pool**. Variation in the gene pool leads to variations in individual phenotypes in the population.

A measure of the genetic variation in a population is the **allele frequency** of a gene. The allele frequency of a gene is the number of copies of an allele divided by the total number of copies of the gene in the population. For example, if a population contains 1000 diploid cats, of which 100 are homozygous for a recessive allele b involved in ear shape, 400 are heterozygous bB and another 500 are homozygous BB for the dominant allele, what is the allele frequency of the b allele? The total number of alleles is 1000 individuals × 2 copies = 2000 total alleles. The number of b alleles is $(100 \times 2) + 400 = 600$ copies (the number of b alleles in homozygotes and heterozygotes). The allele frequency is $600/2000 = 0.3$. The total frequency of all alleles in a population must be 1.0 (=100%).

Sexual reproduction generates variation in the gene pool of a population. Meiotic recombination, independent segregation, and matching of new alleles from different parents in zygotes all contribute to the variability found in a population. Mutation is another important contributor to variation in the gene pool. Changes in allele frequency over time are the basis of changes in individual phenotypes found in the population.

Hardy-Weinberg and Population Dynamics

The allele frequencies in a population determine the genotypes and phenotypes of offspring in each generation. If nothing acts to change the allele frequencies, the population will remain the same from one generation to the next over time. This idea is the foundation of the description of population genetics called the **Hardy-Weinberg rule**. If the allele frequencies in a population do not change between generations, the population is at Hardy-Weinberg equilibrium and evolution is not occurring. If the allele frequencies do change over time, the population is not at equilibrium, and evolutionary forces must be acting on the population. The following conditions must be met for a population to remain at Hardy-Weinberg equilibrium for allele frequencies in a population:

1. Random mating occurs.
2. The population is large enough to avoid random statistical fluctuations in allele frequencies.
3. There is no mutation.
4. There is no migration into or out of the population.
5. There is no natural selection.

Under the above conditions, there is a free flow of genes between members of the same species, while the total gene content is continually being shuffled. We can perform a theoretical cross of the whole population to demonstrate mathematically that the gene pool frequencies remain constant generation after generation. Let us assume for the original gene pool that the gene frequency of the dominant allele for tallness, (T), is p = 0.80 , and the gene frequency of the recessive allele, (t), is q = 0.20. Thus p = 0.80 and q = 0.20. The parents are crossed and their offspring frequencies are shown with a Punnett square below.

Punnett Square

		Possible Sperm		allele frequency in gene pool
		0.8T	0.2t	
Possible Ova	0.8T	.64 TT	.16 Tt	
	0.2t	.16 Tt	.04 tt	

allele frequency in gene pool

The resulting gene frequency in the F_1 generation is: 64 percent TT, 16 percent + 16 percent or 32 percent Tt, and 4 percent tt.

If all of the assumptions to maintain Hardy-Weinberg equilibrium are met, the frequencies of the above F_1 generation are applicable in calculating the frequencies for the F_2 generation.

In working out population genetics problems, we must recall that p = gene frequency of the dominant allele, q = frequency of the recessive allele, and p + q = 1, since frequencies of the dominant and recessive alleles total 100 percent.

Now, using the Punnett square, we may calculate the offspring in the population with alleles p + q. We have:

Punnett Square with pq

Thus, when we cross p + q , we obtain $p^2 + 2pq + q^2 = 1$ (the '1' indicates that the total is 100 percent of the offspring). Note that the result is scientific confirmation of an obvious algebraic identity: $(p + q)(p + q) = p^2 + 2pq + q^2$. The official key for working out the following problem is as follows:

> p = frequency of dominant allele (C)
> q = frequency of recessive allele (c)
> p^2 = frequency of homozygous dominant individuals (CC)
> 2pq = frequency of heterozygous individuals (Cc)
> q^2 = frequency of homozygous recessive individuals (cc)

Another mathematical application of the Hardy-Weinberg principle is explained below:

Problem: In a certain population at Hardy-Weinberg equilibrium, the frequency of dominant homozygous curly hair (CC) is 64 percent. What percentage of the population has curly hair?

Solution: According to the key, p represents the frequency of the dominant allele (C), while q represents the frequency of the recessive allele (c). We are told that the CC frequency is 64 percent. This means that $p^2 = 0.64$ or $p = 0.8$. Since $p + q = 1$, $q = 1 - 0.8$, or 0.2. An individual with curly hair may be either CC or Cc. The frequency of each genotype $= p^2 + 2pq + q^2$.

$p^2 = 0.64$ or 64 percent homozygous curly hair

$2pq = 2(0.8)(0.2)$ or 32 percent heterozygous curly hair

$q^2 = 0.04$ or 4 percent homozygous straight hair

Therefore, the percentage of the population possessing curly hair is as follows:

$p^2 + 2pq = 64$ percent $+ 32$ percent $= 96$ percent

Evolutionary Forces that Disrupt Hardy-Weinberg Equilibrium

The Hardy-Weinberg principle describes the stability of the gene pool. However, no population stays in Hardy-Weinberg equilibrium for very long, because the stable, ideal conditions needed to maintain it do not exist. As conditions change, the changes in the gene pool change the population. Variability within a species also promotes evolutionary trends. Various factors may be introduced into a population to cause change in its gene pool, and other factors influence the variations that are selected for in a given gene pool. Changes in the gene pool induced by forces acting against Hardy-Weinberg equilibrium are the basis of evolution.

Mutation

To maintain equilibrium in the gene pool, there can be no mutation. This condition is not possible to achieve in the real world. Although DNA replication through meiosis and mitosis is extremely accurate, unavoidable errors leading to mutation do still occur at an appreciable rate over time. Cosmic rays and other mutagenic factors are always present. Mutations can be new alleles that did not previously exist, alleles that cause a minor change in activity of a protein, or mutations that lead to significant changes in body plan. Mutations in most cases will be either neutral, having no affect on phenotype, or negative, having a harmful affect on the survival or reproduction of the individual. Mutations are an important source of variation in the gene pool. New phenotypes in a population are the raw material that natural selection acts on to drive evolution. Mutations are the only source of new alleles.

Migration

Different populations can have different allele frequencies in their gene pools. If members of one population mix and interbreed with another population that has different allele frequencies, the gene pool of the population will change. This movement of alleles from one population to another caused by interbreeding is called **gene flow**. Influx from other populations with different gene pools can alter allele frequencies of existing alleles or introduce new alleles.

Population Size

Small populations are statistically more prone than large populations to random events that can alter the gene pool. Changes in allele frequencies in a small population caused by random events are termed **genetic drift**. Although a population may have a large number of alleles, if only a small number of individuals contribute to the next generation, they are not likely to provide a random representation of the diversity in the gene pool. The individuals that do contribute may contain a greater percentage of harmful alleles, or a reduced percentage of favorable alleles than the rest of the population, skewing the gene pool of the next generation. Two circumstances in which genetic drift can occur are bottlenecks and the founder effect.

In a **bottleneck**, a population is reduced to a small number, perhaps through disease, natural disaster, or over-hunting. The individuals left reproduce, but the following generations reflect the gene pool of the individuals that survived the bottleneck, not the larger population that existed before. As an example, if a jar contains a mixture of 1,000 black and 200 white marbles (representing alleles in a gene pool), out of which a person randomly selects 10 marbles, the 10 marbles are not likely to represent the exact proportions in the original pool. If the choice contains 5 black and 5 white marbles, then future marble generations will reflect these new allele frequencies in the marble gene pool. If 10 black marbles are selected, then the white allele will be lost from the marble gene pool. If the white allele represented a disease resistance allele, the bottleneck may cause it to be lost.

Inbreeding is a significant problem following a bottleneck, since the individuals that mate in a small population are more likely to be related and to carry the same harmful recessive genes than in a large population. A major problem in breeding programs for endangered species is that even if the population size is increased, the variability in its gene pool is so reduced that the species may remain vulnerable to infections or other problems that greater diversity in the gene pool might have defended against. Such programs try to prevent mating between relatives to reduce the hazard of harmful recessive homozygous genes emerging from a bottleneck.

The **founder effect** is a similar situation, observed when a small number of individuals of a species migrate into a new habitat. A common example is the colonization of a volcanic island by organisms from another body of distant land. If only a few individuals colonize the island, the new population will reflect only their gene pool and not the larger gene pool of their origin.

Nonrandom Mating

For Hardy-Weinberg to be true, individuals within a population must choose their mate randomly, without respect to their phenotype. If the phenotype influences mate selection, then the genotypes and phenotypes of the population will be changed as well. If the non-random mating selects individuals that are similar for mates, then the offspring will have an increased prevalence of homozygotes than Hardy-Weinberg would predict. Self-fertilization in plants has just this effect, reducing heterozygotes and increasing the abundance of homozygotes.

Natural Selection

Populations of organisms are not all the same. Genetics explains the mechanisms by which offspring inherit traits from parents and the mechanisms by which sexual reproduction introduces variations in the offspring. Mutation is another source of variability in a population. These variations in genotype in a population lead to variations in phenotype. Individuals with different phenotypes, different patterns of physical traits and inherited behaviors, may survive and reproduce in an environment with varying efficiency. The differential production of offspring based on inherited traits is the basis of **natural selection** as described first by Charles Darwin.

Fitness is a key description of natural selection. The fitness of an organism is described by its ability to contribute its alleles and traits to future generations. The factors that will affect fitness include the ability of an organism to survive to reproductive age, to mate and produce offspring, and to raise offspring to maturity. Factors affecting fitness may include the ability to escape predators, the ability to gather food effectively, resistance to disease, ability to attract mates, and care provided to offspring. An individual that has a long life span but few offspring will have poor fitness if other individuals with different traits leave more offspring. An animal that increases the survival of its offspring by caring for them will have greater fitness than an animal that does not provide care for offspring and has a poor rate of survival for offspring. Different organisms tend to increase fitness by either having more offspring that receive little parental care or having fewer offspring that receive more parental care.

Organisms with traits that favor survival, with greater fitness, are more likely to pass on those traits to the next generation than other organisms that lack such traits. From the population genetics point of view, alleles that favor survival and reproduction will be transmitted from one generation to the next more efficiently, and will over time increase in frequency in the population, increasing the frequency of the associated phenotype as well. None of the other forces altering the gene pool and affecting evolution are truly adaptive, but natural selection by definition causes species over time to be better adapted to their environment.

The effect of natural selection on a species depends on the type of selection that takes place. Three types of selective pressure are **stabilizing selection**, **disruptive selection,** and **directional selection**. In a population there is usually a bell-shaped distribution of traits. For example, humans do not exist in a specific set of heights, but over a bell-shaped distribution of heights in the population. Stabilizing selection for a trait does not change the average but tends to sharpen the curve. For example, in a population of bobcats, animals with very large feet may be clumsy and animals with very small feet may sink in snow, so natural selection would be stabilizing toward an average foot size. Disruptive selection tends to select for individuals at either extreme of the distribution for a trait, the opposite of stabilizing selection, leading to two different peaks in the distribution of the population. Directional selection describes a situation in which natural selection drives a change in the average for a trait in the population. If giraffes live in an arid region, for example, and vegetation is accessible only to the tallest members of the population, then selection over many generations will tend to drive evolution of longer necks.

Individual selection acts on an individual and his direct descendants. In social animals, another form of natural selection called **kin selection** takes place. Individuals will share many of the same alleles with close relatives such as siblings and cousins. Natural selection requires the propagation of an organism's alleles in the next generation, but they do not necessarily need to come from the same organism. Lions are social animals, living in prides of related females with their young. If a lion helps to rear and protect her sister's young, she increases her own fitness through kin selection. Insects and other organisms with highly cooperative social systems provide examples of kin selection. Kin selection may be the basis of altruistic behavior observed in nature, in that seemingly selfless acts of sacrifice may increase fitness by increasing the survival of an organism's alleles through his/her kin.

A form of natural selection probably occurred before the origin of the first cell in the development of the first replicating genomes and other prebiotic components. Those molecules that replicated more efficiently had greater fitness than other molecules and came to predominate, carrying their features forward into the first cells and the varieties of life that followed.

Speciation

A **species** is a group of organisms that is able to interbreed productively with the rest of its group and not with other organisms. The appearance of an organism is not the key criteria in deciding whether it belongs to a different species from other similar organisms. The key criteria that defines a species is **reproductive isolation**. If two populations of bird live in different swamps and never mix, that alone does not make them two different species. If the birds are placed together and are still able to breed and produce fertile offspring, they are the same species. The offspring must be fertile for the breeding animals to be considered members of the same species. For example, donkeys and horses are able to breed and produce offspring, mules, but since mules are sterile, donkeys and horses are distinct species. Reproductive isolation between species makes each species a separate unit for evolution to work on. Within a species, genes can be exchanged within the gene pool.

If the gene pools within a species become sufficiently different so that two individuals cannot mate and produce fertile offspring, two different species have developed. No gene flow is possible between two different species. The factors that lead to speciation involve the separation of two gene pools from each other. **Cladogenesis** is the formation of two species from one ancestor species. Cladogenesis can occur when populations occupy the same area or when they are separated geographically from each other. In **allopatric speciation**, populations are separated by a geographic barrier followed by reproductive isolation. The geographic barrier could be a rising sea that divides population, the distance between a series of islands, a river, or a highway. For example, a species of newt that lives in cold moist conditions could be found across a large mountain range (see figure). A long-term change in climate could make the valley regions warmer and drier, leaving only the mountains as an environment for isolated populations of the newt. If the newts in different populations evolve under selective pressure or genetic drift while they are geographically isolated, they may no mate with each other. If the climate cools once again and the newt populations mix in the valleys but do not interbreed, then allopatric speciation occurs, creating distinct species of newt where before there had been one. **Adaptive radiation** occurs when a species enters a series of distinct habitats with open ecological niches and geographic separation permits adaption of different populations to each of the habitats. The evolution of Darwin's finches in the Galapagos Islands is an example of adaptive radiation.

Speciation

A. Original Range—Moist Conditions

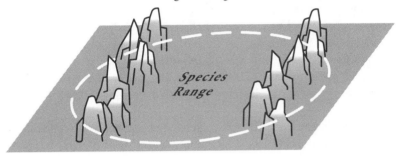

B. Climatic Change—Geographic Isolation

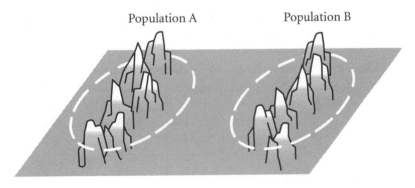

C. Reproductive Isolation—Allopatric Speciation

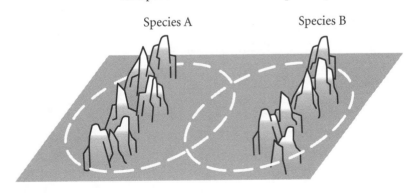

Speciation by populations that occupy the same region is called **sympatric speciation**. For organisms to become a new species through sympatric speciation they must be reproductively isolated while still living in the same region. The most common way for this to occur is through a sudden dramatic genetic change, usually as a result of polyploidy of the genome. Polyploidy can occur within a species if an individual spontaneously (through an accident of meiosis) produces offspring with twice the normal chromosomal number. Polyploidy can also result when a cross between two related species produces a hybrid with the chromosomal complement of both parents. Since the chromosomes don't match, the hybrid cannot go through meiosis to reproduce but if it can reproduce, asexually then it can propagate.

A tetraploid individual cannot produce fertile offspring if it mates with a diploid, but could fertilize itself through self-pollination if it is a plant. Polyploidy is very common in plants and polyploids often appear to have a selective advantage over their diploid precursors.

Mechanisms of reproductive isolation that cause speciation are divided into two groups, called **prezygotic reproductive barriers** and **postzygotic reproductive barriers**. Prezygotic barriers prevent two species from mating. An example of a prezygotic barrier is two species that live in the same region, but in different localities so that they do not contact each other. Another example is different mating rituals, or changes in genitals that prevent two species from mating. If sperm from one species cannot fertilize ova from another, this is another example of a prezygotic barrier. Postzygotic barriers do not prevent mating between species but prevent the formation of fertile offspring.

The rate of speciation and the relationship between speciation and evolution is often debated. One point of view contends that species change slowly and gradually in their form over time. A contending point of view called **punctuated equilibrium** proposes that evolution occurs rapidly in association with speciation in small isolated populations, followed by long periods in which a species changes little.

Chapter 6: Review Questions

1. Which of the following was not present in the prebiotic environment?

   (A) carbon dioxide
   (B) ammonia
   (C) water
   (D) methane
   (E) oxygen

2. The earliest enzymes are thought to have consisted of

   (A) protein
   (B) RNA
   (C) DNA
   (D) saccharides
   (E) triglycerides

3. Which of the following is found in modern life forms?

   (A) L-amino acids
   (B) D-amino acids
   (C) L-sugars
   (D) D-sugars
   (E) more than one of the above

Questions 4–7 refer to the following:

   (A) Cambrian
   (B) Devonian
   (C) Jurassic
   (D) Cretaceous
   (E) Quaternary

4. Mammals evolved during this geologic period

5. Gymnosperms developed during this geologic period

6. Chordates developed during this geologic period

7. One method of dating organic material involved finding the ratio of _____ .

   (A) carbon dioxide to carbon monoxide
   (B) ^{18}O to ^{16}O
   (C) water to carbon dioxide
   (D) ^{14}C to ^{12}C
   (E) carbon monoxide to oxygen

8. _____ structures have _____ functions/features and the same evolutionary origin while _____ structures have _____ functions/features but different evolutionary origins.

   (A) Analogous; similar; homologous; different
   (B) Analogous; similar; homologous; similar
   (C) Homologous; similar; analogous; different
   (D) Homologous; similar; analogous; similar
   (E) None of the above

9. Which of the following is NOT a condition of Hardy-Weinberg equilibrium?

   (A) No natural selection
   (B) Large population size
   (C) No mutation
   (D) No migration into or out of the population
   (E) Non-random mating

10. What is meant by an organism's "fitness," in an evolutionary sense?

   (A) The organism uses oxygen efficiently
   (B) The organism survives to adulthood
   (C) The organism survives to adulthood and reproduces
   (D) The organism survives to adulthood and finds a mate
   (E) The organism is not killed before adulthood

Please see the answers and explanations to these review questions in Section V of this book.

ORGANISMS AND POPULATIONS

The great diversity of living things is the product of billions of years of evolution. Scientists have attempted to sort out and discern the relationships among the vast number of different types of organisms. From this grew the science of the classification of living things, taxonomy, and a new nomenclature for biological science. Scientists also investigated the structure of plants and animals, and the functions of their organs and tissues. To understand living things, biologists classified and studied cells, tissues, and organs, breaking organisms down into their fundamental units. These organisms also live in relationship to each other and the environment; these relationships and interactions are the focus of ecology.

The chapters of this section will explore our knowledge about the diversity of organisms: their biological structure, how they function individually and in populations, and their relationship with their biotic and abiotic environments.

Diversity of Organisms

The science of classification of living things, and the nomenclature it utilizes, is known as **taxonomy**. The modern classification system, originated by Carolus Linnaeus in the 18th century, seeks to group organisms in a hierarchical basis. Early efforts relied heavily on similar anatomy and morphology to group organisms together. The modern classification system relies on evolutionary relationships between organisms, known as **phylogeny**. The bat, whale, horse, and human are placed in the same class of animals (mammals) because they are believed to have descended from a common ancestor. When possible, the taxonomist classifies all species known to have descended from the same common ancestor within the same taxonomic group, or **taxon**.

The methods used to discover the evolutionary relationships of organisms include all of the methods described in the chapter about evolution, including studying the relative DNA sequence of genes, comparative anatomy and embryology, and the fossil record. By describing all of these factors and comparing them to other organisms, scientists can construct **phylogenetic trees** (see figure), depicting the position of the last common ancestor shared by two species. All of the animals that share a common trait such as an anatomical feature or a DNA sequence may be thought of as sharing a common ancestor that also shared the same trait. In some cases, based on analysis of the rate of mutation of a gene or the age of a fossil, the age as well as the position in the phylogenetic tree of an ancestor can be determined.

Phylogenetic Tree

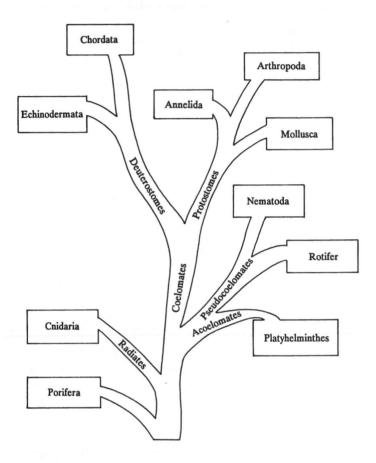

7.1   TAXONOMIC CLASSIFICATION

Classification and Subdivisions

The evolutionary classification scheme groups organisms in increasingly broader taxonomic groups, starting with kingdom as a very broad grouping. Each kingdom has several major phyla that are further divided into classes, orders, families and genera. Finally, the species is the smallest subdivision. A species is a set of organisms that is able to reproduce and produce fertile offspring. In the Linnaean naming system, a species is indicated in a binomial system by its genus, a group of closely related species, and its species printed in italics, with the genus capitalized and the species name in lower case. For examples, the species name of humans in the Linnaean system is *Homo sapiens*, where *Homo* is the genus name and *sapiens* is the species.

The order of classificatory divisions is as follows:

KINGDOM

PHYLUM

CLASS

ORDER

FAMILY

GENUS

SPECIES

The complete classification of humans is:

Kingdom:	Animalia
Phylum:	Chordata
Subphylum:	Vertebrata
Class:	Mammalia
Order:	Primates
Family:	Hominidae
Genus:	*Homo*
Species:	*Sapiens*

One current system of taxonomy has derived a grouping with three domains as the broadest category: bacteria, archaebacteria, and eukaryotes. The classification of an organism or even the structure of the phylogenetic tree can change over time as scientists learn more about species and use new techniques like molecular biology to characterize them. The major groups in this current classification will be described in a survey of the diversity of life on Earth.

7.2   PROTISTS

Protists are a broad group encompassing a variety of organisms that differ in many ways. Protists are single-celled organisms, with some exceptions. Some protists are photosynthetic, while others are heterotrophic. Most live in aqueous environments. Some, but not all, protists are able to move, through the use of cilia, flagella, or amoeboid motion. Reproduction in protists is also diverse, with both sexual and asexual mechanisms of reproduction found.

Examples of Protists

Amoebas

Amoebas have a undefined body shape that changes over time with streaming of cytoplasm in different directions to form extensions of the cell called pseudopods. Pseudopods form with the assistance of cytoskeletal elements and are involved in movement. When a pseudopod forms, cytoplasm can flow into it, slowly moving the cell in the direction of the pseudopod. Amoebas move slowly along a surface, feeding as heterotrophs through phagocytosis of food material they find, surrounding the food with pseudopods and ingesting it into the cytoplasm in a digestive vacuole.

Ciliates

Ciliates include some of the largest and most complex protists. A familiar example of a ciliate is the protist *Paramecium*. Ciliates move through coordinated movement of cilia that cover most of the cell surface and can respond to stimuli like food or pH to alter their motion. A semi-rigid covering of the cell called the pellicle helps to give ciliates specialized forms. Ciliates have contractile vacuoles that maintain osmotic balance, pumping water out of the cell to prevent the cell from swelling with osmotic influx of water. An oral groove in *paramecia* is an indentation in the cell surface that is lined with cilia. The cilia beat food particles like yeast cells down the oral groove to a vacuole that surrounds the food. This forms a food vacuole that pinches off from the cell membrane to digest the contents of the vacuole.

Ciliates have nuclei, but in two different types, **macronuclei** and **micronuclei**, both of which contain copies of the genome. The micronuclei are involved in recombination that occurs during a form of genetic exchange called conjugation. *Paramecia* and other ciliates reproduce mitotically, but through conjugation between two cells can exchange genetic information.

Flagellates

Many protists use flagella to propel themselves through water. Flagellates include parasites, including the organism that causes sleeping sickness, and *euglena*, photosynthetic protists. *Euglena* can function as either autotrophs through photosynthesis or as heterotrophs, by feeding on external material. With a flagellum and the ability to detect light, *euglena* are able to move toward light to increase photosynthesis. *Euglena* reproduce asexually through mitosis.

Slime Molds

Slime molds were previously considered to be fungi, partly on the basis of their mode of nutrition. Slime molds are heterotrophic, feeding by surrounding food particles, and are found as either acellular or cellular species. Acellular slime molds are multinucleated, with many nuclei found inside one very large plasma membrane. The multinucleated acellular slime mold can move by cytoplasmic streaming despite the fact that acellular slime molds can be quite large. When conditions do not favor feeding, the mold can specialize to form **fruiting bodies**, extensions on raised stalks that form haploid spores.

Cellular slime molds are similar except that they are not multinucleate. Individual cells remain distinct. Cells may live independently when they are feeding, but when food is scarce, the cells aggregate to form fruiting bodies that create spores.

Algae

Algae are protists that perform photosynthesis, some of which are single-celled while others are multicellular, such as the giant kelp. **Dinoflagellates** are algae with flagella. **Diatoms** are algae with intricate symmetric shells of silicon. **Brown algae** are multicellular and can be quite large, and include the kelps. Algae can have life cycles in which the organism alternates between a diploid spore-forming stage called a **sporophyte** and a haploid gamete forming stage called the **gametophyte**. Some of the algae are often grouped with plants, and it is likely that the plants arose from green algae.

7.3   FUNGI

All true fungi are heterotrophic, parasitic, or saprophytic; none is autotrophic. Fungi have a cell wall made of chitin, distinct from the cell wall of plants (cellulose) or bacteria (peptidoglycan). Fungi are often involved in breaking down organic material, providing a key service to ecosystems. They do this by secretion of enzymes on dead material for digestion of nutrients that are then absorbed. Some varieties utilize extracellular digestion from which nutrients are absorbed. Notable types are mushrooms, molds and yeasts. Yeasts are often studied as a simple example of a eukaryote. The yeast cell has much in common at a basic molecular level with a human cell. Lichens are a symbiotic joining of fungi and algae. Multicellular fungi have filaments called **hyphae** and the collective fungus as a whole is called the **mycelium**.

Fungi can often reproduce both sexually and asexually. Imperfect fungi do not have a sexual reproduction stage. Asexual reproduction takes place in a few ways, including the formation of haploid spores in either **sporangia** or **conidia** or by splitting of a piece of the fungus to reproduce asexually. Sexual reproduction in fungi does not involve distinct male and female gametes but mating types. Fungi are usually haploid, except for a brief diploid stage after fertilization between haploid gametes. The zygote goes through meiosis usually quickly, to produce haploid spores that can develop into mature haploid organisms.

7.4   PLANTS

Plants are generally multicellular eukaryotes that derive energy and nutrition through photosynthesis, using the energy of the sun to produce glucose. Plants have a cell wall made of cellulose and a life cycle that alternates between diploid **sporophytes** and haploid **gametophytes**. This type of life cycle is called **alternation of generations** (see figure). Both the sporophyte and gametophyte can be complex multicellular stages and the sporophyte and gametophyte often look very different. The sporophyte produces haploid single-celled spores that grow into a complete haploid plant, the gametophyte. The gametophyte produces haploid gametes that through fertilization form a diploid zygote that will grow to form the mature diploid sporophyte once again. The extent and dominance of the sporophyte and the gametophyte in the life cycle of plants changed through evolution. The earliest plants had large gametophytes while more modern plants exist mostly in the sporophyte stage with a transient small gametophyte. The alternation of generations life cycle with an extended haploid phase is in contrast to animals, where a brief haploid phase, the gamete, is observed.

**Alternation of Generations
in a Conifer (Gymnosperm)**

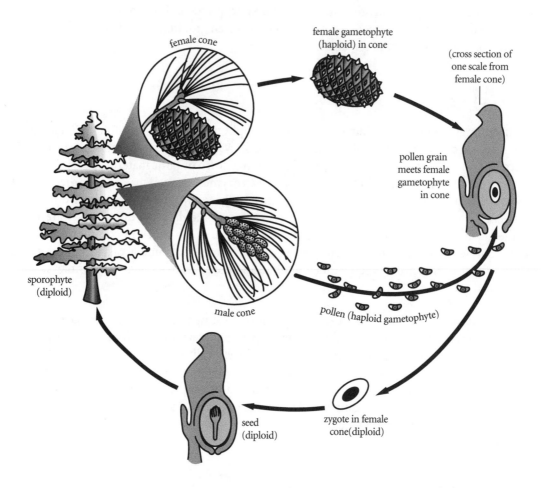

Plants probably arose from green algae growing in moist soil around shallow water. In colonizing land, plants needed to deal with several challenges. These include:

1. Mechanical support against gravity.
2. Different means of reproduction.
3. Prevention of drying out from exposure to air.
4. Movement of nutrients and water from one part of the plant to another.

The cellulose cell wall of plants provides mechanical support. Wood is formed of many cell walls that, when combined together, form a very strong yet light structure. The pressure of water in each cell also presses against the cell wall, helping to provide structure and strength as well. When a plant does not have enough water, the cells shrink and provide less pressure, and the plant wilts. To prevent drying on land, many plants cover leaves with waxy coatings.

To move water and nutrients, most modern plants have tubelike structures involved in moving water from the roots to the leaves, and moving sugars from the areas of photosynthesis in the leaves to the rest of the plant. The tubelike structures make these plants vascular plants, or **tracheophytes**. Vascular plants include ferns, seed plants and flowering plants. The **nontracheophytes** were the first plants. Their size is limited by the lack of a vascular system, so they are usually quite small. Examples of nontracheophytes are mosses. As plants, nontracheophytes do alternate between the haploid and diploid forms in their life cycle. In nonvascular plants, the gametophyte is usually larger and dominant in the life cycle, while in vascular plants the larger and dominant form is usually the sporophyte.

Vascular Plants—Tracheophytes

Most modern plants are vascular plants, or tracheophytes. Tracheophytes have specialized cells called **tracheids** that form tubes and fibers for the movement of fluid in the plant. Vascular plants developed roots and leaves. Tracheophytes can further split into those that form seeds and those that do not. Ferns and horsetails are vascular nonseed plants. The sporophyte of ferns is the large and dominant stage of the fern life cycle, with spores often visible under the leaves. The gametes of ferns need water for fertilization, which is one reason why ferns are usually found in cool wet environments. The gametophyte of this group is usually a smaller structure, but separate and independent of the sporophyte.

The seed plants are vascular plants that include the **gymnosperms**, which produce "naked seeds" and **angiosperms**, the flowering plants. The diploid sporophyte is even more dominant in the seed plants than in the nonseed vascular plants. The gametophyte is usually small and dependent on the sporophyte. Seed plants have separate male and female spore-forming structures (sporangia). The gametophytes develop inside the sporangia. Pollen grains are male gametophytes that are spread by various means to the female sporophyte, where it forms a pollen tube that grows to the female gametophyte. Sperm from the pollen tube fertilize the female gamete to form a diploid zygote. In seed plants, the diploid zygote forms an embryo that halts development at an early stage, forming a **seed**. When the seed finds favorable conditions, the embryo will begin growing again to form the mature sporophyte stage of the plant.

Gymnosperms include the pines and other conifers that often dominate ecosystems in Northern or mountain climates. Wind carries pollen from the male sporophyte to the female cone, where seed development occurs.

Angiosperms are the flowering plants, and have a double fertilization in their life cycle. When angiosperm pollen forms a pollen tube, one sperm fertilizes an egg to form a diploid zygote. Another sperm, however, fertilizes two haploid female nuclei to form a triploid cell that grows to form a tissue in the seed called the **endosperm**. The endosperm provides nutrition to the zygote during development of the sporophyte. Flowers are involved in angiosperm reproduction (see figure). The **stamens** are the male component of the flower where pollen is produced and have a **filament** for support and **anthers** where the pollen is released. The female part of the flower is the **pistil** and includes a **stigma** where pollen is captured, and an ovary at the base. The petals and colors of the flowers are often useful in attracting insects and birds involved in carrying pollen from flower to flower. Angiosperms produce fruits from the ovary of the flower.

Flower Structure

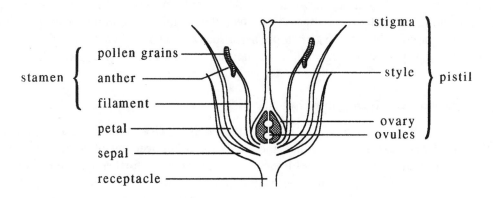

7.5 ANIMALS

Animals are multicellular heterotrophs. As heterotrophs, animals actively acquire food from the environment. Many of the adaptations of animals have evolved for more efficient food capture, including movement and the evolution of complex nervous systems to respond to the environment. Escape from predation in animals is another force driving evolution. Animals have evolved specialized cells and tissues to perform distinct functions.

One of the characteristics of different groups of animals is the body plan. The symmetry of one body plan is organized around a central point radiating outward; this is called **radial symmetry**, and is seen in the sea urchin. An alternative body plan is **bilateral symmetry** in which the body has two sides that are mirror images of each other (see figure). Humans have bilateral symmetry, with the left side of the body reflecting the image of the right side of the body. The head end is called the anterior of the animal and the rear is the posterior. The front, the side with the mouth, is the ventral, and the back is the dorsal (like the dorsal fin).

Body Plans, Symmetry

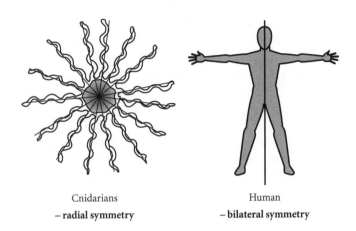

Cnidarians
– radial symmetry

Human
– bilateral symmetry

The development of animals often reveals evolutionary relationships through similarities at different embryonic stages. In early embryonic development, for example, some animals have only two types of cell layers, the endoderm and ectoderm, while others have a third layer, the mesoderm. Cell cleavage during early development is also used to characterize animals. In animals called **protostomes**, cells in the early embryo have spiral cleavage, while in animals called **deuterostomes**, cells are cleaved in a radial pattern. The protostomes include annelids, roundworms, mollusks and arthropods, all of which have bilateral symmetry. Deuterostomes, which have a dorsal nerve cord, include echinoderms and chordates.

An important development has been the evolution of the body cavity. Vertebrates and many other animals have a body cavity between the digestive tract and the exterior of the animal called a **coelom** that is lined with muscle and a membrane (see figure). The coelom is filled with fluid, contains internal organs, and helps in movement by forming a hydrostatic skeleton that the muscles can act against. Animals with a coelom are called **coelomate** and animals without a coelom are called **acoelomate**, in which case the space between the gut and the exterior is filled with tissue.

Coelomate   **Acoelomate**

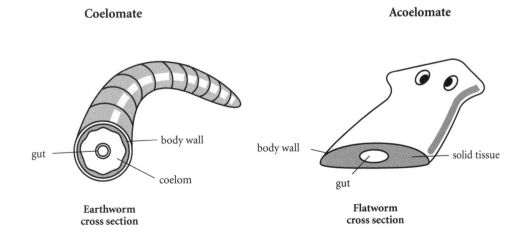

gut — body wall
— coelom
**Earthworm
cross section**

body wall — — solid tissue
gut
**Flatworm
cross section**

Phylum Porifera (Sponges)

The simplest animals are the sponges, placed together in a phylum called porifera. Sponges live in water and resemble colonial organisms. If broken apart, sponge cells can reassemble into a sponge or be split to form several smaller sponges. Sponges do not have organs, although they do have specialized cells. Sponges form sac-shaped structures with specialized cells inside the sac that have flagella to make the water move. The flagella draw water into the sponge through pores and out through separate openings, allowing cells to gather food for the sponge by filtering it from water that is drawn past. Sponges are **sessile**, meaning they do not move, but rely instead on the movement of water to obtain food. Sessile filter feeders represent a highly successful a way of life that is found in several other phyla, including cnidarians, annelids, mollusks, and chordates. Little energy is required to acquire food in this way, although competition for living space and predation are issues that a filter feeder must deal with.

Phylum Cnidaria

Cnidarians evolved early in evolution and include jellyfish, sea anemones and corals. They are aquatic animals with radial symmetry, including tentacles arranged in a circle around a simple gut opening. Cnidarians have a simple digestive cavity with only one opening and two cell layers (see figure). With only two cell layers, cnidarians do not require a respiratory system or a circulatory system. Gases exchange directly between cells and the environment. The tentacles contain structures called **nematocysts** that act as microscopic harpoons laced with toxin the animal can fire to capture prey. These animals have a nerve network that coordinates the movements of the tentacles and body. The life cycle of a cnidarians can go through both an asexual **polyp** stage (anemone shape) as well as a **medusa** stage (jelly-fish shape) that reproduces sexually. Cnidarians often have symbiotic relationships with algae that help provide nutrition. This is the reason that anemones, for example, are often green.

Cnidarian Body Plan

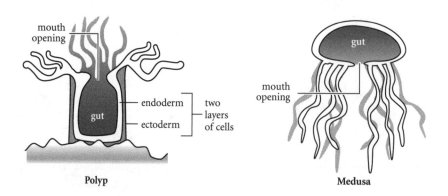

Polyp   Medusa

Phylum Platylhelminthes

Flatworms are ribbonlike and bilaterally symmetrical. They possess three layers of cells, including a solid mesoderm, but are acoelomate protostomes. They have a digestive cavity with only one opening, do not have a circulatory system, and their nervous system consists of eyes, an anterior brain ganglion, and a pair of longitudinal nerve cords. Flatworms such as the well-known planaria are capable of extensive regeneration. Parasites such as flukes and tapeworms are flatworms. Flatworm movement is accomplished through cilia covering the bottom of the animal.

Phylum Nematoda

The roundworm nematodes are protostomes, with a pseudocoelom, lined with muscle on one side. Movement of this muscle layer allows nematodes to wiggle, although not move in a specific direction very effectively. They have digestive tract with both a mouth and an anus. A solid mesoderm is present; a circulatory system is not. They possess nerve cords and an anterior nerve ring. The nematode *C. elegans* is a popular subject of study in molecular biology and developmental biology, since it has a small number of cells (about 950) that can be observed throughout development. The worm shaped body plan has evolved in several different animal groups, probably since the long slender shape allows for better movement through mud or soil than other shapes.

Phylum Annelida

Annelida (for example, earthworms and leeches) are segmented worms that possess a coelom contained in the mesoderm between layers of muscle lining the gut and the body wall. They have well defined systems, including nervous, circulatory, and excretory systems. The segmentation of the coelom forms a hydroskeleton. Movement is coordinated by the nervous system. The circulatory system has several tube-shaped hearts that pump blood through a closed circulatory system. Annelids do not have a respiratory system, but absorb oxygen through their moist surface. Each segment has separate excretory organs.

Phylum Mollusca

The mollusks include snails (gastropods), octopus, squid (cephalopods), and clams (bivalves), animals with diverse structure. Complex protostomes, mollusks have a muscular foot, and a **mantle** that produces the shell found in most mollusks. Some mollusks have a rough rasping **radula** structure that is involved in feeding. Gills for feeding and food gathering are often found as well with cilia on the gills to move water. Gills are located in a space called the **mantle cavity** partially covered by the mantle. Mollusks were probably evolved from segmented protostomes, since some mollusks are still partially segmented, but most mollusks lost segmentation. Mollusks are mostly aquatic except for some snails and slugs.

Phylum Arthropoda

Arthropods have jointed appendages, chitinous exoskeletons, and open circulatory systems. The exoskeleton of arthropods provides protection and support and is the anchor for internal muscles for movement of jointed appendages. For arthropods to grow, they must emerge from their exoskeleton in a process called molting and then grow a new covering. Arthropods have a ventral nerve cord and a complex digestive tract. The three most important classes of arthropods are insects, arachnids, and crustaceans. Insects possess three pair of legs, spiracles, and tracheal tubes designed for breathing outside of an aquatic environment. Tracheal tubes lead from openings in the surface to carry air into the interior tissues. The insect body has three main sections, the head, the thorax and the abdomen. Arachnids have four pair of legs and "book lungs." Examples include the scorpion and the spider. Lastly, crustaceans have segmented bodies with a variable number of appendages. Crustaceans like the lobster, crayfish, and shrimp also possess gills. Arthropods were the first terrestrial animals, and insects with wings were the first flying animals.

Deuterostomes: Phylum Echinodermata

The echinoderms, including sea stars and sea urchins, are spiny and radially symmetrical, contain a water vascular system, and possess the capacity for regeneration of parts. The water vascular system is a hydrostatic system that moves tube feet. Tube feet are involved in movement and capturing food.

Phylum Chordata

Chordates differ from the previous phyla in that a stiff, solid dorsal rod called the notochord is present during embryonic development, as are paired gill slits. Chordates also have dorsal hollow nerve cords and tails extending beyond the anus at some point in their development. The origin of the chordates was probably an organism similar to tunicates, which are sessile filter feeders as adults but have a free-swimming larval stage with a notochord and a dorsal nerve cord.

Vertebrates represent a subphylum of chordates in which the notochord is not found in the adult; it is replaced in the adult by a bony segmented structure called the vertebral column. Vertebrates have two pairs of appendages, and are in general characterized by increasingly complex nervous systems.

The evolution of jaws from gill structures led to the evolution of the modern fishes, including chondrichthyes (cartilaginous fishes such as sharks and rays) and osteichthyes (bony fishes such as trout and perch). The regulation of buoyancy with a swim bladder was another important adaptation in fishes.

About 400 million years ago, a group of fish evolved with simple lungs and lobe-shaped fins that allowed them to occasionally move about on land. With time, these animals evolved stronger limbs that could support their weight on land, evolving into the amphibians, the first terrestrial vertebrates. Amphibians lay and fertilize eggs in water and amphibian embryos develop in water, but adults can move onto land. Examples of amphibians include frogs and salamanders. Gas exchange occurs through the moist thin skin as well as simple lungs or gills for aquatic species.

Reptiles became independent of water and expanded their range by developing hard eggs that can be laid on land without drying out. The egg allows gas exchange and allows embryonic development to occur surrounded by several layers, including the amnion for protection and yolk for nutrition. Reptile lungs are better developed than amphibian lungs and the reptile heart is also more efficient than is found in amphibians. Reptile skin is thicker than amphibian skin allowing reptiles to live on land without drying out.

Birds evolved from reptiles with the development of wings, feathers, and light bones for flight. Birds lay eggs, and most young birds have extended care provided after hatching, unlike most reptiles. Birds have a very rapid metabolic rate compared to reptiles. A famous evolutionary intermediate is Archaeopteryx, a winged creature from 150 million years ago with many features of modern birds as well as of dinosaurs.

Mammals also evolved from the reptiles about 200 million years ago, with mammary glands to nourish young. After the extinction of the dinosaurs 65 million years ago, mammals diversified greatly. Mammals have rapid metabolism and a four-chambered heart, and actively regulate their internal temperature, with hair and sweat glands to regulate temperature. One class of mammals, the monotremes, lays eggs, but the vast majority of mammals give birth to live young. Marsupials give birth to immature young that complete development in a pouch in contrast to the placental mammals, which nourish the embryo through the placenta for a longer period of time before birth.

Primates are a group of mammals that have opposable thumbs adapted for grasping. Primates also have binocular vision for depth perception, a characteristic shared by other vertebrates. Many primates live in trees, but the ancestors of humans such as the australopithecines probably evolved to living on the ground and also to walking upright on two feet for better vision in their ground environment. Fossils reveal the move to upright bipedal locomotion and an increase in brain size in these evolutionary ancestors of humans. Fossils more similar to modern humans such as those of *Homo habilis* and *Homo erectus* reveal tool use as far back as 2 million years ago by hominids.

Chapter 7: Review Questions

1. Which of the following correctly represents classification categories, from largest to smallest?

   (A) kingdom, phylum, class, order, genus, family, species
   (B) kingdom, class, phylum, order, family, genus, species
   (C) kingdom, phylum, class, order, family, genus, species
   (D) kingdom, phylum, order, class, family, genus, species
   (E) kingdom, phylum, class, order, family, species, genus

2. Scientists study the evolutionary relationships of organisms to determine how closely related they are. The evolutionary relationships between organisms are called

   (A) taxonomy
   (B) embryology
   (C) physiology
   (D) genetics
   (E) phylogeny

3. This organism is heterotrophic and can exist in cellular form or multinucleated acellular form which is capable of amoeba-like motion. In unfavorable conditions, fruiting bodies are formed which produce spores.

   (A) slime mold
   (B) sponge
   (C) fern
   (D) algae
   (E) violet (angiosperm)

4. Which organism has radial symmetry?

   (A) frog
   (B) sea urchin
   (C) rabbit
   (D) kangaroo
   (E) crab

Questions 5–7 refer to the following:

   (A) endosperm
   (B) endoderm
   (C) gymnosperm
   (D) angiosperm
   (E) tracheids

5. This group of plants includes the pine

6. This tissue nurtures the developing plant zygote

7. These cells produce tubes and fibers which function in fluid transport in plants

Questions 8–10 refer to the following:

   (A) protostomes
   (B) deuterostomes
   (C) radial symmetry
   (D) bilateral symmetry
   (E) coelom

8. Organisms in phylum *Platyhelminthes* have this type of symmetry

9. Organisms classified in this category show radial cleavage during early development and have a dorsal nerve cord

10. This fluid-filled body cavity is lined with muscle

Please see the answers and explanations to these review questions in Section V of this book.

Structure and Function of Plants and Animals

8.1 ANIMAL REPRODUCTION

One of the fundamental properties of life is that living organisms reproduce themselves, either sexually or asexually. Asexual reproduction produces genetically identical copies of an organism. Asexual reproduction requires minimal energy and can be advantageous when a population is growing rapidly. A population that reproduces asexually, however, will lack genetic variability. In the extreme case, a population produced through asexual reproduction of a single individual would be clonal and homogeneous, with only mutation acting to introduce new alleles in the gene pool over time.

Sexual reproduction increases diversity in the gene pool, allowing organisms to adapt to changing conditions. Sexual reproduction involves the union of haploid cells from two different parents, the male and female, to produce a single diploid cell, the **zygote**, that develops into a new organism. The union of these two haploid cells is called **fertilization**. The production of sex cells through gametogenesis occurs through meiosis (chapter 4). Meiosis changes combinations of alleles through independent assortment and recombination.

Fertilization

In fertilization, the egg nucleus (containing the haploid number, or n chromosomes) unites with the sperm nucleus (also containing n chromosomes). This union produces a zygote of the original diploid or 2n chromosome number. In this way, the normal (2n) somatic number of chromosomes is maintained.

When the egg and the sperm join, the union triggers a cascade of events. These events, which are part of the process of fertilization, may occur either externally or internally. One of the first obstacles to fertilization is for the sperm to encounter the egg. In **external fertilization**, chemical signals aid the sperm in finding an egg of the correct species. In **internal fertilization**, the sperm must pass through the female reproductive tract to encounter the egg. The motility and viability of sperm is affected by the environment it encounters.

Once the sperm encounters the egg, the sperm must penetrate through several layers that surround the egg. In humans, these layers include a layer of follicle cells that surround and support the egg, and a jellylike layer called the **zona pellucida**. In invertebrates such as the sea urchin, the sperm must penetrate the jelly coat and the **vitelline envelope**, which is similar to the zona pellucida. Enzymes contained in the acrosome at the tip of the sperm are released when the sperm contacts the outer layer to digest the glycoproteins and polysaccharides that form this layer. An acrosomal process extends the head of the sperm, allowing the head of the sperm to contact the next layer. Specific recognition of sperm and egg proteins is required at this step. If successful, the sperm and egg membranes fuse and the sperm nucleus enters the egg.

Within seconds after fertilization of the egg by a sperm, the egg reacts to prevent fertilization by additional sperm. The reaction includes a change in membrane potential, a release of calcium in the egg cytoplasm, and a change in the egg membranes that blocks further entry of sperm. The increase in cytoplasmic calcium stimulates the beginning of egg development. The egg is initially paused in the second meiotic division, and completes meiosis only after fertilization, expelling a polar body.

External Development

External development occurs outside the female's body, in water or on land. The eggs of fish and amphibians, for example, are fertilized externally in water. The embryo then develops inside the egg, feeding on the yolk in the egg. Such embryos are given very little parental care; therefore, many embryos must be born to ensure survival of at least some, because of threats from predators, environmental extremes, and other hazards.

Aquatic animals often release eggs and sperm into the environment where fertilization takes place. Sessile species will often release huge numbers of gametes at a specific time of year, leaving it to chance to determine whether or not fertilization occurs. The large number of gametes released helps improve the chances that the correct gametes will meet and fertilization will occur. External fertilization is not an option on land, so animals on land use internal fertilization, carrying the male gametes into the body of the female. The penis, reproductive structure of the male, carries sperm into the female reproductive tract through the vagina. The number of eggs produced depends upon a number of factors. One of these factors is the type of fertilization employed. Because very few sperm actually reach the egg during external fertilization, this process requires large quantities of eggs to ensure at least some degree of fertilization success. Finally, the less care the parents provide, the more eggs are required to guarantee survival of enough offspring to continue the species.

External development on land occurs in reptiles, birds, and a few mammals, such as the duck-billed platypus. There are many adaptations for embryonic development within eggs and on land. One of these is a hard shell for protection, which is brittle in birds and leathery in reptiles. Extraembryonic membranes also help to provide a favorable environment for the developing embryo.

Types of extraembryonic membranes include the **chorion**, which lines the inside of the egg shell (see figure). This moist membrane permits gas exchange through the shell. The **allantois** carries out functions such as respiration and excretion. The allantois layer has many blood vessels to take in O_2 and give off CO_2, water, salt, and nitrogenous wastes. A third embryonic membrane, the **amnion**, encloses the amniotic fluid. Amniotic fluid provides a watery environment for the embryo to develop in, and provides protection against shock. Finally, the **yolk sac** encloses the yolk. Blood vessels in the yolk sac transfer food to the developing embryo.

Egg

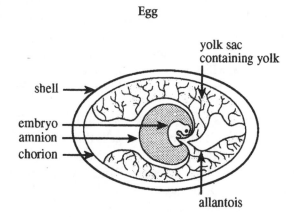

Internal Development

In animals that develop internally, fertilization and embryonic development occur within the mother. This internal development can take a number of different forms, depending on whether or not a placenta is utilized in sustaining the embryo. In some nonplacental animals, development occurs inside the mother, but the embryo lacks a placenta so there is no region of exchange of materials between the blood of the mother and the embryo. Eggs must therefore be relatively large, as their yolk must supply the developing embryo's needs. Marsupials such as opossums are examples of nonplacental animals. They develop inside the oviduct, obtaining food from the yolk of the egg. Marsupials give birth to live young.

The **placenta** includes tissues of both the embryo and the mother. It is the site at which exchange of food, oxygen, waste, and water can take place. In placental animals, there is no direct contact between the bloodstreams of the mother and the embryo. Transport between the maternal and embryonic circulatory systems is accomplished by diffusion and active transport in the placenta between intermingled blood vessels of the mother and embryo. The eggs of placental animals are very small, since the embryo requires only a small amount of yolk for nourishment until a placental connection is made. Humans, for example, have no yolk, but they do have a yolk sac. The umbilical cord that attaches the embryo to the placenta is composed completely of tissues of embryonic, not maternal, origin. This cord contains the umbilical artery and vein that carry embryonic blood into and out of the placenta. As in birds and reptiles, the amnion of placental mammals provides a watery environment to protect the embryo from shock.

Early Development

From fertilization to birth, the single-celled zygote must divide, grow and differentiate into all of the tissues and organs of the organism. Embryonic development begins with rapid mitotic cell division, called the cleavage stage. In cleavage, the cell does not grow in size, but rapidly divides the existing RNA and protein in the egg cytoplasm into smaller cells. Cell divisions start in the oviduct within hours after fertilization with the completion of meiosis and the fusion of sperm and egg nuclei. In the earliest stages, mitotic divisions result in one cell producing two cells, which produce four cells, which produce eight cells, and so on. This ultimately creates what is known as a **morula**, or a solid ball of cells. This morula continues to divide and form the **blastula**, which is a hollow ball of cells. The central cavity is filled with fluid secreted by the cells, and is referred to as the **blastocoel**.

In mammals, the blastula consists of an outer layer of cells called the trophoblast that will develop to form the placenta and an inner mass of cells that will develop into the embryo. As the human embryo passes through the early cell divisions, the developing embryo travels down the oviduct, and, within five to ten days, implants itself in the uterine wall. The uterus responds to the implanted embryo with an increased blood supply, allowing the embryo to receive nutrition from the uterus as the placenta develops.

In the very early embryo, every cell has the potential to develop into any tissue in the fully developed animal. Such unrestricted developmental fate is called **totipotency**. As the embryo proceeds through development, cells become **determined**, or committed to specific developmental fates. Cells that specialize in form to perform a specific function such as nerve cells or muscle cells are **differentiated**. Cells can be determined to a specific developmental path without changing their external appearance to reflect a differentiated state. Development and differentiation of cells is affected in part by their environment—the cells they are surrounded by. **Induction** is a developmental process in which one cell causes a neighboring cell to take a specific developmental path. Transplantation of cells from one part of the developing embryo to another can indicate if cell fate is determined or is induced by surrounding tissues. If cell fate is determined, cells will differentiate in the same way regardless of location in the embryo. As development of the organism progresses, the developmental fate of the cell becomes progressively narrowed. The genome is not altered in most cells as they take on a specific differentiated state in a tissue, but the pattern of gene expression appears to be responsible for the different fates of cells. Expression of specific genes is often regulated by transcription factors.

Gastrulation

In mammals, the cells in the blastula are not yet determined to a developmental fate as a specific tissue. In **gastrulation**, cells from the blastula form three germ layers that will give rise later in development to specific types of tissues (see figure). The three germ layers formed during gastrulation are the **endoderm**, the **mesoderm**, and the **ectoderm**. To begin gastrulation, rapid division of cells at one end of the blastula causes an inpocketing or involution known as the two-layer gastrula. Two germ layers, ectoderm and endoderm, are initially present, endoderm on the inside and ectoderm on the exterior. **Mesoderm** cells move between the endoderm and ectoderm to form a third cell layer in the middle. The opening formed at the point where the endoderm folds inward is called the **blastopore**.

Gastrulation

In embryonic development, the cells of each germ layer differentiate and specialize to form tissues, organs, and organ systems. The ectoderm develops into the epidermis of skin, nervous system, and sweat glands. The endoderm becomes the lining of digestive and respiratory tracts, parts of the liver and the pancreas, and the bladder lining. Finally, the mesoderm develops into the muscles, skeleton, circulatory system, excretory system (except bladder lining), gonads, and the inner layer of skin (dermis).

The pattern of gastrulation differs between different evolutionary groups. In **protostomes**, including annelids, arthropods, and mollusks, the future mouth forms from the blastopore. In **deuterostomes**, including echinoderms and chordates, the blastopore forms the anus of the mature organism, with the mouth forming later in development.

Neurulation

After laying down the basic developmental fate of cells as tissues, the tissues begin to organize themselves into the patterns of future organs. In vertebrates, an early step in development following gastrulation is **neurulation**, in which the first components of the nervous system are determined from the ectoderm. A **notochord** develops from mesoderm along the dorsal side of the embryo, a characteristic shared by all chordates. The notochord induces the ectoderm above it along the exterior surface of the embryo to fold inward and form a **neural tube** of tissue that will become the nervous system, including the spinal cord and brain.

Human Male Reproductive System

Human males, and other male mammals, produce sperm in the **testes**, located in an outpocketing of the abdominal wall called the **scrotum**. Spermatogenesis occurs in the **seminiferous tubules** in the testes (see figure). Primary spermatocytes on the exterior of the tubules produce spermatids that become sperm and move into the center of the tubules as they are continuously produced. **Sertoli cells** support the sperm as they develop and **Leydig cells** produce testosterone. FSH and LH are two pituitary hormones that are involved in spermatogenesis as well as in the female reproductive system. In the **epididymis** on the exterior of testes, sperm finish maturation and are stored. The **vas deferens** is a duct that carries sperm from the testes to the urethra, which also connects the bladder to the exterior through the penis. Glands along the path secrete a liquid (semen) that carries and provides nutrients for the sperm. The glands include the **seminal vesicles** and the **prostate gland**. During sexual intercourse, smooth muscle along the ducts and glands propel sperm and glandular secretions together as semen that is ejaculated from the erect penis into the vagina.

Male Reproductive Tract

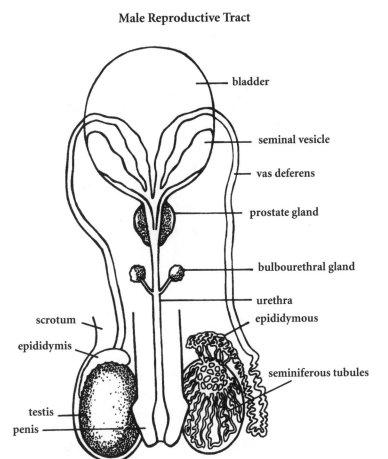

As gonads, the testes have a dual function; they produce both sperm and male hormones (such as testosterone). These hormones regulate secondary sexual characteristics of the male, such as facial and pubic hair and deepening of the voice. Increased testosterone production during puberty induces these changes.

Human Female Reproductive System

Ovaries are paired structures in the lower portion of the abdominal cavity. Through meiosis, ovaries produce eggs in follicles. During ovulation, the eggs leave the follicle and are carried into the upper end of the **oviducts**, also called the **fallopian tubes** (see figure on next page). At birth, all the eggs that a female will ovulate during her lifetime are already present in the ovaries, paused in meiotic prophase I. These eggs develop and ovulate at a rate of one every 28 days (approximately), starting in puberty, in response to hormonal regulation. The ovaries also produce the female sex hormones, estrogens. Estrogens regulate the secondary sexual characteristics of the female, including the development of the mammary (milk) glands and the widening of the hip bones (pelvis). They also play an important role in the menstrual cycle, which involves the interaction of the pituitary gland, ovaries, and uterus.

Female Reproductive Tract

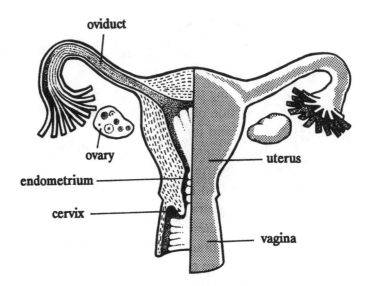

The Menstrual Cycle

The menstrual cycle involves a complex interplay of hormones secreted by the hypothalamus (GnRH: gonadotropin-releasing hormone), the pituitary (LH and FSH) and the ovary (estrogen and progesterone). In general, each of these hormones stimulates the secretion of the next hormone down the regulatory path, and inhibits the secretion of the hormones that precede it. GnRH stimulates FSH and LH, which stimulate estrogen and progesterone secretion. Estrogen and progesterone generally inhibit secretion of GnRH, LH and FSH.

There are four stages in the menstrual cycle:

1. Follicular stage
2. Ovulation
3. The corpus luteum (luteal) stage
4. Menstruation

In the follicular stage, primary oocytes in the ovary along with follicular cells that support the oocyte form a follicle. FSH (follicle stimulating hormone) from the anterior pituitary gland stimulates a follicle to mature. As the follicle matures, it produces estrogen, which promotes thickening of the uterine lining to support an embryo. This stage lasts approximately nine to ten days. When the follicle is mature, a surge in LH secretion by the pituitary causes the egg to burst out of the follicle and the ovary during ovulation (see figure).

STRUCTURE AND FUNCTION OF PLANTS AND ANIMALS

Menstrual Cycle

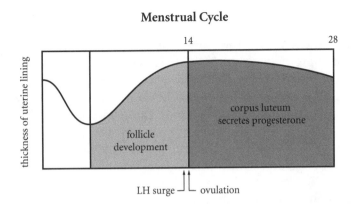

After ovulation, LH acts on the remaining follicular cells in the ovary to form the **corpus luteum**, which secretes further estrogen and progesterone to support further uterine thickening and development of secretory tissue in the uterine lining. If fertilization does not occur, the corpus luteum degrades after a few days, and without the secretion of estrogen and progesterone, the uterine lining regresses as well, leading to menstruation.

If fertilization occurs, the embryo goes through a small number of cell divisions in the first few days as it moves down the oviduct into the uterus. A few days after fertilization the embryo is a **blastocyst** (a hollow ball of cells), and implants in the uterine lining to proceed through embryonic development. The tissues associated with the implanted embryo produce **hCG** (human chorionic gonadotropin), a hormone that takes the place of LH to maintain the corpus luteum. The corpus luteum then continues to secrete progesterone and estrogen, to maintain the uterine wall. As pregnancy continues, the placenta develops and takes over the production of estrogen and progesterone and the corpus luteum degenerates.

If sperm are present in the oviduct at the time of ovulation, fertilization takes place in the oviduct. If there are two or more eggs released by the ovaries, it is possible that all of them will be fertilized, due to the extremely high number of sperm. The result of multiple fertilizations will be fraternal (dizygotic) twins, which are produced when two separate sperm fertilize two eggs. These twins are related in the same way that any two siblings are. If there is only one fertilized egg, twins may still result. This may occur through separation of identical cells during the early stages of cleavage (for example, the two-, four-, or eight-cell stage) into two or more independent embryos. These develop into identical (monozygotic) twins, triplets, and so forth, since they all came from the same fertilized egg.

In the first three months of embryonic development, the first trimester, the embryo goes through organogenesis and the majority of the body plan is formed. Later in development, when the embryo resembles an infant in form, it is called a fetus. As the second and third trimester pass, the fetus grows and its major organ systems prepare for life outside of the womb. The growing pressure of the fetus in the uterus and changes in hormonal levels make the uterine muscles more reactive to stretching reflexes. A combination of the hormone oxytocin and reflexes caused by pressure induce labor to begin, in which smooth muscle in the uterine wall contracts against the baby to press it downward against the cervix. Stretching of the cervix allows the uterine muscles to press the baby through the cervix and the vagina.

8.2 NUTRITION

All organisms require energy to grow, survive and reproduce. The chemical energy that drives many of these processes in the cell is ATP. Different organisms get the energy to produce ATP from different sources. Plants are photoautotrophs that use the energy of the sun to produce ATP through photosynthesis. This ATP is used to produce glucose, which can be burned in plant mitochondria to supply energy. Excess carbohydrate energy that the plant does not consume through respiration can be stored as starch. Animals are heterotrophic, which means that they must obtain food, which provides the raw material for energy, repair, and growth of tissues. Digestion is the process by which animals extract nutrients from food.

The process of digestion encompasses the events that take place after food material is ingested, preparing nutrients for use by the cells of the body. Digestion involves mechanical and chemical breakdown of food and absorption of digested nutrients. Digestion may be defined as the breakdown of large molecules into small molecules. Digestion can be intracellular, occurring through the action of intracellular enzymes. It can also be extracellular; in this process, enzymatic secretions in a cavity such as the intestine break down nutrients into simpler compounds, which are in turn absorbed by cells lining the gut to undergo further processing.

In many organisms, mechanical digestion, or the physical breakdown of large particles of food into small particles, occurs through cutting and grinding in the mouth and churning in the digestive tract. The molecular composition of these food particles is unchanged, but by breaking food into small pieces, the surface area enzymes act on is increased. Vertebrate teeth are specialized structures involved in the mechanical digestion of food, breaking it into smaller pieces. Teeth, like other parts of the digestive system, are adapted to fit the source of nutrition used by a species. Herbivores, plant eaters, have flattened broad teeth (molars) that grind tough plant material. Carnivores, animals that eat other animals, have sharp incisors that cut and slash the tissue of their prey.

Chemical breakdown of molecules in digestion is often accomplished through enzymatic hydrolysis. Food in large part consists of large biological polymers, including proteins, nucleic acids and polysaccharides such as starch. Organisms first hydrolyze these with enzymes into their building block components. The smaller digested nutrients (glucose, amino acids, fatty acids, and glycerol) are absorbed by cells lining the gut to be metabolized or transported to other parts of the body.

Let's take a look at how different organisms ingest and digest their food.

Protozoans utilize intracellular digestion. In amoebas, pseudopods surround and engulf food (via phagocytosis) and enclose it in food vacuoles. Lysosomes containing digestive enzymes fuse with the food vacuole and release their digestive enzymes, which act upon the nutrients, breaking down macromolecules like proteins, nucleic acids, and polysaccharides. The resulting simpler molecules then diffuse into the cytoplasm. The unusable end products are eliminated from the vacuoles.

STRUCTURE AND FUNCTION OF PLANTS AND ANIMALS

In the paramecium, cilia sweep microscopic food such as yeast cells into the oral groove where a food vacuole forms around food. Eventually, the vacuole breaks off into the cytoplasm and progresses toward the anterior end of the cell. Enzymes are secreted into the vacuole and the products diffuse into the cytoplasm. Solid wastes are expelled at the anal pore.

Hydra (phylum Cnidaria) employ both intracellular and extracellular digestion. Tentacles bring food to the mouth (ingestion) and release the particles into a cuplike sac called the gastrovascular cavity. The endodermal cells lining this gastrovascular cavity secrete enzymes. Thus, digestion principally occurs outside the cells (extracellularly). However, once the food is reduced to small fragments, the gastrodermal cells engulf the nutrients and digestion is completed intracellularly. Undigested food is expelled through the mouth. Every cell is exposed to the external environment, thereby facilitating intracellular digestion.

Annelid Digestive Tract

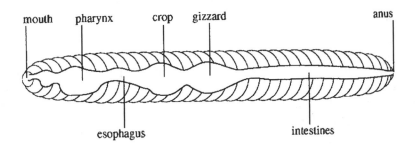

Since the earthworm's body is many cells thick, only the outside skin layer contacts the external environment (see figure). For this reason, this species requires a more advanced digestive system and circulatory system. Like higher animals, earthworms have a complete one-way, two-opening digestive tract. This enables specialization of different parts of the tract for mechanical and chemical digestive processes and absorption of the food that has been ingested. These parts include the mouth, pharynx, esophagus, crop (to store the food), gizzard (to grind the food), intestine (which contains a large dorsal fold that provides increased surface area for digestion and absorption), and anus (where undigested food is released). The digestive tract of arthropods is similar, with the addition of complex feeding appendages and salivary glands (see figure).

Arthropod Digestive Tract

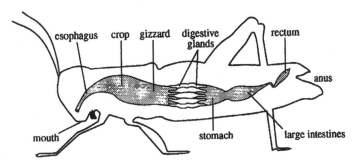

Digestion in Humans

The human digestive system consists of the alimentary canal and the associated glands that pour secretions into this canal (see figure). The alimentary canal consists of the entire path of food through the body: the oral cavity, pharynx, esophagus, stomach, small intestine, large intestine, and rectum. The vertebrate gut is lined with many layers. The interior of the digestive tract is called the **lumen**. Lining the lumen is a layer called the mucosa, which absorbs material from the lumen as well as secretes material into the lumen. Beneath the mucosa, the submucosa layer contains blood vessels that absorb material into the circulation and nerves that regulate digestion. Layers of smooth muscle line the gut in layers that run along the length of the digestive tract or in circles around its circumference. These muscles work together to move material through the tract in contractile waves called **peristalsis**. The action of smooth muscle lining the digestive tract also mixes food for better digestion and absorption.

Many glands line this canal, such as the gastric glands in the wall of the stomach and intestinal glands in the small intestine. Other glands, such as the pancreas and liver, are outside the canal proper, and pour their secretions into the canal via ducts. In the mouth, the salivary glands produce saliva, which both lubricates food and begins starch digestion. Saliva contains **salivary amylase**, an enzyme that digests starch to maltose (a disaccharide). As food leaves the mouth, the esophagus conducts it to the stomach via the cardiac sphincter by means of peristalsis. Rings of muscles called **sphincters** regulate the flow of material through the digestive tract at key control points such as the junction between the esophagus and the stomach and between the stomach and the small intestine.

The Stomach

In the stomach, gastric glands produce **hydrochloric acid** and the enzyme **pepsin**. The acidity destroys ingested microorganisms, and helps to break down ingested food. The acidity of the stomach (pH 1-2) also provides the low pH environment necessary for the optimum enzymatic activity of pepsin, a protease enzyme that begins the digestion of proteins into amino acids (see figure on page 200). Pepsin is initially secreted by gastric glands as pepsinogen, an inactive precursor. Many proteases involved in digestion are secreted as precursors called **zymogens** that lack protease activity. The protease zymogens are activated after secretion to prevent injury to the tissue that synthesizes the protease. The acid in the stomach helps to activate pepsinogen to pepsin after secretion. The stomach lining secretes mucus to protect the stomach lining from acid and pepsin. If the protection of the stomach lining fails, ulcers can result.

Human Digestive System

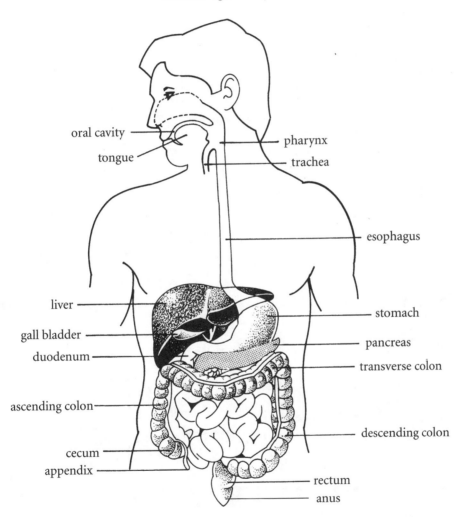

oral cavity

tongue

pharynx

trachea

esophagus

liver

gall bladder

duodenum

ascending colon

cecum

appendix

stomach

pancreas

transverse colon

descending colon

rectum

anus

Digestion

A. Fats

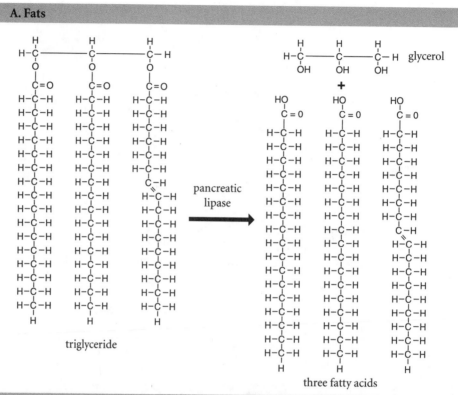

triglyceride

pancreatic
lipase

glycerol

+

three fatty acids

B. Proteins

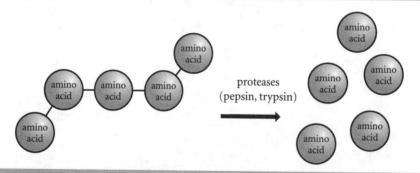

proteases
(pepsin, trypsin)

C. Carbohydrates

sucrose

amylase

glucose + fructose

The stomachs of vertebrate herbivores called **ruminants**, such as cows and deer, are modified to be able to use plants for food. Vertebrates cannot digest cellulose, the main component of the cell wall in plants, so they rely on symbiosis with microorganisms that live in special modified stomachs to digest the cellulose. Termites also require a similar symbiosis with a microorganism to digest cellulose in their digestive tract.

The Small Intestine

Aborption of nutrients from food as well as most of the digestion of food takes place in the small intestine, which has three sections, the **duodenum**, **jejunum**, and **ileum**. Smooth muscle in the stomach wall contracts to propel **chyme** (partially digested food in the stomach) into the small intestine through the **pyloric sphincter** which closes the connection between the stomach and the small intestine. The small intestine is very long and has a very large surface area to optimize the absorption of nutrients. **Villi** are finger-shaped projections that line the wall of the small intestine, greatly increasing its surface area for absorption, and each villi is lined with **microvilli** as well (see figure). Villi contain capillaries and lacteals (projections of the lymphatic system) that carry absorbed nutrients from the intestine to the rest of the body.

Absorption in Humans

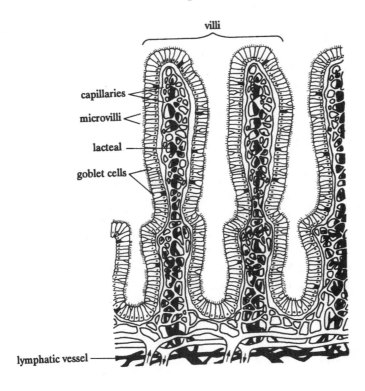

The pancreas and the liver possess glandular tissue that secretes material through ducts into the lumen of the small intestine. The liver produces **bile**, which is essential to the digestion of fats. Bile is stored in the gall bladder prior to its release into the small intestine through the bile duct. Bile is composed of steroid-based salts that are modified with highly polar groups to create detergent-like molecules that can emulsify fats, breaking them up into smaller fat droplets so that enzymes from the pancreas can digest the fats more effectively.

The pancreas secretes **proteases, lipase, amylase,** and **bicarbonate** into the small intestine through the pancreatic duct. Pancreatic lipase, with the assistance of bile, breaks down dietary fats, triglycerides, into glycerol and free fatty acids. The proteases produced by the pancreas include **trypsin** and **chymotrypsin**, both of which are secreted initially as inactive zymogens that are activated by proteolytic cleavage after secretion into the intestine. Once activated, these proteases continue to digest ingested proteins into smaller peptides. Amylase from the pancreas digests starch into maltose.

The digestive enzymes secreted by the pancreas have optimal activity at an alkaline pH. Also, the thin wall of the small intestine that is best for absorption is vulnerable to the strongly acidic pH of the stomach. The necessary alkaline environment is created by the release of large quantities of bicarbonate ion (HCO_3^-) by the pancreas. Bicarbonate neutralizes the chyme acidity in the duodenum.

Before nutrients can be absorbed, digestion must reduce proteins and sugars into very small subunits. The intestine only absorbs monosaccharides. Individuals who cannot digest the disaccharide lactose found in milk to its component monosaccharides cannot digest dairy products. Additional digestive enzymes, disaccharidases for digestion of maltose, lactose, and sucrose, in the lining of the microvilli complete the digestion of disaccharides to monosaccharides, simple sugars like glucose and fructose. Peptidases in the microvilli also complete protein digestion to single amino acids or very small peptides with 2-3 amino acids that can be absorbed.

Absorption

Active transport mechanisms are involved in the absorption of glucose and amino acids into the villi. A gradient of sodium across the intestinal wall is used to drive cotransport of these nutrients into epithelial cells that line the villi. From the epithelial cells, the nutrients are then transported into the capillaries that pass through the villi, for transport to the rest of the body. Blood flow from the intestine first flows to the liver through the hepatic portal system before passing to the rest of the body, allowing the liver to metabolize many of the ingested nutrients. Fatty acids and glycerol from digested triglycerides are passively absorbed by epithelial cells, then resynthesized into triglycerides within the cells and packaged into lipoprotein particles called **chylomicrons** that contain mostly triglycerides. Chylomicrons pass into lymphatic lacteals that pass through the villi, and through the lymphatic system they eventually pass into the circulatory system (see figure).

Chylomicrons

The liver plays an important role in processing chylomicron. Chylomicrons are absorbed by the liver and the triglycerides are repackaged to deliver triglycerides to other tissues, including fat cells. The residual particles left behind, LDL, can accumulate in the cells lining arteries, contributing to coronary heart disease if they are overly abundant in the blood stream.

Many essential vitamins are absorbed in the intestine as well. These vitamins often act as cofactors for enzymes or other proteins. Vitamin A is a precursor for an essential cofactor for a pigment in the eye involved in vision. Niacin is a precursor for NADH and NADPH. Riboflavin (vitamin B_2) is the precursor for $FADH_2$. Vitamin C is required for the proper synthesis of collagen that is secreted in the extracellular matrix. Its absence causes scurvy, whose symptoms involve weakening of connective tissue. Inadequate amounts of other vitamins can also cause diseases in individuals with insufficient food supplies or with poor diets. Vitamin D is required for calcium absorption, and its absence can cause rickets, resulting in weak bone formation. Vitamin D can be synthesized in the skin through exposure to sunlight, but is commonly supplemented in milk for those who do not receive abundant exposure to sunlight. Minerals are also essential nutrients. Iodine is required for the synthesis of thyroid hormone, and its absence can cause thyroid insufficiency and goiter, enlargement of the thyroid. For this reason, table salt commonly contains iodine.

The Large Intestine

The next organ in the human digestive tract is the large intestine. Most of the absorption of digested nutrients occurs in the small intestine. Meanwhile, the large intestine is mainly devoted to activities like water and vitamin K absorption. Vitamin K, an essential nutrient involved in blood clotting, is not abundant in foods. Humans receive this essential nutrient through a symbiotic relationship with an intestinal bacteria, *Escherichia coli*. The large intestine also absorbs the bicarbonate ions from digested material. The final indigestible material forms feces that are stored in the rectum prior to their elimination through the anus.

Regulation of Digestion and Energy Stores

Hormones and the nervous system regulate the digestive process by controlling secretion by glands and the action of smooth muscle in the walls of the digestive organs. Digestion is controlled in part by local reflexes within the nervous system. For example, filling of one section of the intestine can cause nerves in the intestine wall to initiate peristalsis of the muscles in the region, thus moving the food further along through the intestine. The sympathetic nervous system sends signals from the brain to inhibit hormone secretion and muscle movement associated with digestion while the parasympathetic nervous system stimulates digestion.

Many hormones regulate the action of digestive glands and smooth muscle. When food fills the stomach, cells in the stomach wall secrete the hormone **gastrin** that increases secretion in the stomach as well as the activity of stomach muscle. Cells in the intestinal wall also respond to the presence of food by secreting hormones that regulate digestion. When the chyme in the small intestine becomes acidic because of stomach emptying, intestinal cells secrete the hormone **secretin**, which stimulates the pancreas to release bicarbonate ions to neutralize the acid. The presence of fats or protein in the small intestine causes intestinal cells to secrete the hormone **cholecystokinin** (CCK) that stimulates the secretion of pancreatic enzymes and the release of bile.

Long Term Storage and Use of Energy Molecules

The body is continuously burning glucose for energy, and must maintain a constant glucose supply in the blood. Even a momentary lapse in blood glucose levels can be disastrous for tissues such as the brain, which burns a large amount of glucose, and can cause coma. The body must continuously produce glucose to maintain the supply. After a meal containing starch or sugars, the glucose supply will be abundant. The body stores some of this supply as glycogen, mainly in the liver and skeletal muscle, to use between meals. More energy is stored as fats, and proteins can also be burned for energy during conditions of starvation, leading to loss of tissues such as muscle.

The liver helps regulate blood glucose levels. Glucose and other monosaccharides absorbed during digestion are delivered to the liver via the hepatic portal vein. Glucose-rich blood is processed by the liver, which converts excess glucose to glycogen for storage. If the blood has a low glucose concentration, the liver converts glycogen into glucose and releases it into the blood, restoring blood glucose levels to normal. In addition, the liver synthesizes glucose from noncarbohydrate precursors such as pyruvate and from amino acids via the process of **gluconeogenesis**. Fatty acids are burned for energy through a process called **beta-oxidation**.

The primary hormones that regulate the use of glucose are **insulin** and **glucagon**. After a meal when blood glucose is high, insulin is secreted by pancreatic endocrine cells, beta-cells in the Islets of Langerhans. Insulin acts on the liver and other tissues to take up glucose out of the blood. In the liver, insulin stimulates glycogen formation, represses gluconeogenesis, and stimulates fat production in cells. When glucose levels fall, insulin levels fall and glucagon levels rise, reversing the above pattern. Glycogen is then broken down in the liver to produce glucose that is released into the blood. **Epinephrine**, the hormone secreted by the adrenal gland during a fight-or-flight response, also stimulates the release of glucose from the liver to allow for rapid action if necessary.

8.3   THE CIRCULATORY SYSTEM

Cells must exchange material with the external environment. During cellular respiration, all aerobic organisms consume oxygen and produce carbon dioxide and must exchange these gases with the environment. Cells often produce nitrogenous waste that must be removed, and cells must obtain nutrients. In simple organisms such as single-celled protozoans, cnidarians with only two cell layers, and sponges, the small body size and close proximity of all cells to the external aquatic environment allows all cells to directly exchange material with the external environment. The same is true for roundworms and for flatworms.

However, large multicellular organisms, particularly those in a terrestrial environment, cannot rely on passive diffusion of materials between cells and the environment, and have adapted circulatory systems to move fluids within the body, carrying nutrients to the tissues and wastes away. The circulatory system includes the blood, the heart, and the vessels that carry blood through the body. Circulatory systems can be either closed, in which the blood is always contained within blood vessels, or open, in which the blood in at least part of the body mixes directly with the tissues in open sinuses.

Invertebrate Circulatory Systems

Most cells in annelids (earthworms) are not in direct contact with the external environment. An internal closed circulatory system indirectly brings materials from the external environment to cells. Blood travels toward the head (anterior) through dorsal blood vessels. Five aortic arches or 'hearts' force blood down to the ventral vessel, which carries blood to the posterior and up to complete the circuit (see figure). The blood carries oxygen and carbon dioxide between cells and the skin of the earthworm where gas exchange occurs. The blood also circulates nutrients from the digestive tract to the rest of the body.

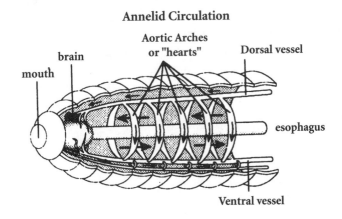

Annelid Circulation

Arthropods and mollusks utilize an open circulatory system, in which blood flows through a dorsal vessel and then out into spaces called sinuses. In these sinuses, blood is not enclosed in blood vessels but directly bathes cells, and exchange of food takes place. Air exchange, meanwhile, is accomplished through a tracheal system of air tubes. Blood then reenters blood vessels. The heart consists of a simple beating tube. In general, closed systems are more efficient at moving material within the body.

Vertebrate Circulatory Systems

The vertebrate circulatory system is closed, with a chambered **heart** that pumps blood through **arteries** that lead away from the heart to capillaries. **Capillaries** are extremely small vessels in tissues where exchange of material between the circulation and the tissues actually occurs. From the capillaries, blood is carried back to the heart through **veins**. Valves in the heart and the veins help to prevent blood from flowing backward through the system. **Atria** in the heart receive blood from the body and **ventricles** are muscular chambers that pump blood back out through arteries to the body.

In fish, the heart has two chambers: one atrium and one ventricle. Blood is pumped from the ventricle to capillaries in the gills to receive oxygen, then collected in a large artery to pass directly to the rest of the body before returning to the heart in veins. Amphibians have a three-chambered heart. The amphibian ventricle pumps blood to both the lungs and the rest of the body at the same time through two different major arteries. This arrangement allows oxygenated blood from the lungs and deoxygenated blood from the body to mix in the ventricle before it is delivered back to the body. It also allows higher arterial pressure in blood pumped to the tissue.

Pumping blood through a capillary bed with a large number of small vessels creates resistance. The heart creates pressure in the circulatory system when it contracts to force blood through the system. The highest pressure in the blood vessels is found in the arteries leading away from the heart to the capillaries. On the other side of the capillaries, very low pressure remains to push blood back through the veins to the heart.

The greater metabolic needs of birds and mammals require efficient circulation, with complete separation of blood flow to the lungs and the other tissues of the body. To accomplish this, both birds and mammals evolved hearts with four chambers: two atria and two ventricles. The mammalian heart is really two pumps in one, with one atrium and ventricle devoted to pumping blood to the lungs through the pulmonary circulation, and another atrium and ventricle devoted to pumping blood to the rest of the body through the systemic circulation. This organization avoids the mixing of oxygenated and deoxygenated blood found in the amphibian heart and allows the high arterial pressure required for rapid delivery of material to metabolically active tissues.

Human Circulation

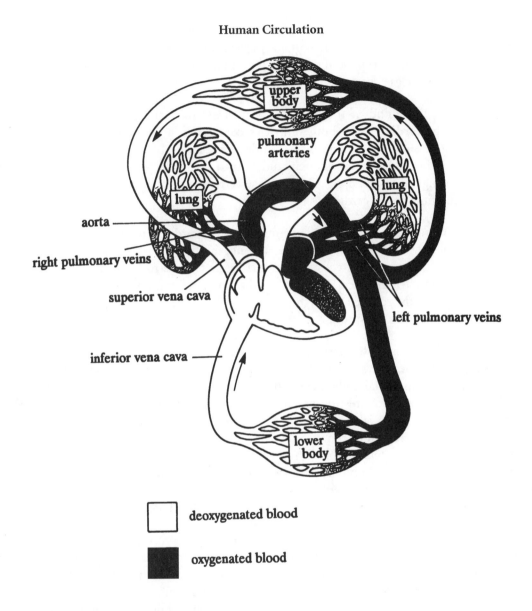

The Human Circulatory System

The Heart

The human heart is a four-chambered pump, with two collecting chambers called atria and two pumping chambers called ventricles. The right ventricle pumps deoxygenated blood to the lungs through the pulmonary artery. Oxygenated blood returns through the pulmonary vein to the left atrium. From there it passes to the left ventricle and is pumped through the aorta and arteries to the rest of the body. Valves in the heart are essential to efficient pumping of the heart. When blood is pumped out of the ventricles, valves close to prevent the blood from flowing backward into the heart after the contraction of the ventricles is complete.

The aortic valve blocks backflow from the aorta into the left ventricle and the pulmonary valve prevents backflow of blood from the pulmonary artery into the right ventricle. Another set of valves, one between each atrium and ventricle, prevent blood from flowing backward into the atria when the ventricles contract.

The heartbeat is the sound of the atria and ventricles contracting in a regular repeated pattern called the **cardiac cycle**. Both ventricles contract in a portion of the cycle called **systole**. During **diastole**, the ventricles relax and are filled with blood by contraction of the atria. The sound of the heartbeat is often studied by a doctor to listen to the action of the heart valves during the cardiac cycle. The electrical activity of the heart can also be studied to look for irregularities in the cardiac cycle.

The heart is composed of a special type of muscle called cardiac muscle. Cardiac muscle is striated like skeletal muscle. Cardiac muscle cells are not multinucleated like skeletal muscle, but the cytoplasm of cardiac muscle cells is connected between cells to form electrical synapses from one cell to another. Electrical impulses called action potentials, similar to those found in neurons, spread directly from cell to cell throughout the heart muscle to trigger the contraction of the atria and ventricles in unison.

The heart beat is regulated by nervous control, but is not initiated by a signal from the nervous system. The heart initiates each heartbeat internally at a special pacemaker region in the heart called the **sinoatrial node**. Cells maintain a negative membrane potential across the plasma membrane called the resting potential. In neurons, an action potential is triggered when the membrane potential becomes sufficiently depolarized (less negative) to trigger the opening of voltage-gated ion channels (see Section 8.10, Nervous System), that then propagate the action potential through the cell. In skeletal muscle a similar wave of depolarization by an action potential triggers muscle contraction. In the sinoatrial node, pacemaker cells spontaneously depolarize their membrane potential at a steady rate on their own, causing voltage-gated channels in the pacemaker cells to open.

When the action potential is initiated in the pacemaker cells, it spreads rapidly from cell to cell throughout both atria to cause the atria to contract together. The cardiac action potential cannot pass directly from the atria to the ventricles, however, since there are no direct cell to cell connections between the atria and ventricles. The impulse is carried through the **atrioventricular node** from the atria to the ventricles, then through the **Bundle of His** and the **Purkinje fibers**, all of which carry the action potential to the ventricles, where it will spread throughout the cardiac muscle rapidly from cell to cell. The passage of the impulse through the atrioventricular node delays the impulse so that the timing of contraction by the ventricles coincides with the completion of atrial contraction.

The rate of the heart beat is regulated by the sympathetic and parasympathetic nervous systems, components of the autonomic nervous system. Action of the sympathetic system directly or indirectly through the hormone epinephrine causes the heart rate to increase by acting on the sinoatrial node pacemaker. The parasympathetic system is more important in the regulation of the heart rate. The vagus nerve of the parasympathetic system directly innervates the sinoatrial node, and when stimulated, slows the heart rate.

STRUCTURE AND FUNCTION OF PLANTS AND ANIMALS

In coronary heart disease, the arteries that supply the heart with blood become blocked by the accumulation of plaques in the lining of the arteries. A diet high in fat raises the level of LDL, a blood lipid. High levels of LDL and low exercise favor plaque accumulation. If the plaque is broken, a blood clot called a thrombosis can form, which can completely block the coronary artery, causing a myocardial infarction (a heart attack). During a heart attack, the tissues of the heart are deprived of oxygen and can become damaged, hindering heart function. Reduction of fat in the diet, exercise, stress reduction and medication can all reduce plaque formation and the risk of heart attacks.

Blood Vessels

The **arteries** carry blood from the heart to the tissues of the body. They repeatedly branch into smaller arteries (arterioles),which supply blood to the tissues via the capillaries. Arteries are thick-walled, muscular, and elastic, conducting blood at high pressure. Arterial blood is oxygenated, except for the pulmonary artery, which carries deoxygenated blood from tissues to the lungs to renew the oxygen supply.

Veins, on the other hand, carry blood back to the heart from the capillaries. Veins are relatively thin-walled, conduct blood at low pressure and contain many valves to prevent backflow. Veins have no pulse and usually carry dark red, deoxygenated blood (except for the pulmonary vein, which carries recently oxygenated blood from the lungs). Skeletal muscle contraction around veins can help to move blood back through the veins to the heart.

Capillaries are thin-walled vessels that are very small in diameter. In fact, their walls are only one endothelial cell thick. Capillaries, not arteries or veins, permit exchange of materials between the blood and the body's cells. Fluid, which contains water with solutes such as nutrients and hormones, seeps from capillaries, driven by pressure out of the thin-walled vessels. Cells and proteins are retained in the capillaries, drawing most water back into the capillaries by osmosis. Fluid that remains in tissues can enter the lymphatic system to be filtered and cycled back to the circulatory system.

Regulation of Blood Flow

The needs of tissues for blood flow can increase and decrease over time. The flow of blood is regulated locally in the tissues to match the supply of blood to metabolic needs at any time. Smooth muscle in the walls of arterioles can constrict to reduce the flow of blood to capillaries in a tissue. The smooth muscle relaxes when the blood leaving the capillaries is low in oxygen, allowing more blood to flow through the arteriole and through the capillaries, increasing the oxygen supply to the tissue. Other factors such as carbon dioxide level or lactate level can also cause relaxation of arteriole smooth muscle to increase blood flow. Local control allows the blood flow in each tissue to be finely controlled to match its needs.

For the body as a whole, the nervous system regulates blood flow via the autonomic nervous system. The sympathetic nervous system causes constriction of arteries in many tissues such as those of the digestive tract, but causes dilation of arteries in skeletal muscle. The central control of blood flow occurs in the medulla of the brain, which receives information from sensors in the aorta about stretching, and from oxygen sensors in other arteries. If the aorta is stretched, this indicates that the arterial blood pressure is high, and triggers the control center in the medulla to inhibit the sympathetic nervous system, relaxing the arteries in the periphery and slowing the heart rate. If blood is lost, decreasing arterial pressure, the stretch sensors can trigger a response from the sympathetic nervous system to constrict vessels in the periphery and increase the heart rate to compensate for the lost blood. If the oxygen sensors in the arteries detect an oxygen level that is too low, they trigger a signal through the medulla to increase the heart rate and arterial pressure to deliver more blood and oxygen to the tissues.

Stimulation of the sympathetic nervous system results in the secretion of epinephrine by the adrenal glands; epinephrine increases the heart rate and constricts arteries to increase arterial pressure. Two additional hormones that regulate blood pressure are vasopressin and angiotensin. When blood pressure in the arteries falls, the kidney secretes an enzyme called renin that activates the peptide hormone angiotensin. Angiotensin acts on the smooth muscle in the arterioles and arteries to cause constriction and increase the central pressure. Vasopressin is secreted by the posterior pituitary in response to the stretch sensors and has similar effects of constricting arteries and increasing central pressure.

Inflammation of local tissues causes arterioles to expand, increasing the flow of fluid into capillaries; the flow of fluid from the capillaries into the tissues causing swelling. Histamine is a signaling molecule that is released in an allergic reaction, causing increased blood flow and permeability of capillaries in a tissue, leading to swelling. Generalized swelling called **edema** can occur if the osmotic balance of plasma is low or the venous pressure is too high due to heart disease.

Blood

Circulation would not be possible without circulatory fluid, the blood. The liquid component of blood that contains dissolved nutrients, wastes, proteins, hormones, and fibrinogen is the **plasma.** The fraction of blood without cells left after clotting is called **serum**, containing glucose, lipids, salts, hormones, and albumin. Several materials are suspended in plasma and transported throughout the body, including red blood cells, white blood cells, and platelets. If blood is placed in a centrifuge, the cells can be precipitated and the volume of cells in blood measured. The percent of blood volume occupied by cells measured this way is called the **hematocrit** and is about 40% of the total blood volume.

Red blood cells, also called **erythrocytes,** are the most common cells found in blood, and their primary function is to transport oxygen. After they are formed in the bone marrow, human red blood cells lose their nuclei and become biconcave discs. Red blood cell production is stimulated by the hormone **erythropoietin** produced by the kidneys. Erythrocytes survive in circulation for about four months. Through repeated passage through narrow capillaries and eventual damage to the cell, red blood cells wear out and are destroyed in the spleen and liver. Red blood cells contain hemoglobin (the red pigment containing iron), which unites with oxygen to form oxyhemoglobin. Hemoglobin binds oxygen in a cooperative manner to create a sigmoidal binding curve. In the tissues, the partial pressure of oxygen is low and hemoglobin releases oxygen. Acidic conditions such as those generated by lactic acid in muscle can also stimulate hemoglobin to release oxygen in tissues. Carbon monoxide can bind to hemoglobin irreversibly, preventing hemoglobin from binding oxygen and preventing oxygen from reaching tissues.

During maturation, red blood cells also lose mitochondria, which renders them incapable of performing aerobic respiration. If they were able to carry on aerobic respiration, they would themselves use up the oxygen that they were assigned to carry to the tissues of the body. Instead, they produce ATP without using oxygen, through glycolysis (see Section 8.5, Respiratory Gas Exchange).

Red blood cells manufacture two major types of antigens, antigen A (associated with blood type A) and antigen B (associated with blood type B). In any given individual, one, both, or neither antigen may be present. The same pattern appears in every red blood cell.

The plasma of every individual also contains antibodies for the antigens that are not present on the individual's red blood cells. If an individual were to produce antibodies against his or her own red cells, they would agglutinate and the blood would clump. Type A individuals have anti-B antibodies, and type B individuals have anti-A antibodies. Type O individuals, who have neither A nor B antigens, have both anti-A and anti-B antibodies. Type AB individuals have neither type of antibody. These relationships are depicted in the graph below.

	Blood Type	Antigen on Red Blood Cell	Antibodies Found in Plasma
	A	A	anti-B
	B	B	anti-A
Universal Recipient	AB	A, B	none
Universal Donor	O	none	anti-A, anti-B

White blood cells, also called **leukocytes**, are involved in immune functions in the body (see Section 8.4, Immune System). White blood cells include **phagocytes**, which engulf bacteria via amoeboid motion, and various types of lymphocytes, such as B and T cells which are involved in the immune response. B cells produce antibodies, or **immunoglobins**, which are secreted proteins specific to foreign molecules such as viral or bacterial proteins. Helper T cells coordinate the immune response and killer T cells directly kill cells that are infected with intracellular pathogens or cells that are aberrant (such as malignant cells).

Phagocytes consist of **neutrophils**, which are the first cells to arrive at a site of inflammation to eat bacteria and other foreign particles. They are the primary component of pus. Macrophages and monocytes are also phagocytic cells that engulf through phagocytosis and present foreign components, such as bacteria and viruses, to specific arms of the immune system.

Platelets are not truly cells but are cell fragments produced in the marrow as pieces of **megakaryocyte cells**. At the site of a bleeding injury, activation of the protease **thrombin** cleaves **fibrinogen** protein in the blood to make fibrin that polymerizes to form a net across the wound, trapping more cells and blocking the flow of blood.

The Lymphatic System

The fluids of the blood and extracellular fluids are very similar in composition, except that the extracellular fluids do not contain cells and have very little protein. These are retained in the capillaries. Fluids that pass into the tissues would accumulate if a mechanism did not return them to circulation. One means to accomplish this is the lymphatic system. Lymph vessels are a system of tubes which are semi-independent of the blood system. Lymph vessels permeate the tissues, collecting extracellular fluid (at this stage known as lymph) and carrying it back to the circulatory system. The lymph is moved through the system without direct pumping, but mainly by the pressure of skeletal muscle acting against the lymph vessels to move liquid forward past valves that prevent backflow. The liquid inside the vessels is at very low pressure, essentially negative pressure, helping to draw liquids out of tissues and into the lymph system. The system ultimately returns lymph to the blood system via the largest lymph vessel, the thoracic duct, which empties lymph back into circulation shortly before it enters the heart. Blockage of the system can cause swelling as liquid accumulates in the tissues.

One of the functions of the lymph system is to assist in immune surveillance. As the lymph moves through the lymphatic system, it passes through lymph nodes. The lymph nodes and immune cells inside filter material out of the lymph including bacteria, extracellular proteins, or cancerous cells. This filtration system cleanses the extracellular fluids.

8.4   THE IMMUNE SYSTEM

The body presents a potential breeding ground for many foreign organisms, including bacteria, viruses, fungi, protists, and parasites. The challenge for the immune system is to prevent harmful organisms from colonizing the body. Some of the body's defenses are aimed against specific organisms, while others are nonspecific deterrents to all potential infectious agents.

Non-Specific Immune Defenses

Nonspecific agents against infection can act as barriers to prevent entry into the body or they can actively destroy potential pathogens. The skin is an important barrier to infection covering most of the body. Most infectious agents cannot penetrate through an unbroken layer of skin. More common points of entry are a wound in the skin or the mouth. Mucus secretions in the digestive tract, the nose, the respiratory tract or the reproductive tract can trap microorganisms. Cilia lining the respiratory tract move the trapped organisms out of the system up the trachea. Tears contain an antibacterial enzyme called lysozyme that can digest the peptidoglycan cell wall found in many bacteria. The acidity of the stomach is a barrier against entry of organisms through the mouth.

Another defense against infection is often the community of organisms found living in the body. Microorganisms are found in association with humans on the skin and in the digestive tract, particularly the large intestine. The microorganisms such as *E. coli* that inhabit the intestine are symbiotic in their relationship, causing no harmful consequences for the host human, and actually offering advantages to the host as well as the *E. coli*. In addition to providing the essential nutrient Vitamin K, the bacteria compete for the intestinal environment with potential infectious agents, preventing them from colonizing and causing disease. One argument against taking unnecessary antibiotics is that it may destroy the indigenous *E. coli* of the intestine and lead to infection by pathogenic organisms.

Additional non-specific defenses include complement proteins and some white blood cells. Proteins in the blood and tissues can act as nonspecific protections against infection. The complement system consists of a series of proteins secreted into the blood that defend against infection in a general manner. Complement proteins bind to the surface of many pathogens, attracting white blood cells to the infectious agents to destroy them. Complement proteins can also work together when bound to a bacterium to form a tube-shaped opening through the bacterial membrane, causing the cell to burst. Phagocytes, including neutrophils and macrophages, are white blood cells that seek out and ingest infectious agents through phagocytosis. Once internalized, the invading agents are destroyed inside the phagocyte.

Specific Immune Response

The immune system blocks infections nonspecifically in many ways, but on occasion organisms evade the non-specific defenses. The next level of defense is immune responses against specific organisms. B cells and T cells are types of white blood cells, also called lymphocytes, that recognize and eliminate specific foreign agents. The main challenge for these cells is to recognize and remove foreign materials that do not belong in the body (non-self) while leaving unscathed all of the normal tissues of the body (self).

Immune cells recognize not the whole infectious agent, but specific molecules that are part of the agent. **Antigens** are the agents or molecules that the immune system responds against. Each B or T cell responds to a specific antigen from a pathogen, such as a protein in the bacterial cell wall or a viral envelope protein. The B cell response is sometimes called the **humoral response** and the T cell response is sometimes called the **cellular response**. B cells respond to antigens by making **antibodies**. Antibodies are proteins, also called **immunoglobulins**, that recognize and bind to specific antigens. All antibodies have the same general structure (see figure), with one end called the constant region and variable regions at the other end that differ from one antibody to another. It is the differences in the variable regions that determine the antigen specificity of an antibody.

Antibody Structure

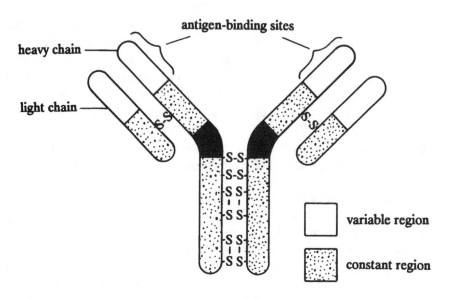

There are five different classes of antibodies that perform different immune functions depending on where they are found in the body. The constant regions determine the class of an antibody, including IgA, IgE, IgM, IgG and IgD. IgM is expressed first during an infection and is found on the surface of the B cells, along with IgD. Most antibodies in blood are in the IgG class. IgE are involved in allergic responses and IgA are often secreted from body linings such as mucus membranes.

The sequence of variable regions determines antibody specificity. Antibodies are composed of protein subunits joined by disulfide bonds (see figure). The variable regions join together between polypeptide chains to fold into a complex quaternary structure that forms the binding site for antigens. The variable regions contain a large number of potential amino acid sequences, producing a vast potential repertoire of antigen specificity. A broad range of variable regions is created during differentiation of B cells. The immunoglobulin genes contain many different potential sections that can be pieced together through a specific type of immune cell recombination to form complete immunoglobulin genes that can be expressed in the mature B cell. Changes in the sequences of the rearranged immunoglobulin gene introduce additional variability in the encoded variable chains and increased variety in the range of antigens that antibodies can recognize. These changes in the immunoglobulin genes are one of the few exceptions that are known to the general rule that differentiated cells have the same genome as all other cells.

There is a huge number of potential pathogens and antigens that B cells must be prepared to respond against as well as a large number of self antigens that B cells should not respond against. This problem is apparently solved through clonal selection of lines that recognize non-self but not self. During development, B cells produce a vast range of immunoglobulin genes and antibodies. The antibodies are first expressed on the surface of the B cells when they develop. B cells that respond against antigens in the body, self antigens, are eliminated by inducing the cells to die. This leaves many different clonal lines of cells that each express only one type of antibody. The population of many different clones includes cells that have many different antigen specificities in the antibodies expressed on their surface. When a pathogen invades, it may contain antigens that a B cell clone recognizes. If so, this B cell will proliferate, through clonal selection, and begin to secrete antibody into the blood. Antibody responses by B cells are selected for out of a large range of possibilities.

When B cells encounter an antigen that binds to the antibody they express on their surface, they become activated. An activated B cell will proliferate and differentiate into two forms, plasma cells and memory cells. **Plasma cells** are synthetic factories that produce and secrete antibody. The plasma cells continue to produce antibody until they die after several days, and the antibody may circulate for a short period longer. The other type of cells that are produced are memory cells. **Memory B cells** do not produce antibody, but express antibody on their surface. The memory cells may remain in circulation for years, and if they encounter antigen, they will differentiate to produce new plasma cells that secrete antibody. Since the memory cells are amplified by clonal selection along with the plasma cells during the initial infection, they create protection against future infections by the same agents. The amplification of memory cells is the reason that people don't usually get the same disease twice and also why vaccinations are often repeated over a long period of time.

T cells play a different role in the specific immune response, the cellular response. T cells include cytotoxic T cells and helper T cells. Cytotoxic T cells kill cells that are infected by a pathogen that the T cell recognizes, and helper T cells coordinate the immune response of other cells against specific antigens. Unlike the humoral response of B cells which involves secreted factors (antibodies), the cellular response occurs by direct interaction between cells of the immune system. Like B cells, T cells produce proteins with a broad range of antigen specificity that are selected for by clonal selection out of a random population. The antigen recognition proteins expressed by T cells are **T cell receptors**, which are not secreted. The T cell receptor is expressed in the membrane of T cells and does not recognize antigen in solution, but only recognizes antigen on the surface of other cells in a specific context.

For antigen to stimulate a T cell receptor, it must be presented to the cell as part of a complex of proteins called the **MHC** (major histocompatibility complex) or **HLA** (human leukocyte antigens) that are found in the plasma membrane of cells. Two types of MHC are mainly involved in the T cell response, **MHC Class I** and **MHC Class II**. MHC I are present on the surface of all cells and MHC II are only present on immune cells, including macrophages, B cells and T cells. MHC present antigen from inside the cell at the cell surface where T cells can recognize it. For the cellular immune response to occur, the T cell receptor of a cytotoxic T cell must recognize antigen presented in the MHC I on the surface of an infected cell. If it does, this cytotoxic T cell clone will proliferate, as occurs in the clonal selection of B cells. In future interactions of cytotoxic T cells from this clone with infected cells, the T cell receptor will recognize antigen presented in the MHC I, and the cytotoxic T cell will kill the infected cell (see figure).

MHC Class I

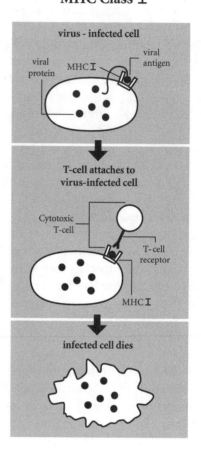

T cells are also involved in the response of B cells to antigen. Helper T cells are required for B cells to respond to antigen stimulation. Immune cells such as macrophages will digest antigen and present it in MHC Class II at the cell surface. Helper T cells with T cell receptors that recognize that antigen will be stimulated and proliferate and will stimulate B cells that respond to the antigen to proliferate into plasma cells. Many **cytokines** are secreted by helper T cells to communicate with other cells to coordinate the immune response.

MHC molecules play a key role in self/non-self recognition in the immune system. There are many different forms of the MHC molecules in the human gene pool, and if two people do not match in their MHC genes, a transplant of an organ between them is likely to be rejected by the immune system of the recipient. The T cells of the recipient will recognize the MHC of the donor organ as foreign, non-self, and will mount an immune reaction against it. Occasionally, B cells and T cells that respond to self-antigens elude elimination and result in autoimmune disorders such as multiple sclerosis, in which an immune response against myelin damages the nervous system. Rheumatoid arthritis appears to involve an immune response against proteins found in the joints. These conditions may be caused by defects in T cell selection. For a T cell to reach circulation, it must be able to respond to MHC and not to MHC with self-antigen bound.

8.5 RESPIRATORY GAS EXCHANGE

A vital function that living organisms must maintain is that of gas exchange, or respiration. With the advent of the oxygen rich atmosphere found on earth today, most organisms rely on the oxidation of glucose through aerobic respiration to generate energy. Oxygen is used as the final electron acceptor in electron transport as part of aerobic respiration to produce ATP, and CO_2 is produced from burning glucose in the Krebs cycle. The respiratory gases oxygen and carbon dioxide diffuse freely across plasma membranes, and so in all organisms these molecules move in and out of cells by simple diffusion down a concentration gradient.

In aquatic organisms including prokaryotes, protists, sponges and cnidarians, every cell is directly exposed to the external environment and respiratory gases can be easily exchanged by direct diffusion of these gases through the cell membrane. Complex multicellular organisms require more complex mechanisms to exchange gases between cells and the exterior atmosphere.

Mucus secreted by cells at the external surface of the annelid's body provide a moist surface for gaseous exchange from the air to the blood through diffusion. The annelid's circulatory system then brings O_2 to the cells and waste products such as CO_2 back to the skin, excreting them into the outside environment. The arthropod respiratory system consists of a series of respiratory tubules called **tracheae**. These tubules open to the outside in the form of pairs of orifices called **spiracles**. Inside the body, the tracheae subdivide into smaller and smaller branches, enabling them to achieve close contact with most cells. In this way, this system permits the direct intake, distribution, and removal of respiratory gases between the air and the body cells. Specialized cells for transporting oxygen are not found. Since a blood system does not intervene in the transport of gases to the body's tissues, this system is very efficient and rapid, enabling most arthropods to produce large amounts of energy relative to their weights. The direct diffusion of air through tracheae is one factor that limits body size in arthropods.

Respiration in Fish

Water contains less oxygen than air, making respiration more of a challenge in water. Limited quantities of oxygen can often limit life in a body of water, while oxygen is seldom the limiting resource on land. As a result, aquatic organisms require a large surface area in contact with water to maximize gas exchange. The gills of fishes are divided into numerous thin-walled, threadlike gill filaments that are well fed by capillaries. The walls of the gills are very thin, to maximize the diffusion of gases between the blood and water by minimizing the distance that must be traveled. Fish gills are protected from the outside by an **opercular flap** to prevent other organisms from eating them. As water passes over these gill filaments, O_2 diffuses into the blood, while CO_2 leaves the blood to enter the water. Arteries then transport the oxygenated blood through the body. The water, meanwhile, passes out of the body through openings on either side of the head, taking the discarded carbon dioxide with it.

Respiration in Fish

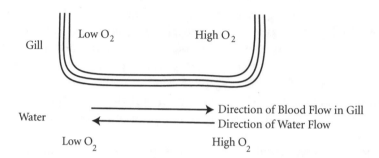

A key feature of the gills in fish is the use of **countercurrent exchange** to maximize the exchange of gases between the blood inside the gills and the water flowing over the gills (see figure). The blood flowing through the capillaries in the gills moves in the opposite direction of the water moving across the gills outside. This arrangement maximizes gas exchange by maximizing the concentration gradient of gases in blood and water.

Respiration in Humans

Amphibians evolved lungs as they moved onto land, simple air sacs with very little surface area. The small surface area of the amphibian lung requires that they supplement gas exchange in the lungs with exchange across the thin moist skin. For organisms such as mammals, however, significant gas exchange across the skin does not occur, requiring that the lungs and respiratory system evolve to supply their greater metabolic needs.

Humans have developed a complex system of respiration to transport oxygen to their cells and to rid their bodies of waste products like carbon dioxide. The air passages involved in respiration consist of the nose, **pharynx, larynx**, the **trachea** which carries air down the throat, **bronchi** that lead into each lung, and bronchioles that branch throughout the lungs to end in tiny saclike compartments called **alveoli** that are the site of gas exchange (see figure). The great number of alveoli in the lungs creates a very large surface area for gas exchange.

STRUCTURE AND FUNCTION OF PLANTS AND ANIMALS

Human Respiratory System

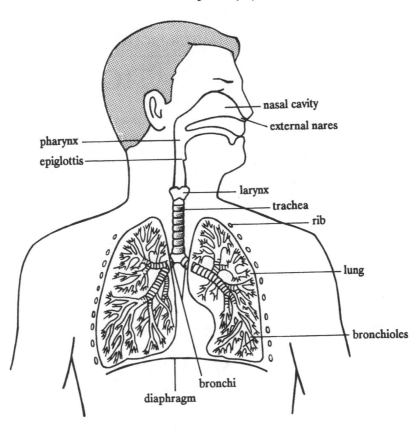

For gas exchange to occur, the lungs must move air in and out to bring external air in contact with the gas exchange surfaces in the alveoli. The lungs are found in a sealed cavity in the chest, the **pleural cavity**, sealed by a pleural membrane bound by the ribs and chest wall and by the muscular **diaphragm** on the bottom. Negative pressure in the pleural cavity keeps the pleural membrane drawn tightly outward against the walls of the chest cavity, keeping the lungs inflated. If the pleural cavity is punctured, allowing it to contact the external environment, the lung can collapse. The diaphragm is curved upward when relaxed, and flattens when contracted, expanding the chest cavity. Chest muscles move the ribs up and out as the diaphragm moves working together to create a larger chest cavity and a vacuum that draws air into the respiratory passages (inhalation). When the diaphragm and intercostal rib muscle relax, the reverse process decreases the size of the chest cavity and forces air out of the lungs (exhalation). Exhalation is usually passive, meaning that the elastic tissues of the lung draw the chest inward during exhalation and that muscle contraction is not involved in exhalation. During strenuous exercise, however, additional muscle groups can cause active exhalation.

The rate of breathing is regulated in the medulla to supply tissues with the correct levels of oxygen and the correct rate of CO_2 removal. Excess CO_2 in the blood stimulates the medulla to send messages to the rib muscles and the diaphragm to increase the frequency of respiration. The respiratory center is less sensitive to changes in the level of oxygen in the blood.

Since air moves in and out of the lungs through the same tube, the trachea, not all of the air in the lungs is exchanged with every breath. The total amount of air in the lungs is about six liters, but the average breath at rest only exchanges about 0.5 liter of air. Even the strongest exhalation leaves air behind in lungs, called the **residual volume**. The air that is breathed in mixes with the air already in the lungs, and diffuses down to the alveoli. The air in the alveoli has a different composition from atmospheric air: it has more water vapor, more carbon dioxide and less oxygen than the atmospheric air outside of the lungs.

Gas Transport and Exchange

The alveoli, located in the lungs, have thin, moist walls and are surrounded by thin-walled capillaries. The walls of the alveoli are moist to allow for gases to dissolve in the thin layer of fluid and then diffuse across the respiratory membranes. Oxygen diffuses from the alveolar air into the blood through the alveolar and capillary membranes and CO_2 and H_2O diffuse out in the same manner. Since passive diffusion drives gas exchange, both in the lungs as well as the tissues, gases always diffuse from higher to lower concentration. In the tissues, O_2 diffuses into tissues and CO_2 leaves, while in the lungs this is reversed due to high oxygen pressure and low CO_2.

The liquid inside the alveoli is largely water, which has surface tension. This surface tension tends to draw the walls of the alveoli together along with the liquid, potentially causing them to collapse and lose the ability to exchange gases. What prevents this from happening in most adults is the secretion of detergent-like surfactants in the alveoli to reduce surface tension. Premature infants may be born before they are able to secrete such surfactants, causing difficulty in breathing.

Oxygen is transported in the blood by red blood cells. Red blood cells contain the oxygen transport protein hemoglobin. The cooperative nature of oxygen binding to hemoglobin allows the blood to efficiently pick-up and deliver oxygen to tissues. Hemoglobin is saturated with oxygen under the conditions normally found in the lung. As blood travels to the tissues, the low oxygen in tissues allows some of the oxygen bound to hemoglobin to release and diffuse into the tissues. The cooperative nature of oxygen binding allows small changes in the tissue oxygen level to have large changes in the delivery of oxygen to the tissue when metabolic needs are increased. Increased pH in tissues such as muscle during exercise will also decrease the affinity of hemoglobin for oxygen, causing more oxygen to be delivered to the muscle.

CO_2 is carried in blood mainly as dissolved carbonate ions and does not use a specific protein carrier for the most part. When CO_2 dissolves in blood in tissues, it enters red blood cells, and is converted by the enzyme carbonic anhydrase into bicarbonate ions to be transported back into the blood. In the lungs, carbonic anhydrase converts the bicarbonate back into carbon dioxide to be exhaled. Bicarbonate ions also play an important role as a pH buffer in the blood and are regulated by the kidney to maintain the plasma pH within a narrow range.

8.6   HOMEOSTASIS

Life requires that the conditions inside the body are maintained in a fairly constant state despite changes in the external environment. Homeostasis is the process by which a stable internal environment within an organism is maintained. Some important homeostatic mechanisms include the maintenance of a constant internal body temperature (thermoregulation), the removal of metabolic waste products (excretion), and the maintenance of water and solute balance (osmoregulation).

Thermoregulation

Warm-blooded animals, **homeotherms**, regulate their internal body temperature. Mammals are homeotherms. Cold-blooded animals such as lizards are called **poikilotherms** and do not have a constant internal temperature. Poikilotherms compensate for their lack of physiological temperature regulation through behavior: If it is hot out, they stay out of the sun, or risk overheating.

In humans, the skin (see figure on next page) protects the body from microbial invasion and from environmental stresses, such as dry weather and wind. Specialized epidermal cells called melanocytes synthesize the pigment melanin, which protects the body from ultraviolet light. The skin is a receptor of stimuli, such as pressure and temperature. The skin is also an excretory organ (removing excess water and salts from the body) and a thermoregulatory organ helping control both the conservation and release of heat.

To cool off, mammals can pant or sweat. Sweat glands secrete a mixture of water, dissolved salts, and urea via sweat pores. As sweat evaporates, the skin is cooled by the aborption of heat that occurs during the evaporation of water. Thus, sweating has both an excretory and a thermoregulatory function. Sweating is under autonomic (involuntary) nervous control. Panting allows cooling by evaporative heat loss as well.

To warm up, a mammal can increase its metabolic rate and insulate itself from the environment. Animals in cold environments like the north and south poles tend to have a large size and rounded shapes—adaptations that reduce the surface area presented to the frigid environment. Subcutaneous fat in the hypodermis insulates the body. Hair or fur traps and retains warm air at the skin's surface. Hormones such as epinephrine can increase the metabolic rate, thereby increasing heat production. Thyroid hormone can increase the long-term metabolic rate to increase the generation of metabolic heat in a cold environment. In addition, muscles can generate heat by contracting rapidly (shivering). Heat loss can be inhibited through the constriction of blood vessels (vasoconstriction) in the dermis, moving blood away from the cooling atmosphere. Likewise, dilation of these same blood vessels (vasodilation) dissipates heat.

Human Skin

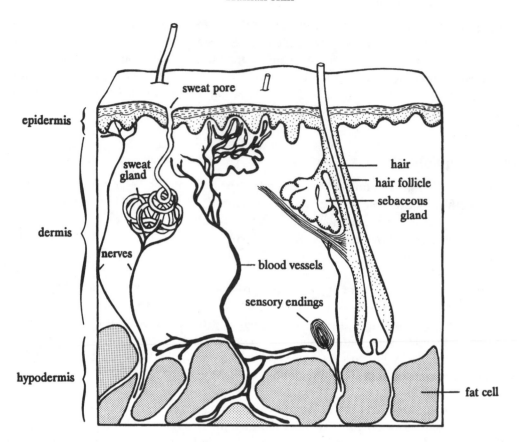

Most mammals have a layer of fur which traps and conserves heat. Some mammals exhibit varying states of torpor in the winter months in order to conserve energy; their metabolism, heart rate, and respiration rate greatly decrease during these months. Hibernation is a type of intense or extreme torpor during which the animal remains dormant over a period of weeks or months with body temperature maintained below normal.

Mammalian body temperature is controlled by the hypothalamus. The hypothalamus tries to adjust the core body temperature according to a temperature set-point. The hypothalamus uses many of the mechanisms indicated above to regulate body temperature, including dilation or constriction of blood vessels, sweating, and shivering. Information from the skin affects the set-point, and the temperature of the hypothalamus itself is used to determine the response of the body. Fevers are a response to infection that alters the set-point to a higher level than normal, to slow replication of the microorganisms responsible for the infection.

8.7   EXCRETION

Cells require water and specific concentrations of salts. Most organisms have a salt concentration of the extracellular fluids that is roughly the same as the concentration of salt in the oceans. Cells must regulate the salt and water content of their bodies to function. Also, cells produce nitrogenous wastes from protein metabolism that can be toxic if not removed. The excretory system removes nitrogenous wastes from the body and also regulates the salt and water balance of the body.

Invertebrate Excretion

Simple organisms such as cnidarians or sponges have all cells in contact with the external, aqueous environment. Water-soluble wastes such as carbon dioxide and the highly toxic ammonia can therefore exit via simple diffusion through the cell membrane. Some freshwater protozoa, such as the paramecium, possess a contractile vacuole, an organelle specialized for water excretion by active transport. Excess water, which continually diffuses into the hyperosmotic cell from the hypo-osmotic environment (in this case, fresh water), is collected and periodically pumped out of the cell. This permits the cell to maintain its volume and pressure. Flatworms such as planarians that live in freshwater have tubules that end in specialized excretory flame cells. Cilia projecting from flame cells drive water out of excretory pores, removing excess water from the body. Annelids have two pairs of **nephridia** (or metanephridia) in each body segment to excrete water, mineral salts, and nitrogenous wastes. In arthropods, nitrogenous wastes are excreted in the form of solid uric acid crystals. The use of solid nitrogenous wastes is an adaptation that allows arthropods to conserve water. Mineral salts and uric acid accumulate in the **Malphigian tubules** and are then transported to the intestine to be expelled along with solid wastes of digestion.

Ammonia is a highly toxic compound if present in significant levels in tissues, so many organisms convert ammonia to either urea or uric acid that are then excreted from the body.

Vertebrate Excretion

The principal organ of excretion in humans is the **kidney**. The kidney forms urine that passes to the **urinary bladder** through the **ureter** (see figure on next page). From the bladder, urine passes to the exterior through the **urethra**. The kidneys filter the blood to remove harmful metabolic waste such as urea while retaining cells, proteins, salt, glucose and other essential factors in the blood. In addition to removing wastes, the kidneys regulate the volume and salt content of the extracellular fluids, including blood. To accomplish this, the kidneys filter the blood, then secrete material into the filtrate and reabsorb material from the filtrate to form urine.

Human Excretory System

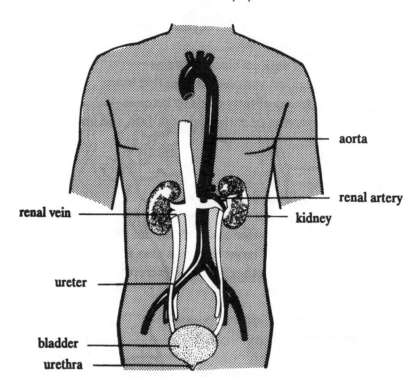

One of the main metabolic waste products that the kidneys remove is urea. When excess amino acids are present, or during a period of starvation when other energy sources are depleted, the body will break down amino acids from proteins and burn them for energy via the Krebs cycle. During this process, nitrogen is enzymatically removed from the amino acid and released as ammonia, which is highly toxic. In humans, the liver converts the ammonia to urea, which is much less toxic than ammonia. The kidneys then remove the urea from the bloodstream.

Bony marine fish live in a very salty environment, much saltier than their body fluids, and must contend with osmotic loss of fluids. To prevent water loss, marine bony fish swallow water and actively transport salt ions either from their gills or their kidneys. They cannot produce urine that is saltier than their body fluids to save water but they produce very small volumes of urine.

Cartilaginous fish such as sharks and rays must also contend with the same salty ocean environment, but do so by accumulating urea in their tissues to very high levels to have the same osmotic potential as seawater. The urea prevents osmotic water loss. Amphibians and fresh water fishes have more salt in their extracellular fluids than the environment, and must contend with the tendency of water to flow into their body through osmosis. These animals tend to excrete large quantities of dilute urine to remove excess water from their bodies.

STRUCTURE AND FUNCTION OF PLANTS AND ANIMALS

Terrestrial animals must contend with water loss through evaporation of body fluids to the environment. Reptiles and birds contend with this problem through excretion of uric acid as a solid, so they do not lose water in urine. Birds and mammals have both developed the ability to conserve water by making urine that is more concentrated than their extracellular fluids.

The basic unit of the kidney that performs these functions is a small tubelike structure called the **nephron** (see figure) that filters blood and modifies the filtrate to produce urine. The blood that is to be filtered enters each nephron in a ball-shaped cluster of capillaries called the **glomerulus**. The pressure of blood in the glomerulus squeezes the liquid portion of the blood out of the glomerulus through a sievelike filtering structure. Blood cells are too large to pass through the sieve and most proteins are retained in blood by their size and charged nature. Other small molecules such as salts, amino acids, glucose, water, and urea pass easily into the filtrate. The filtrate that leaves the blood enters a cup-shaped structure around the glomerulus, **Bowman's capsule**, the starting end of the tubelike nephron.

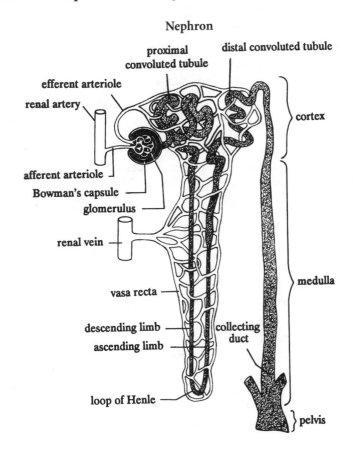

Nephron

From Bowman's capsule, the urinary filtrate will move down the nephron tubule, becoming increasingly modified as it progresses. Before urine is ready to be excreted, it passes through 1) the proximal convoluted tubule, 2) the loop of Henle, 3) the distal tubule, and 4) the collecting ducts. The first step in modifying the urinary filtrate occurs in the **proximal convoluted tubule**. In this region, active transport pumps glucose, amino acids, sodium, and proteins out of the filtrate. Water follows by osmosis, concentrating the urine and greatly reducing the volume of filtrate. The reabsorbed material reenters the blood in capillaries that surround the nephron. This step conserves necessary metabolites that would otherwise be wasted in urine at the same time that it concentrates the urinary filtrate and conserves water.

From the proximal tubule, the filtrate passes to the **loop of Henle**. While the glomerulus and Bowman's capsule of each nephron are located in the outer region of the kidney (the cortex), the loop of Henle dips down into the inner kidney region called the **medulla**. The medulla has a very high concentration of extracellular sodium. As the filtrate passes down the loop of Henle, water is drawn out of the filtrate due to osmosis, passing from the low ion concentration in the filtrate to the high ion concentration of the extracellular fluid in the medulla. When the filtrate passes back up the loop of Henle, sodium is pumped out into the medulla. These two steps further reduce the volume of the urinary filtrate, drawing out the water along with the sodium, and help to preserve the high concentration of sodium in the medulla.

After passing through the distal tubule, the filtrate must pass through the collecting duct before passing out to the ureter and the urinary bladder. The collecting duct passes back down through the high ion concentration in the medulla. To make concentrated or dilute urine, the hormone **ADH** (antidiuretic hormone) regulates the permeability of the collecting duct walls. The high salt concentration created this way in the medulla is required later in the urine formation process to create urine in the collecting ducts that is more concentrated than the extracellular fluids.

When a person needs water, they will secrete ADH and excrete more concentrated urine. ADH is secreted by the posterior pituitary gland when stretch sensors in the arteries detect a drop in blood pressure. ADH acts on the walls of the collecting ducts to make them more permeable to water. Since the fluid of the medulla is very concentrated with ions, water will flow out of the collecting ducts if the walls of the collecting duct are water permeable and allow osmosis, saving water and creating concentrated urine. If no ADH is present, the walls of the collecting ducts do not permit osmosis, and the urine will remain dilute.

Another hormone that regulates urine formation is the steroid hormone **aldosterone**. Aldosterone is secreted in response to low extracellular sodium and acts on the distal tubule to increase the resorption of sodium from the urinary filtrate. The resorption of sodium causes water to be removed from the filtrate by osmosis, reducing the urine volume and increasing the volume of the extracellular fluids of the body, increasing blood pressure. Thus, secretion of aldosterone is another means the body can use to conserve water.

8.8   ENDOCRINE SYSTEM

The body has two communication systems to coordinate the activities of different tissues and organs. One communication system is the nervous system and the other is the endocrine system. The endocrine system is the network of glands and tissues that secrete **hormones**, chemical messengers produced by cells to act on other cells. The target cells that a hormone acts on must have receptors that specifically recognize the hormone and then transmit a response in the cell. Some hormones act on nearby cells; this is called a **paracrine** signal. Some hormones act on the same cell that makes them; this is called an **autocrine** signal. In some cases endocrine cells are alone and in other cases they are organized into endocrine glands that produce and secrete hormones.

Compared to the nervous system, the signals conveyed by the endocrine system take much more time to reach their destination and cause a response. A nervous impulse is produced in a millisecond and travels anywhere in the body in less than a second. Hormones require time to be synthesized, can travel no more quickly than the blood can carry them, and often cause effects by inducing protein synthesis or transcription that require time. However, hormone signals will tend to be more long lasting than nerve impulses. When the nerve impulse ends, a target such as skeletal muscle returns quickly to its starting state. When a hormone induces protein synthesis, the proteins remain long after the hormone is gone. Often the two systems work together. The endocrine glands, such as the pancreas or the adrenal cortex, can be the direct targets (effectors) of the autonomic nervous system. The hormone adrenaline acts in concert with the sympathetic nervous system to produce a set of results similar to those produced directly by sympathetic neurons.

Endocrine glands secrete hormones directly into the bloodstream. This is in contrast to **exocrine** secretions that do not contain hormones and are released through ducts into a body compartment. An example of exocrine secretion is the secretion by the pancreas of digestive enzymes into the small intestine through the pancreatic duct. Both endocrine and exocrine functions can be found in the same organ. The pancreas simultaneously produces exocrine secretions such as digestive enzymes and endocrine secretions such as insulin and glucagon.

Two Types of Hormones: Steroid Hormones and Peptide Hormones

Based on their chemical structure, hormones can be classified into two groups: steroid hormones and peptide hormones. In both cases, the hormone must bind to a protein receptor on the target cell to affect that cell. The types of receptors they bind to are different, however. Peptide hormones are large, hydrophilic, charged, and cannot diffuse across the plasma membrane. As a result, the receptors they bind to are located on the cell surface. When a peptide hormone binds to its receptor on the surface of a target cell, it activates the receptor and causes it to transmit a signal into the cellular interior. This signal can be to turn on a protein kinase that phosphorylates certain proteins and change their activity, or to activate enzymes that create **second messengers** in the cell such as cyclic AMP, that amplify the signal and alter many different cellular activities. Other second messengers include increased cytoplasmic calcium released in response to some hormones that alters the activity of many enzymes, or inositol phosphate, produced through hormonal activation of an enzyme that hydrolyzes a membrane lipid to release diacyl glycerol as well. This form of indirect signaling by a hormone is called a **signal transduction cascade** because of the amplification of the signal by downstream signaling factors. One receptor protein with hormone bound can activate thousands of enzyme molecules, each of which can produce many thousands of molecules of second messenger, leading to yet further amplification downstream through more pathways of interacting signaling molecules (see figure).

Cyclic AMP—Second Messengers

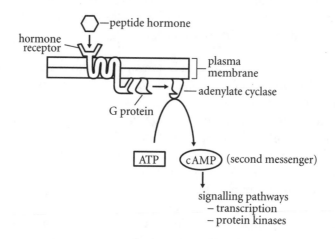

One large class of peptide hormone receptors is the G-protein coupled receptors. These receptors are found in the plasma membrane with seven transmembrane domains that pass through the membrane and anchor the receptor at the cell surface. When these receptors bind hormone, they activate another class of signaling molecules called G proteins that interact with the receptor. Activated G proteins bind GTP and activate downstream signaling proteins such as adenylate cyclase, the enzyme that produces the second messenger cyclic AMP to amplify the signal.

Peptide hormones can be small peptides such as ADH, with just a few amino acid residues, or large complex polypeptides like insulin. Since peptide hormones are secreted proteins, they are synthesized on the rough ER in the cell, then packaged and processed in the Golgi before they are delivered to the plasma membrane for secretion. The release of hormones is usually regulated. The hormones are stored in secretory vesicles in the cytoplasm, waiting for the signal that fuses the vesicle with the plasma membrane, dumping the hormones into the extracellular fluid and blood and transporting them through the circulatory system to distant target tissues in the body.

Steroid hormones are small and hydrophobic, and most hormones of this class are derived from cholesterol, including estrogen, progesterone, testosterone, and cortisol. Since they are small and hydrophobic, they can diffuse through the cell membrane. These hormones bind steroid hormone receptors after they have diffused into the cell through the plasma membrane. The receptors with hormone enter the nucleus and bind to the promoters of genes to regulate transcription, turning genes on or off. The signals created by steroid hormones are changes in gene transcription and protein expression.

STRUCTURE AND FUNCTION OF PLANTS AND ANIMALS

Since steroid hormones freely diffuse through membranes, they are not stored after production. They are usually secreted at a rate equal to their production. It is the rate of production of these hormones that is highly regulated, by controlling the activity of the enzymes that produce the hormones.

Endocrine Glands

Hormones are secreted by a variety of endocrine glands, including the hypothalamus, pituitary, thyroid, parathyroids, adrenals, pancreas, testes, ovaries, pineal gland, kidneys, gastrointestinal glands, heart, and thymus (see figure). Some hormones regulate a single type of cell or organ, while others have more widespread actions. The specificity of hormonal action is determined by the presence of specific receptors on or in the target cells.

Human Endocrine System

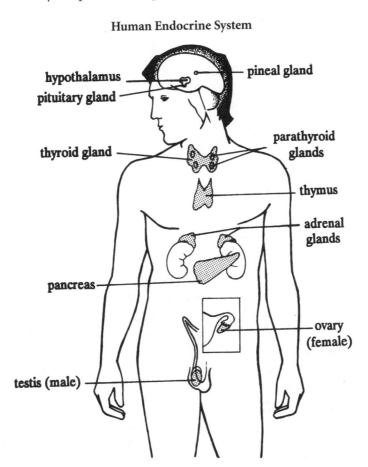

A common principle that regulates the production and secretion of many hormones is the **feedback loop**. Often, several hormones regulate each other in a chain. For example, the hypothalamic hormone corticotropin acts on the anterior pituitary to release ACTH, which acts on the adrenal cortex to release cortisol. In a feedback loop, the level of the last hormone regulates the production of earlier hormones in the loop. When cortisol blood levels increase, cortisol acts on the pituitary to decrease further ACTH secretion, which leads to a decrease in cortisol production. Acting in this way, feedback loops act to maintain hormones at a relatively constant level.

Hypothalamus and Pituitary Gland

The **hypothalamus**, a section of the posterior forebrain, is located above the pituitary gland and is intimately associated with it. The pituitary has two parts with different functions, the **anterior pituitary** and the **posterior pituitary**. The posterior pituitary makes vasopressin and oxytocin. These hormones are synthesized in the hypothalamus and delivered by nerves that pass directly into the posterior pituitary where they are secreted. **Vasopressin** is also called **ADH** (antidiuretic hormone) and acts on the kidney to conserve water (see Section 8.7, Excretion). **Oxytocin** plays a role during childbirth to stimulate labor.

The anterior pituitary secretes several hormones and is regulated by the hypothalamus through a **portal blood circulation** that carries blood directly from the hypothalamus to the pituitary. In most parts of the circulatory system, blood flows directly back to the heart from capillaries, but in a portal system blood flows from capillaries in one organ to capillaries in another. When the hypothalamus is stimulated (by feedback from endocrine glands or by neurons innervating it), it releases hormonelike substances called **releasing factors** into the anterior pituitary-hypothalamic portal circulatory system. These hormones are carried directly to the pituitary by the portal system. In their turn, these releasing factors stimulate cells of the anterior pituitary to secrete the hormone indicated by the releasing factor.

The anterior pituitary secretes the following hormones:

- **Growth hormone** fosters growth in a variety of body tissues.
- **Thyroid stimulating hormone** (TSH) stimulates the thyroid gland to secrete thyroxine.
- **Adrenocorticotrophic hormone** (ACTH) stimulates the adrenal cortex to secrete corticosteroids including cortisol in response to stress.
- **Prolactin** is responsible for milk production by the female mammary glands.
- **Follicle-stimulating hormone** (FSH) spurs maturation of seminiferous tubules in males and causes maturation of ovaries in females. It also encourages growth of follicles in the ovaries.
- **Luteinizing hormone** (LH) induces interstitial cells of the testes to mature by beginning to secrete the male sex hormone testosterone. In females, a surge of LH stimulates ovulation of the primary oocyte from the follicle. LH also helps to create the corpus luteum.
- **Endorphins** are peptide hormones that act in the CNS to block pain signaling by binding to the opiate receptors. Opiates are naturally occurring mimics of the endorphin hormones.

STRUCTURE AND FUNCTION OF PLANTS AND ANIMALS

Thyroid Gland

The thyroid hormone, **thyroxine**, is a modified amino acid that contains four atoms of iodine. It accelerates oxidative metabolism throughout the body. An abnormal deficiency of thyroxine causes goiter, an enlargement of the thyroid gland, decreased heart rate, lethargy, obesity, and decreased mental alertness. Hyperthyroidism, caused by too much thyroxine, is characterized by profuse perspiration, high body temperature, increased basal metabolic rate, high blood pressure, loss of weight, and irritability. Thyroxine secretion is stimulated by cold and environmental cues that act on the hypothalamus to stimulate the secretion of thyrotropin-releasing hormone, which acts on the anterior pituitary in turn to secrete thyrotropin. Thyrotropin from the pituitary regulates thyroxin production.

The thyroid gland also produces **calcitonin**, a hormone that regulates calcium ion concentration in the blood. Calcitonin stimulates bone cells called osteoblasts that build bone and depresses the activity of osteoclasts that break down bone. The net result of increased calcitonin is to decrease the amount of calcium in blood by increasing bone formation. Calcitonin is opposed by the action of parathyroid hormone.

Parathyroid Glands

The parathyroid glands are small pealike organs located on the posterior surface of the thyroid. They secrete **parathyroid hormone**, which regulates the calcium and phosphate balance between the blood and other tissues. Increased parathyroid hormone increases bone resorption and elevates plasma calcium. Decreased calcium in blood causes secretion of parathyroid hormone, which increases the activity of osteoclast cells in bone that remodel bone to release calcium.

Pancreas

The pancreas is a multifunctional organ. It has both an exocrine and an endocrine function. The exocrine function of the pancreas secretes enzymes through ducts into the small intestine. The endocrine function, on the other hand, secretes hormones directly into the blood stream.

The endocrine function is centered in the **islets of Langerhans**, which contain alpha and beta cells that secrete glucagon and insulin, respectively. **Insulin** stimulates the muscles and other cells to remove glucose from the blood. It is also responsible for spurring both muscles and liver to convert glucose to glycogen, the storage form of glucose. The islets of Langerhans also secrete **glucagon**, which responds to low concentrations of blood glucose by stimulating the breakdown of glycogen into glucose.

Adrenal Glands

The adrenal glands are situated on top of the kidneys and consist of the adrenal cortex on the exterior of the gland and the adrenal medulla on the outside of the gland. In response to stress, ACTH stimulates the adrenal cortex to synthesize and secrete the steroid hormones, which are collectively known as corticosteroids. The corticosteroids, derived from cholesterol, include glucocorticoids, mineralocorticoids, and cortical sex hormones. Glucocorticoids such as cortisol are involved in glucose regulation and protein metabolism, and their presence is important in stress responses. Glucocorticoids raise blood glucose levels and decrease protein synthesis. They also reduce the body's immunological and inflammatory responses. ACTH from the pituitary induces cortisol production. Elevated cortisol represses ACTH expression and lowers cortisol levels, acting as a feedback loop to maintain relatively constant cortisol levels.

Mineralocorticoids, particularly aldosterone, regulate plasma levels of sodium and potassium, and consequently, the total extracellular water volume. Aldosterone causes active reabsorption of sodium and passive reabsorption of water in the nephron. The adrenal cortex also secretes small quantities of androgens (male sex hormones) in both males and females.

The secretory cells of the adrenal medulla can be viewed as specialized sympathetic, postganglionic nerve cells that secrete hormones into the circulatory system. This organ produces **epinephrine** (adrenaline) and **norepinephrine** (noradrenaline), both of which belong to a class of amino acid-derived compounds called catecholamines. Epinephrine increases the conversion of glycogen to glucose in liver and muscle tissue, causing a rise in blood glucose levels and an increase in the basal metabolic rate. Both epinephrine and norepinephrine increase the rate and strength of the heartbeat, and dilate and constrict blood vessels. These in turn increase the blood supply to skeletal muscle, the heart, and the brain, while decreasing the blood supply to the kidneys, skin, and digestive tract. These effects are known as the "fight or flight response," and are elicited by sympathetic nervous stimulation in response to stress. Both of these hormones are also neurotransmitters.

Ovaries and Testes

The gonads are important endocrine glands, with testes producing testosterone in males and ovaries producing estrogen and progesterone in females. The sex steroids first play a role in early development in sex determination. The early embryo of males and females is the same, including gonads that appear similar in both sexes. Around the seventh week of development, the gonads of males, in response to information on the Y chromosome, start to produce testosterone. This testosterone induces the differentiation of testes and the other structures of the male reproductive system. If the Y chromosome is not present, or the genes involved in testosterone signaling are defective, the default pathway for sexual development results in development of the female reproductive system.

During puberty, expression of GnRH by the hypothalamus acts on the pituitary to induce FSH and LH production. In females, FSH and LH act on the ovaries to cause estrogen production and the development of the secondary sexual characteristics of women. The onset of the menstrual cycle also results. At menopause, estrogen and progesterone production drop dramatically. In men, FSH and LH production during puberty act on the testes to induce testosterone production and the development of secondary sex characteristics of men.

8.9   MOTOR SYSTEMS

Animals have almost universal desires to gather food, escape predators, and find mates. The ways that animals meet these needs are varied. One mechanism is through the use of cilia and flagella, projections that extend from eukaryotic cells. Cilia and flagella in eukaryotes are produced from microtubules organized in a ring with nine groups of two microtubules, and sometimes more microtubules in the center of the assembly. Motor proteins join the microtubules together and consume ATP to drive the sliding movement of the microtubules past each other, bending the flagella or cilia back and forth, producing a whiplike motion that drives movement. In a protist or a free-swimming cell such as a sperm, flagella movement can drive the cell through water. If cilia are found on the surface of a mollusk gill or in the human respiratory tract, the cilia moves water over the surface.

Within cells, the cytoskeleton drives movement of cytoplasmic contents from one part of the cell to another. Microtubules can be used as tracks for the movement of organelles in the cell, for example, with motor proteins attaching to the organelle and the microtubule hydrolyzing ATP to drive movement. Actin microfilaments can also drive movement of cytoplasmic contents and change the shape of a cell by interacting with myosin. Amoebas, protists, and amoeboid cells in multicellular organisms, such as macrophages, use actin microfilaments for cellular motion. These cells form extensions called pseudopods, and through polymerization of actin and motion of actin and myosin filaments past each other, move cytoplasm into the pseudopod. This changes the cell's shape and causes the cell to move in the direction of the pseudopod. The motion of muscle cells in multicellular animals is also based on the motion of actin and myosin filaments past each other.

Hydrostatic Skeletons

The muscles that run the length of the cnidarian body wall of animals such as anemones can contract and shorten the body if the inner water chamber is sealed by closing the gut opening. Sealing off the inner water compartment allows the muscles to press against the water to cause movement.

The same type of hydrostatic skeleton is involved in annelid locomotion. Each segment of annelid worms can expand or contract independently based on contraction of muscles that run around the worm and muscles that run the length of the worm. Contraction of circular muscles makes the worm segments longer and narrower, and contraction of the muscles along the length of the worm makes the segments fat and wide. Annelids advance principally through the action of these muscles against a hydrostatic skeleton divided into compartments in each segment, as well as through bristles in the lower part of each segment. These bristles, called **setae**, anchor the earthworm temporarily in earth while muscles push the earthworm ahead.

Exoskeleton

The exoskeleton is the hard skeleton that covers all the muscles and organs of some invertebrates. Exoskeletons found in arthropods are composed of chitin. In all cases, the exoskeleton is composed of noncellular material secreted by the epidermis. Although it serves the function of protection, an exoskeleton imposes limitations on growth. Thus, periodic molting and deposition of a new skeleton are necessary to permit body growth. Muscles attached to the interior of the exoskeleton between jointed sections contract, allowing movement.

Vertebrate Endoskeleton

The vertebrate endoskeleton serves as a framework for movement and support within all vertebrate organisms. In this framework, skeletal muscles are attached to bones, permitting movement when a muscle contracts by bringing two bones together. The endoskeleton also provides protection, since bones surround delicate vital organs. For example, the rib cage protects the thoracic organs (heart and lungs), while the skull and vertebral column protect the brain and spinal cord.

Cartilage

Cartilage is firm and flexible, but is not as hard or as brittle as bone. It makes up the skeletons of lower vertebrates, such as sharks and rays. In higher animals, cartilage is the principal component of embryonic skeletons, and is replaced during development by the aptly termed replacement bone. Cartilage is often found at joints in humans, such as the hips and knees. Because cartilage has no vessels or nerves, it takes longer to heal than bone.

Bone

Bone makes up the skeleton of mature higher vertebrates, including humans. Bone arises through the replacement of cartilage or through direct ossification. Bone produced through the latter process is called dermal bone; the bones of the skull are examples of this. In replacement bone, on the other hand, cells called osteoblasts replace the cartilage that has already formed. A hollow cavity is formed within each bone; this cavity is subsequently filled with bone marrow, the site of formation of blood cells.

Mature bone is continually remodeled by osteoblast cells that lay down new bone, and osteoclast cells form tunnels through bone to break it down. This remodeling of bone responds to hormones and is involved in the regulation of the amount of calcium found in blood.

The division between spongy and compact bone is based on function and internal structure. Spongy bone is located in the central portions of bone. It consists of a network of hard spicules separated by marrow-filled spaces in which blood cells are produced. The low density and ability to withstand lateral stress that are characteristic of bone may be attributed to this type of spongy bone.

Meanwhile, compact bone, located on the outer surfaces and articular surfaces, is responsible for the hardness of bone and its ability to withstand longitudinal stress. It consists of cylindrical units called **Haversian systems**. Each unit consists of cells embedded in a matrix of inorganic material (calcium phosphate) which gives bone its hardness. These cells radiate around a central blood vessel within a Haversian canal.

Bones are held together by sutures or immovable joints, such as those in the skull, or by movable joints such as the hip joint. In the latter type, ligaments serve as bone-to-bone connectors, while tendons attach skeletal muscle to bones and help bend the skeleton at the movable joints. In the vertebrate skeleton, the **axial skeleton** is the midline basic framework of the body, consisting of the skull, vertebral column, and the rib cage. The **appendicular skeleton**, on the other hand, includes the bones of the appendages and the pectoral and pelvic girdles.

Attachments and Joints

A skeletal muscle originates at a point of attachment to the stationary bone. The insertion of a muscle is the portion attached to the bone that moves during contraction. An **extensor** extends or straightens the bones at a joint, like the straightening of an arm at the elbow. A **flexor** bends a joint to an acute angle, as in bending the elbow to bring the forearm and upper arm together.

Joints are held together by ligaments, connective tissue joining two bones together, and tendons, connective tissue joining muscle to bone. There are many different types of joints, including the ball and socket joints found in the hip and joints that form hinges in the knees.

Types of Muscles

Vertebrates possess three different types of muscle tissues: smooth, skeletal, and cardiac. Muscle contraction in all three types of muscle depends on the movement of actin and myosin filaments past each other, using ATP.

Skeletal muscles, or voluntary muscles, produce intentional physical movement. The somatic nervous system innervates skeletal muscle. Each fiber is multinucleated (that is, it has more than one nucleus in each cell) and is crossed by alternating light and dark bands called striations. The striations are formed by overlapping strands of the contractile proteins actin and myosin. The thin filaments in striations consist of actin and the thick filaments consist of myosin. The contractile unit in each array of actin and myosin filaments is the **sarcomere**, containing myosin in the middle and overlapping actin filaments on each end. Muscle contraction results in myosin filaments sliding over the actin filaments, causing each sarcomere to shrink in length.

The contraction of skeletal muscle is regulated in the muscle cell by **troponin-tropomyosin** proteins that bind to the actin filaments. In the absence of calcium, troponin-tropomyosin block myosin from binding to actin. When a motor neuron fires an action potential that reaches the synapse of the nerve with a skeletal muscle cell, the **neuromuscular junction**, the muscle has an action potential that spreads across the whole cell, and causes the cell to release calcium from internal stores in the sarcoplasmic reticulum. The calcium cause troponin-tropomyosin to change its conformation and allow myosin to bind actin, moving the actin and myosin filaments past each other.

The strength of muscle contraction is not determined by the strength of an action potential that reaches it or the strength of the action potential in the muscle cell. The action potential in both nerves and muscle cells is an all-or-none response. Increased frequency of action potentials can increase the strength of muscle contraction. If the action potentials fire quickly enough, the muscle cell cannot relax between contractions.

Smooth muscle, or involuntary muscle, is generally found in visceral systems and is innervated by the autonomic nervous system. Each muscle fiber consists of a single cell with one centrally located nucleus. Smooth muscle is nonstriated. Examples are the muscles found in the walls of the arteries and veins, the walls of the digestive tract, the bladder, and the uterus. Smooth muscle cells have only one nucleus, and do not have a striated appearance under the microscope. Smooth muscle cells release calcium in the cytoplasm to trigger contraction, but do not use the troponin-tropomyosin system used in skeletal muscle to regulate actin-myosin contraction.

Cardiac muscle is the tissue that makes up the heart. It has characteristics of both skeletal and smooth muscle. Cardiac muscle is both striated and mononucleate. The nervous system modulates the inherent heart beat of cardiac muscle.

8.10 NERVOUS SYSTEM

The body has two communication systems that coordinate the actions of different parts of the body and responses to the environment. The endocrine system uses chemical messengers called hormones to transmit signals between cells. Rather than chemical messengers, the nervous system uses electrochemical impulses as messengers that travel through a network of cells to their destination. The information carried in these impulses enables organisms to receive and respond to stimuli from their external and internal environments. Your brain and spinal column control your breathing, your movement, and your ability to feel, see, touch, hear, and taste, as well as your ability to understand complex emotions and more abstract concepts.

Neurons: The Functional Unit of the Nervous System

All nervous systems start with specialized cells called neurons as their most basic functional unit. The **neuron** is a specialized cell that is designed to transmit information in the form of electrochemical signals called **action potentials**, or **nerve impulses**. These signals are generated when the neuron alters the voltage across its plasma membrane. The property that neurons have that allows them to carry an action potential is an excitable membrane that can vary its potential. The basic parts of the neuron's cell structure are the cell body, the dendrites, and the axon (see figure). The **cell body** contains the nucleus and most of the organelles and is the site of most protein synthesis and energy production in neurons. The **dendrites** project from the cell body to receive chemical information from other neurons and carry this information to the cell body. The axon is a very long, slender projection of the neuron that transmits the action potential from the cell body to the target cells. The axon can be as long as a meter when the cell body is located in the central nervous system and the axon must carry the action potential to a target in the extremities.

Neuron

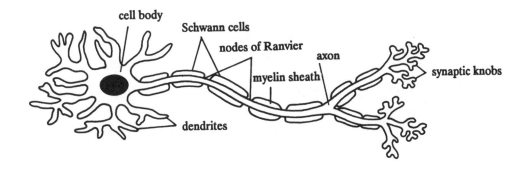

Resting Potential

The action potential involves manipulating the voltage across the plasma membrane to carry information. The voltage across the plasma membrane is altered by moving ions back and forth across the membrane. All cells have a voltage across their plasma membrane that is generated through the actions of a protein called the **Na$^+$/K$^+$ ATPase**. Using the hydrolysis of ATP for energy, this protein pumps sodium ions out of the cell and potassium into the cell. This activity is essential to maintain the osmotic balance of cells. Some of the potassium leaks back out of the cell through the **potassium channel**, an ion channel that selectively allows potassium ions to flow down a K$^+$ concentration gradient established by the Na$^+$/K$^+$ ATPase. With more positive ions on the outside of the cell, a net negative voltage of about −70mVolts exists across the plasma membrane of most cells and is called the resting potential.

The Action Potential

All cells have a resting potential, but not all cells have an excitable membrane. Most cells maintain the membrane potential at the resting potential. Cells with an excitable membrane alter their membrane potential to carry information via the action potential. The difference is that neurons, as well as muscle cells, have ion channel proteins in their plasma membrane that open to allow ions through in response to a decrease in the membrane potential. This protein in neurons is the **voltage-gated sodium channel**. When the membrane potential is more negative than the resting potential, moving from −70 mVolts to −90 mVolts, the membrane is **hyperpolarized**. When the membrane potential is less negative than usual, changing from −70 mVolts to −50 or 0 mVolts, the membrane is **depolarized**. The voltage-gated sodium channels are closed at the resting potential and do not let ions through the membrane. A change in membrane potential is what causes the voltage-gated sodium channels to open to allow sodium through (see figure).

Action Potential

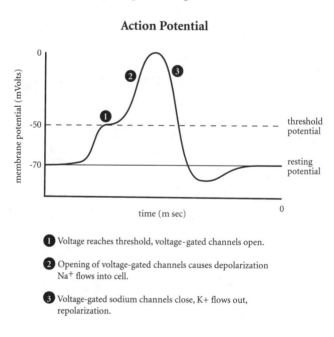

❶ Voltage reaches threshold, voltage-gated channels open.

❷ Opening of voltage-gated channels causes depolarization Na$^+$ flows into cell.

❸ Voltage-gated sodium channels close, K+ flows out, repolarization.

To understand the action potential, first think of the changes across a single section of membrane. If the membrane voltage becomes less negative than the resting potential, changing from −70 mVolts to −50 mVolts, then the voltage-gated sodium channels in the neuron's plasma membrane will open. The voltage at which the voltage-gated channels open is called the **threshold potential**. When these channels are open, sodium will diffuse freely through the channel to cross the plasma membrane, flowing down a gradient from the outside of the cell into the cytoplasm. The opening of these channels in one region of the membrane and the entry of the sodium through the channels causes membrane depolarization (the membrane is less polarized, moving toward 0 potential). After the voltage-gated sodium channels have opened and depolarization is complete, the channels close rapidly again, allowing the membrane voltage to return to its normal potential within a millisecond. The return of the voltage to its normal negative state is called **repolarization**. As a section of membrane depolarizes, this triggers the threshold for the voltage-gated channels in the adjacent section of membrane to depolarize. In this way, the action potential moves along the length of the axon in a wavelike manner until it reaches the end of the neuron at the synapse. The action potential can travel very rapidly in this manner, traversing the length of an axon at 120 meters per second.

In some neurons, the action potential moves along the entire length of an axonal membrane as a continuous wave of ion movement. In other vertebrate neurons, a substance called **myelin** surrounds the axons to allow action potentials to travel more quickly. Myelin is formed by **glial cells** that wrap their plasma membrane around axons, insulating the axon. Small spaces between the myelin coating are called **nodes of Ranvier** (see the figure of the neuron). In a myelinated neuron, the action potential jumps from node to node, bypassing the insulated myelin regions where no ions cross the membrane. The action potential can travel much more rapidly this way in myelinated neurons, since it can jump forward instead of traversing the plasma membrane of the entire length of the axon. This method of jumping forward is called **saltatory conduction**. Another factor that affects the speed of an action potential is the size of the neuron—larger neurons carry action potentials more quickly.

Size and Frequency of Action Potentials

There are two important characteristics of action potentials affecting the way nerves carry information. One factor is that every action potential in a neuron is the same size. Once the neuronal membrane reaches the threshold for depolarization, it will fully depolarize. This makes the action potential an all-or-nothing response. Either a neuron fires an action potential or it does not. The strength of a stimulus does not change the size of the action potential depolarization or the duration of the depolarization.

The second factor differentiates the strength of a stimulus. A touch of a finger is different from a blow from a hammer, and the organism must recognize this. Each action potential may be the same size, but the neuron can carry action potentials more frequently to indicate a stronger stimulus. Thus, a weak signal may trigger one action potential in a second, while a strong stimulus may trigger ten in the same period.

When an action potential has just passed through a section of membrane, that section of membrane cannot carry an action potential again immediately. First it must finish depolarization and repolarization. This limit to the frequency of action potential firing in a neuron is the **refractory period**. The refractory period places an upper limit on the number of action potentials that can pass through a neuron in a unit of time. The refractory period and the directional nature of the neuron also mean that neurons only carry action potentials in one direction, from the cell body out to the end of the axon. Action potentials do not move back up the axon to the cell body.

The Synapse – Communication Between Cells

When an action potential reaches the end of a neuron, chemicals called **neurotransmitters** are released to communicate with the next cell across a small gap between cells. The junction between cells that communicate is called the **synapse**. There are two types of synapses: chemical synapses and electrical synapses.

At a chemical synapse, when the action potential reaches the end of the axon, it encounters a rounded terminal filled with vesicles. The vesicles contain neurotransmitter. Important neurotransmitters include **seratonin**, **dopamine**, **acetylcholine**, and **glutamate**. In response to the action potential, voltage-gated calcium channels open and allow calcium into the cell. The presence of calcium causes some of the vesicles to fuse with the plasma membrane and release their contents into the gap between cells. The gap between the neuron and the target cell is called the **synaptic cleft**. Neurotransmitter diffuses across the synaptic cleft and binds to receptors on the target cell plasma membrane. Once bound, these receptors will open ion channels, allowing specific ions through the membrane in response to neurotransmitter. These ions can then elicit a response in the postsynaptic cell (see figure).

Synapse

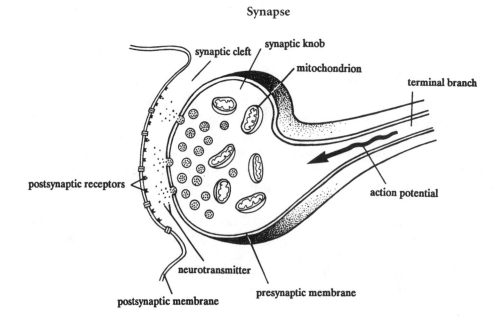

Each neurotransmitter has specific receptors that it interacts with at the synapse, and each receptor opens a channel that allows a specific ion through. An **excitatory neurotransmitter** is one that binds to a receptor that depolarizes the membrane of the postsynaptic cell. For example, the neuromuscular junction is the synapse formed between somatic motor neurons and skeletal muscle. Action impulses from the CNS travel down the axon of motor neurons until they reach the neuromuscular junction. Acetylcholine is the neurotransmitter released at these synapses. Acetylcholine diffuses across the synaptic cleft, where it binds to acetycholine receptors in the muscle cell membrane. These receptors are ligand-gated ion channels that bind acetycholine and open to allow sodium ions to diffuse into the cell. When the sodium enters, it depolarizes the plasma membrane of the target muscle cell. If the depolarization of the target reaches the threshold to open voltage-gated sodium channels, an action potential will be initiated in the muscle cell membrane by voltage-gated channels, and will propagate throughout the muscle cell membrane, triggering contraction of the muscle cell. The more action potentials that reach the muscle and the more muscle cells involved, the stronger the muscle contraction.

A neurotransmitter can also bind to a receptor that opens to allow chloride to enter the postsynaptic membrane, causing a hyperpolarization, with the membrane potential moving away from the threshold for triggering an action potential. This neurotransmitter is inhibitory, since it makes it more difficult for an action potential to start in the target cell. The most common inhibitory neurotransmitter in the CNS is GABA (gamma-aminobutyric acid).

A neuron can form synapses with many neurons, all of which can release neurotransmitter to alter the membrane potential of the target cell. The information from all of the synapses a neuron interacts with is combined in the cell body of a neuron in a process called **summation**. Summation is the means that a single neuron uses to process information from all of its stimulating neurons and to decide whether or not to initiate an action potential itself. If all of the combined changes in the potential of the neuron cause it to reach the threshold depolarization to open voltage-gated channels, then it will fire an action potential. If not, the neuron will remain silent (no action potential).

As important as turning on the signal caused by release of neurotransmitter is the mechanism used to turn off the signal. At the neuromuscular junction, for example, if the acetylcholine is not removed, the muscle cell will continue to contract. Once a neurotransmitter is released into the synaptic cleft, it will continue binding to the postsynaptic receptors unless it is removed from the synapse in some way. There are several ways to remove the neurotransmitter from the synapse. One way is for the neurotransmitter to diffuse into the surrounding fluid. The synapse is small, making diffusion fairly rapid, but in most cases other mechanisms are involved. Another way to remove neurotransmitter from the synapse is with an enzyme that degrades the neurotransmitter. At the neuromuscular junction, **acetylcholinesterase** is an enzyme that acts on acetylcholine to degrade and inactivate it. Pesticides or nerve gas that inactivate this enzyme can be deadly by causing uncontrollable muscle contraction. A third way to remove neurotransmitter is to take it back up into cells at the synapse. This occurs for norepinephrine and seratonin at synapses.

Organization of the Nervous System

As organisms evolved and became more complex, their nervous systems also underwent corresponding increases in complexity. Simple organisms can respond only to simple stimuli, while complex organisms like humans can discern subtle variations of a particular stimuli, such as a particular shade of color.

Invertebrate Nervous Systems

Protozoa are single celled and have no nervous system, although they do have receptors that respond to stimuli such as heat, light, and chemicals. Sponges are multicellular but have almost no response to the environment and have no nerves. Cnidarians have a nervous system consisting of a network of cells, the nerve net, located between the inner and outer layers of the cells of their bodies. The nerve net is decentralized, and allows coordinated motion of muscle fibers to contract in response to a stimuli. Annelids possess a primitive central nervous system consisting of a ventral nerve cord and an anterior 'brain' of fused ganglia. The more complex behavior of annelids requires more centralized information processing, as well as distributed processing of local movement by nerve ganglia in each segment. Arthropods have a better developed, more refined nervous system than annelids. They have more specialized sense organs, including simple or complex eyes and a tympanum for detecting sound.

Human Nervous System

The human central nervous system (CNS) is composed of the brain and the spinal cord. The brain contains all functions that require more information processing than simple reflexes. It consists of an outer portion containing neuronal cell bodies (gray matter) and an inner portion containing axons (white matter). Information flows from sensory neurons back to the CNS in afferent nerves. The CNS processes the information and sends a response out to the body through efferent neurons.

Human Brain

The human brain is composed of the following regions:

Cerebral cortex. The cortex controls all voluntary motor activity by initiating the responses of motor neurons present within the spinal cord. It also controls higher functions, such as memory and creative thought. The cortex is divided into hemispheres, left and right, with some specialization of function between them.

Olfactory lobe. This serves as the center for reception and integration of olfactory input.

Thalamus. Nervous impulses and sensory information are relayed and integrated en route to and from the cerebral cortex by this region.

Hypothalamus. Such visceral drives as hunger, thirst, pain, temperature regulation, and water balance are controlled by this center.

Cerebellum. Muscle activity is coordinated and modulated here.

Pons. This serves as the relay center for cerebral cortical fibers en route to the cerebellum.

Medulla oblongata. This influential region controls vital physiological functions such as breathing, heart rate, and gastrointestinal activity. It has receptors for carbon dioxide; when carbon dioxide levels become too high, the medulla oblongata forces you to breathe.

Human Brain

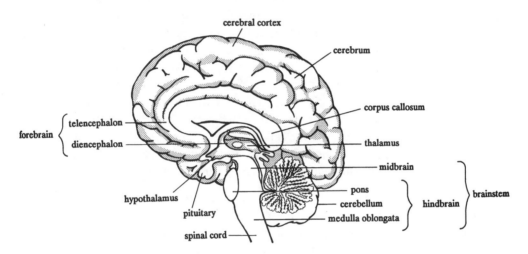

The **spinal cord** is also part of the CNS. The spinal cord acts as a route for axons to travel out of the brain. It also serves as a center for many reflex actions that do not involve the brain, such as the knee-jerk reflex. The spinal cord consists of two parts. The dorsal horn is the entrance point for sensory nerve fibers whose cell bodies are contained within the dorsal root ganglion. The **ventral horn**, on the other hand, contains the cell bodies of motor neurons. Fibers from the cerebral cortex synapse on the ventral horn motor neurons, thereby initiating muscular contractions.

The lower sections of the human brain perform more "primitive" functions. The spinal cord performs simple reflex actions, the medulla processes simple physiological functions such as breathing and heart beat, the cerebellum coordinates motion, and the forebrain with the cortex handles higher thought. The cortex and the forebrain have grown in size and importance through the evolution of the vertebrates.

The Peripheral Nervous System (PNS)

The peripheral nervous system carries nerves from the CNS to target tissues of the body. The peripheral nervous system consists of 12 pairs of cranial nerves, which primarily innervate the head and shoulders, and 31 pairs of spinal nerves, which innervate the rest of the body. Cranial nerves exit from the brainstem and spinal nerves exit from the spinal cord. The PNS has two primary divisions, the somatic and the autonomic nervous systems.

The Somatic Nervous System (SNS)

This system innervates skeletal muscles and is responsible for voluntary movement. Motor neurons release the neurotransmitter acetylcholine (ACh) onto ACh receptors located on skeletal muscle. This causes depolarization of the skeletal muscle, leading to muscle contraction. In addition to voluntary movement, the somatic nervous system is also important for reflex action. There are both monosynaptic and polysynaptic reflexes. Monosynaptic reflex pathways have only one synapse between the sensory neuron and the motor neuron. The classic example is the knee-jerk reflex. When the tendon covering the patella (kneecap) is hit, stretch receptors sense this and action potentials are sent up the sensory neuron and into the spinal cord. The sensory neuron synapses with a motor neuron in the spinal cord, which, in turn, stimulates the quadriceps muscle to contract, causing the lower leg to kick forward. In polysynaptic reflexes, sensory neurons synapse with more than one neuron. A classic example of this is the withdrawal reflex. When a person steps on a nail, the injured leg withdraws in pain, while the other leg extends to retain balance.

The Autonomic Nervous System

The autonomic nervous system regulates the involuntary functions of the body. The autonomic system innervates the heart and blood vessels, the gastrointestinal tract, urogenital organs, structures involved in respiration, and the muscles of the eye. In general, the autonomic system innervates glands and smooth muscle, but not skeletal muscles. It is made up of the sympathetic nervous system and the parasympathetic nervous system.

The Sympathetic Nervous System

This system utilizes norepinephrine as its primary neurotransmitter. It is responsible for activating the body for emergency situations and actions (the fight-or-flight response), including strengthening of heart contractions, increases in the heart rate, dilation of the pupils, bronchodilation, and vasoconstriction of vessels feeding the digestive tract. One tissue regulated by the sympathetic system is the adrenal gland, which produces adrenaline in response to stimulation. Adrenaline produces many of the same fight-or-flight responses as the sympathetic system alone.

The Parasympathetic Nervous System

Acetylcholine serves as the primary neurotransmitter of the parasympathetic nervous system. One of this system's main functions is to deactivate or slow down the activities of muscles and glands. These activities include pupillary constriction, slowing down of the heart rate, bronchoconstriction, and vasodilation of vessels feeding the digestive tract. The principal nerve of the parasympathetic system is the vagus nerve. Most of the organs innervated by the autonomic system receive both sympathetic and parasympathetic fibers, the two systems being antagonistic to one another.

Sensory Systems of the Nervous System

To move through the environment and respond to it, animals have evolved sensory systems. In humans, sight, hearing, balance, taste, smell, and touch provide an influx of data for the nervous system to assimilate. They also make life more enjoyable. One of the common themes in sensory systems is that the sensory cells receive information and use action potentials to transmit the information to the nervous system. All sensory cells send action potentials. The brain determines the nature of the sensory information based on what cell it comes from and what region of the brain receives the information.

Vision

The eye evolved in invertebrates such as flatworms consists of patches of cells on the surface in a depression that are light sensitive. The worm can move toward or away from light using this type of visual system, but it cannot form an image. Arthropods have compound eyes, with many sections gathering light from different directions. Mollusks also have eyes, with sophisticated eyes found in squid that rival those of vertebrates. The squid eye is an image forming system that is very similar to the vertebrate eye in structure, although it evolved independently, providing an example of convergent evolution.

The human eye detects light energy and transmits information about intensity, color, and shape to the brain. The transparent cornea at the front of the eye bends and focuses light rays. These rays then travel through an opening called the pupil, whose diameter is controlled by the pigmented, muscular iris. The iris responds to the intensity of light in the surroundings (light makes the pupil constrict). The light continues through the lens, which is suspended behind the pupil. This lens focuses the image onto the retina, which contains photoreceptors that transduce light into action potentials. Muscle attached to the lens can stretch its shape to change its focus to view either near or far objects. The lens changes shape to focus on nearby objects. With age the lens becomes less flexible, making focusing on nearby objects difficult.

Human Eye

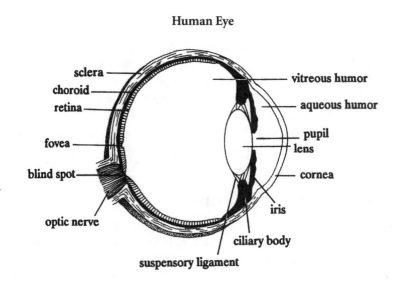

There are two types of photoreceptors in the retina: **cones** and **rods**. Cones respond to high-intensity illumination and are sensitive to color, while rods detect low-intensity illumination and are important in night vision. There are three different types of cones with different color sensitivity depending on the visual pigment they express. Color blind people lack one or more visual pigments. The visual signal is received by the protein **rhodopsin**, which has a retinal group bound to it. This retinal group isomerizes when it is hit by a photon, changing the structure of rhodopsin and starting the signal transduction cascade that changes the membrane potential of the sensory cell and leads to an action potential. The photoreceptor cells synapse onto bipolar cells, which in their turn synapse onto ganglion cells. Axons of these cells bundle to form the optic nerves, which conduct visual information to the brain. Interpretation of these potentials as an image occurs in the cerebral cortex.

Hearing and Balance

The ear transduces sound energy into impulses that are perceived by the brain as sound. It is also responsible for maintaining equilibrium (balance) in the body. Sound waves pass through three regions as they enter the ear. First, they enter the outer ear, which consists of the **auricle** (pinna) and the **auditory canal**. Located at the end of the auditory canal is the **tympanic membrane** (eardrum) of the middle ear, which vibrates at the same frequency as the incoming sound. Next, three bones, or ossicles (**malleus**, **incus**, and **stapes**), amplify the stimulus, and transmit it through the oval window, which leads to the fluid-filled inner ear. This inner ear consists of the **cochlea** and **semicircular canals**. The cochlea contains the **organ of Corti**, which has specialized sensory cells called hair cells. Vibration of the ossicles exerts pressure on the fluid in the cochlea, stimulating the hair cells to transduce the pressure into action potentials, which travel via the auditory nerve to the brain for processing. The frequency of sound is detected by the position in the cochlea of hair cells that are stimulated, and volume is detected by number of signals that are received.

Human Ear

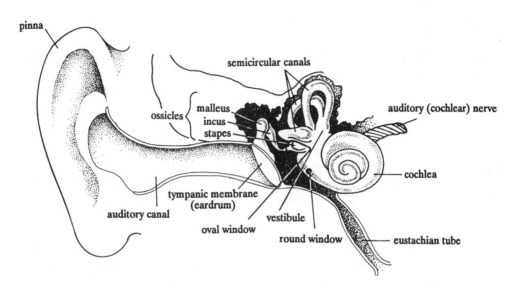

The semicircular canals are used for balance. Each of the three semicircular canals in the inner ear is perpendicular to the other two and filled with a fluid called endolymph. At the base of each canal is a chamber with sensory hair cells; rotation of the head displaces endolymph in one of the canals, putting pressure on the hair cells in it. This changes the nature of the impulses sent by the vestibule nerve to the brain. The brain interprets this information to determine the position of the head.

Taste and Smell

One type of sensory cell, called a chemosensor, detects chemicals in the environment. Taste and smell both involve chemosensors. Taste buds are located on the tongue, the soft palate, and the epiglottis. The outer surface of a taste bud contains a taste pore, from which microvilli, or taste hairs, protrude. Interwoven around the taste buds is a network of nerve fibers that are stimulated by the taste buds, and these neurons transmit impulses to the brainstem. There are four main kinds of taste sensations: sour, salty, sweet, and bitter.

Olfactory receptors are found in the olfactory membrane, which lies in the upper part of the nostrils over a total area of about 5 cm^2. The receptors are specialized neurons from which olfactory hairs project. When odorous substances enter the nasal cavity, they bind to receptors in the cilia, depolarizing the olfactory receptors. Axons from the olfactory receptors join to form the olfactory nerves, which project direction to the olfactory bulbs in the base of the brain.

8.11 ANIMAL BEHAVIOR

Animals must find food and mates and avoid predators. The behaviors of animals help them to accomplish these feats. Some behaviors are determined genetically, while others are learned. The structure of the nervous system, the endocrine system and other aspects of animal physiology can determine the type of behaviors that are observed.

Reflexes

The simplest responses of an organism to its environment are reflexes, automatic responses to simple stimuli. A simple reflex is controlled at the spinal cord level of a vertebrate and involves a two-neuron pathway. In a simple reflex, sensory neurons carry action potentials to the CNS to synapse with a motor nerve going away from the CNS that innervates a muscle or gland that will produce a response.

A slightly more complex reflex is the knee jerk reflex in humans, which involves a few neurons (see figure). The knee jerk is an adaptation to protect the knees. When stretch receptors in the knee are stimulated by the doctor's mallet, action potentials travel in sensory neurons to the spinal cord where they synapse with motor neurons and with **interneurons**. Motor neurons pass directly back to the muscle attached on the muscle on the upper part of the upper leg, causing contraction of this muscle. Contraction of this muscle causes extension of the lower leg—a kicking movement. Interneurons are short neurons in the spinal cord that help to form circuits between sensory and motor neurons. In the knee jerk reflex, the impulse is carried by the interneuron to motor neurons, causing opposing muscle groups on the under side of the leg to relax and allowing the leg to extend. Reflexes often involve coordination of complex sets of muscles required to move the whole arm or leg, all without conscious control of the action.

Reflex Arc

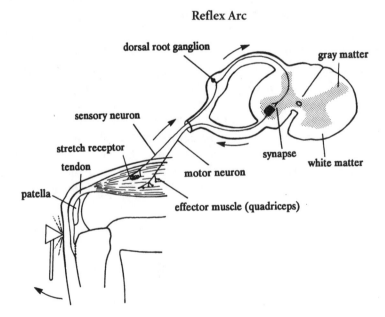

At the same time that the reflex causes the leg to extend, an additional signal to the brain makes a person aware of the situation. When a hand is placed on a hot stove, pain receptors transmit a signal to the spinal cord to initiate a reflex that withdraws the hand even before the signal to the brain is processed to make a person aware of the heat or pain. This type of reflex response is an important adaptation to dangerous situations that require immediate responses.

Simple animals have limited numbers of neurons and only very simple behaviors. Although reflex behavior is important in the behavioral response of lower animals, it is relatively less important in the behavioral repertoire of higher forms of life such as vertebrates due to their increasing dependence on more complex behaviors.

More complex reflex patterns involve neural integration at a higher level, such as the brainstem and even the cerebrum. An example is the **startle response**, the alerting response of an animal to a significant stimulus (for example, a potential danger or hearing one's name called). The startle response involves the interaction of many neurons, including a part of the CNS termed the **reticular activating system**.

Fixed Action Patterns

Along with inheriting physical traits such as feather color and neck length, animals can inherit behaviors rather than learn them. In some circumstances such as avoiding predators, behaviors need to be inherited because the process of learning the trait is likely to be lethal. If animal young are not reared by adults, essential behaviors are likely to be genetically inherited. Behaviors that provide increased fitness will be selected for, and the more important the behavior is to survival or reproduction, the greater the selective pressure for the behavior. Although many behaviors are inherited, they are not usually the product of a single gene that can be followed through Mendelian genetics. It appears more likely that the behaviors are the product of many genes that act on the development of the nervous system, such as growth factors, hormones, hormone receptors, and neurotransmitter receptors.

Fixed action patterns, some of which are amazingly complex, are behaviors that are genetically coded rather than learned. A scientist can determine if a behavior is genetic by raising young in an environment in which they do not have the opportunity to learn the behavior. If a young beaver is placed in an environment with water and wood, it will build a dam, even if it has never seen a dam or seen its parents build a dam. The behavior is a fixed action pattern. Spiders can spin their webs even if they have never seen another web, indicating that this behavior is inherited and is a fixed action pattern. Most of the time, fixed action patterns are resistant to change and are modified very little by learning.

An environmental cue will often trigger the expression of a fixed action pattern. The stimulus that elicits the behavior is referred to as the **releaser**. The releaser is often a very simple stimulus. An example of a fixed action pattern is the retrieval-and-maintenance response of many female birds to an egg of their species. Certain kinds of stimuli are more effective than others in releasing a fixed action pattern. Hence an egg with the characteristics of that species will be more effective in triggering the response than one which only crudely resembles the natural egg. In some cases however, it is not the entire natural object but a simple characteristic of the object that is the true releaser. Baby birds sometimes have a fixed action pattern to peck at the beak of the adult and cause the adult to feed them. The entire adult is not necessary to release the behavior, however. Often only a single key characteristic such as coloration or the beak alone will suffice as an artificial releaser of a fixed action pattern, and can trigger the behavior more strongly than the natural releaser.

Critical Period

Although true fixed action patterns are not learned, there are many factors that can affect their expression. These are specific time periods during an animal's early development when it is physiologically capable of developing specific behavioral patterns such as fixed action patterns. If the proper environmental pattern is not present during this critical period, the behavioral pattern will not develop correctly. The most likely reason for this is that while it is developing, the nervous system must be stimulated to form neural networks in the pattern required for a behavior to be expressed. In some animals there is a visual critical period. If visual stimuli are not present during this critical period, the visual system will not develop properly later on, no matter how much stimulus is given, since the optical processing center in the brain did not receive the stimuli required to develop properly.

Hormones are one influence on critical periods and the expression of genetically influenced behaviors. Fixed action patterns are generally associated with behaviors that have a strong fitness cost associated with them, including mating behaviors. Mating with the wrong species is very costly in terms of fitness since it cannot result in fertile offspring. The songs of songbirds and complicated mating behaviors of many species are designed to both test the fitness of potential mates and prevent mistakes in mating with the wrong species. Many mating behaviors are fixed action patterns designed to prevent mistakes, but they are only expressed if the hormones involved in reproduction, such as estrogens and testosterone, are present during key stages in development. In rats and birds, these hormones have been shown to act on the developing nervous system to alter reproductive behavior, allowing fixed action patterns to be expressed. The expression of male and female reproductive behavior is in part genetic and instinctive, but will not occur if the proper hormonal instructions do not lay the correct nervous framework during development.

In other cases, learning and genetics interact. Many birds cannot perform the correct song without hearing it and learning it when they are young during a critical period soon after birth. However, they may have a genetic template for singing that only allows them to learn the correct song for their species, and not other songs. It is a subject of debate how much of language learned by young humans is influenced by a genetic template.

Imprinting

Imprinting is a process in which environmental patterns or objects presented to a developing organism during a brief critical period in early life become accepted permanently as an element of its behavioral environment. To put it another way, these patterns are 'stamped in' and included in an animal's behavioral response. A duckling, for example, passes through a critical period in which it learns that the first large moving object it sees is its mother. However, if a large, moving object other than its mother is the first thing it sees, the duckling will follow it, as Konrad Lorenz discovered when he was pursued by newborn ducklings that assumed that he was their mother.

Behavioral Cycles

Most organisms, including plants, animals and even protists, have patterns of daily activity, including the timing of eating and sleeping. Underlying these daily cycles in activity are hormonal, nervous and biochemical changes that occur in a daily rhythm. The internal body temperature of humans varies over the course of the day, going down slightly at night during sleep. Daily cycles of behavior are called **circadian rhythms**. Most animals have internal mechanisms that create circadian rhythms even in the absence of external stimuli. However, the circadian rhythm of many organisms does not exactly follow the 24 hour cycle of the sun without external stimuli. Animals that are isolated from the natural phases of light and dark will often follow a pattern of behavior that is either slightly shorter or slightly longer than 24 hours. A human in a cave and allowed to follow a free cycle without external stimuli will usually follow a cycle slightly longer than 24 hours. Each day, external clues such as the sun reset the internal clock, bringing it back to the actual 24 hour daily cycle of the sun. This resetting of the circadian rhythm is called **entrainment**. The cycle is thus initiated intrinsically, but modified by external factors.

Learning and Memory

Learned behavior involves a change in the way an animal behaves based on experience. Learning is a complex phenomenon that occurs, to some extent, in all animals. In lower animals, instinctual or innate behaviors are the predominant determinants of behavior patterns, and learning plays a relatively minor role in the modification of these predetermined behaviors. Higher animals, on the other hand, learn many responses to the environment. Some learned changes, such as habituation and sensitization, are the result of changes within a neuron. Other more complicated types of learning involving memory and long-term changes in behavior are the result of changes in the synaptic connections in networks of neurons. Many forms of learning and memory appear to involve **long-term potentiation** of neurons that causes long-term changes in neurons and stimulates the formation of new neural synapses. Practicing a new activity such as riding a bicycle, for example, causes repeated stimulation of specific neural circuits involved in performing the activity. This stimulation alters the neurons involved, so that the behavior is encoded in a series of connections that make it easier to perform in the future.

Habituation and Sensitization

Habituation is one of the simplest learning patterns, involving the suppression of the normal responses to stimuli. When habituation occurs, repeated stimulation will result in decreased responsiveness to that stimulus. If the stimulus is no longer regularly applied, the response tends to recover over time. If you poke a snail once, it retracts into its shell, but if you poke it repeatedly, habituation occurs and the snail learns to ignore the stimulus. The receptors and sensory neurons involved become down-regulated over time and no longer signal, causing the habituation. Habituation can be due to reduced release of neurotransmitter by sensory neurons, by removal of neurotransmitter receptors from the postsynaptic membrane, or other mechanisms. Habituation also occurs in the CNS, allowing constant stimuli to be ignored. Habituation is reversible: for example, if you leave the snail alone for a while and then touch it again, its retraction response will return.

Sensitization is related to habituation, but is the opposite response to a repeated stimulus. In sensitization, the repeated stimulation of a sensory neuron produces a more sensitive response rather than a reduced response. Like habituation, sensitization is a local response caused by a change in a neuron that is not lasting and does not alter the connections neurons form with each other.

Associative Learning

A simple form of learning called associative learning occurs when an animal links in its mind two events or stimuli that appear to be related to produce the same response. The **conditioned reflex** is a type of associative learning that was studied by Pavlov and involves the association of a normally physical response with an environmental stimulus that is not normally related to the response. Pavlov studied the salivation reflex in dogs. In 1927, he discovered that if a dog was presented with an arbitrary stimulus (e.g., a bell) and then presented with food, it would eventually salivate on hearing the bell alone. He developed the following terminology:

- An **established (innate) reflex** consists of an unconditioned stimulus (e.g., food for salivation), and the response that it naturally elicits is the unconditioned response (e.g., salivation).

- A **neutral stimulus** is a stimulus that will not by itself elicit a response, prior to conditioning (e.g., a bell). During conditioning, the establishment of a new reflex, a neutral stimulus is presented with an unconditioned stimulus. When the neutral stimulus elicits a response in the absence of the unconditioned stimulus, it becomes the conditioned stimulus.

- The product of the conditioning experience is termed the **conditioned reflex** (e.g., salivation at the sound of a ringing bell).

Operant or Instrumental Conditioning

This form of conditioning involves conditioning responses to stimuli with the use of reward or reinforcement. When the organism exhibits a behavioral pattern that the experimenter would like to see repeated, the animal is rewarded, with the result that it exhibits this behavior more often. B. F. Skinner used the "Skinner Box" to show that animals in a cage could be conditioned to push down a lever to release food from a food dispenser. Instrumental conditioning can be performed through positive reinforcement (such as a food reward) or negative reinforcement (such as giving an animal a painful electric shock whenever it exhibits a certain behavior).

Organisms eventually "unlearn" conditioned responses, if they are not reinforced. Extinction is the gradual elimination of conditioned responses in the absence of reinforcement. The recovery of such a conditioned response after extinction is termed spontaneous recovery.

Intraspecific Interactions

Just as an organism communicates within itself via nervous and endocrine systems, it also requires methods to communicate with other members of its species. These methods include behavioral displays, pecking order, territoriality, and responses to chemicals.

Mechanisms of Communication

Many animal behaviors involve mechanisms of communication. Communication takes many different forms. Chemical signals appear to be a very ancient means of communication between organisms, just as hormones in the body allow cells to communicate with each other. In fact, some of the most primitive multicellular organisms, slime molds, use cyclic AMP as a chemical communication that causes the individual amoeboid cells to come together to form a multicellular fruiting body for sporulation. Insects sometimes use wind-born chemical pheromones for attraction of mates, and vertebrates also use chemical signals in urine to mark territory. Auditory signals can travel long distances, tactile communication is useful at close range in crowded conditions, and visual communication is useful in species with highly evolved visual systems like birds and mammals.

A **display** may be defined as an innate behavior that has evolved as a signal for communication between members of the same species. According to this definition, a song, a call, or an intentional change in an animal's physical characteristics is considered a display.

Pecking Order and Territoriality

Frequently, the relationships among members of the same species living as a contained social group become stable for a period of time. When food, mates, or territory are disputed, a dominant member of the species will prevail over a subordinate one. This social hierarchy is often referred to as the **pecking order**. This established hierarchy minimizes violent intraspecific aggressions by defining the stable relationships among members of the group; subordinate members only rarely challenge dominant individuals.

Members of many terrestrial vertebrate species defend a limited area or territory from intrusion by other members of the same species. These territories are typically occupied by a male or a male-female pair. The territory is frequently used for mating, nesting, and feeding. **Territoriality** serves the adaptive function of distributing members of the species so that the environmental resources are not depleted in a small region. Furthermore, intraspecific competition is reduced. Although there is frequently a minimum size for a species' territory, that size varies with population size and density. The larger the population, and the scarcer the resources (e.g., food) available to it, the smaller the territories are likely to be.

Reproductive Behavior

For a behavior or physical trait to evolve, it must be selected for. Natural selection selects for traits that allow better survival and reproduction. One form of selection is **sexual selection**, the selection of traits that are involved in successful mating behaviors.

Males and females exhibit very different reproductive behaviors that have probably evolved in response to the different roles they play in reproduction. Sexual selection in some species results in exaggerated physical traits among males such as coloration in male birds. Males produce large numbers of sperm that have a small cost in production. Females in their lifetime produce a relatively small number of eggs, particularly in birds and mammals, and in most species are primarily responsible for the care and provision of the developing young. Females generally select from among the males who compete for mating opportunities. Mating displays by males often involve defending territories, providing food or otherwise demonstrating fitness to the female. These behaviors cost the male energy, but increase fitness by increasing their chances of reproduction.

Social Behavior

Many animals live together in societies and have evolved social behaviors that increase the fitness of animals in social groups. The evolution of social traits must increase the fitness of the individuals in the social group. Living in a group can allow better hunting by predators such as lions, in which the females hunt together to increase their chances of success. Social behavior can also allow animals to escape predators more effectively. Large flocks of birds or schools of fish are more effective at avoiding predators, seeking safety in numbers.

Animals that live together in a social group are often related and share genes with each other. This is true in insect societies as well as mammalian social groups. In living together and performing social behaviors, individuals increase their own fitness in part through the genes they share with related individuals in the social group. The altruistic behavior demonstrated by social groups, in which an individual appears to sacrifice themselves, can increase their fitness by allowing related individuals to survive. Selective pressure can favor altruistic behavior in a social group in what is called **kin selection**.

8.12 STRUCTURE AND FUNCTIONAL SYSTEMS OF FLOWERING PLANTS

Members of the plant kingdom are photoautotrophs that utilize the energy of the sun, carbon dioxide, water, and minerals to manufacture chemical energy used in respiration and stored in carbohydrates. All plants exhibit alternation of generations in their life cycle between diploid and haploid forms. Within the plant kingdom are several phyla, with one of the key distinguishing characteristics between phyla being the presence or absence of vascular tissue. Within the tracheophytes, which have vascular tissue, two of the important modern phyla are the gymnosperms and the angiosperms. The gymnosperms such as the conifers have "naked seeds" that do not have endosperm and are not located in true fruits. The angiosperms are the flowering plants. They have flowers, true fruits, and a double fertilization to create endosperm to nourish the plant embryo.

Monocots vs. Dicots:
Two Angiosperm Classes

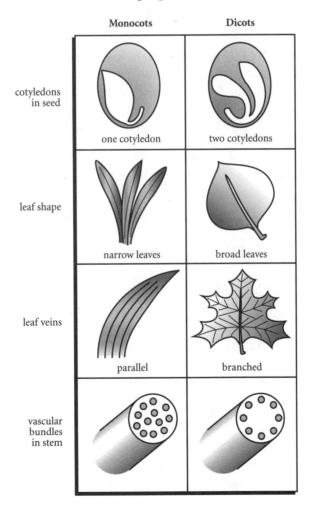

Angiosperms are the dominant plant form in many terrestrial environments today. The discussion of plants in this section will pertain most directly to the angiosperms. Two of the main types (phylogenetic classes) of angiosperms are the monocots and the dicots. These classes are distinguished on the basis of the number of leaves (cotyledons) in the seed embryo, either one (monocot) or two (dicot). Other characteristics of monocots and dicots will be discussed in more detail (see figure).

Structure of Plants

Plants with vascular tissues usually have three types of structures, or organs. These are **leaves**, **roots** and **branches**. The leaves provide most of the photosynthesis of the plant, the roots provide support in the soil, water and minerals, and the branches hold the leaves up to light and convey nutrients and water between the leaves and the roots. Each of these structures can be specialized in many ways.

Plants with taproots have long roots with a single extension deep into the soil while other plants have highly branched roots. Cells on the surface of roots often have long extensions called **root hairs** that increase the surface area of roots. Some plants without root hairs have a symbiotic relationship with fungi that increase the surface area of the root to absorb water and minerals. In legumes, nitrogen-fixing *Rhizobium* species of bacteria infect roots and form root nodules in symbiosis with the plant. The roots of plants often play an important role in preventing erosion. Tropical rain forest that is cleared is highly vulnerable to erosion of the thin soil if the plants and their root systems are absent.

Leaves can have a variety of shapes. Monocot leaves are usually very narrow with veins that run parallel to the length of the leaf, while dicot leaves are broad and with veins that are arranged in a net in the leaf. Modified leaves form thorns in cacti, tendrils in pea plants and petals in flowers. The leaves produce all of the energy of the plant and are specialized to gather sunlight. The broad shape helps to gather sunlight for themselves and in some cases to block sunlight for competing plants. The shape and arrangement of leaves is one of the key features used to distinguish plant species.

Terrestrial plants have broad leaves that maximize the absorption of sunlight but also tend to increase loss of water by evaporation. Leaves of terrestrial plants have a waxy cuticle on top to conserve water. The lower epidermis of the leaf is punctuated by **stomata**: openings that allow diffusion of carbon dioxide, water vapor, and oxygen between the leaf and the atmosphere. A loosely packed **spongy layer** of cells inside the leaf contains chloroplasts with air spaces around cells. Another photosynthetic layer in the leaf, the **palisade layer**, consists of more densely packed elongated cells spread over a large surface area. A moist surface that lines the photosynthetic cells in the spongy layer is necessary for diffusion of gases into and out of cells. Air spaces in leaves increase the surface area available for gas diffusion by the cells. The size of stomata is controlled by **guard cells** around that can open and close the opening. These cells open during the day to admit CO_2 for photosynthesis and close at night to limit loss of water vapor through **transpiration**, the evaporation of water from leaves that draws water up through the plant vascular tissues from the soil. The upper surface layer of cells in leaves has no openings, an adaptation that reduces water loss from the leaf.

Vascular Tissue in Plants: The Stem

The stem is the principal part of the plant involved in transport. Vascular (conducting) bundles run up and down the stem. The vascular bundle at the center of the stem contains **xylem**, **phloem**, and **cambium** tissues. The vascular bundles of monocots are arranged spaced throughout the stem while the vascular bundles of dicots are spaced in a ring in the stems.

Xylem is a tissue that conducts water and minerals from the roots to the rest of the plant. Xylem is located on the inside of the vascular bundle, towards the center of the stem. The thick walls of these cells give the plant its rigid support. In woody dicots, older xylem cells at the innermost layer die, forming wood. The outer layer of xylem in trees is alive and is called the sapwood. Two types of xylem cells have been identified: vessel elements and tracheids that form tubes to draw liquid up into the plant.

The vascular tissue of the xylem contains a continuous column of water from the roots out to the leaves, extending into the veins of the leaves. Leaves regulate the amount of water lost through evaporation. Water is very cohesive, so that as water is lost by leaves, it creates tension that draws the entire column of water upward through the xylem. This process of water movement is called **transpiration** (see figure).

Transpiration

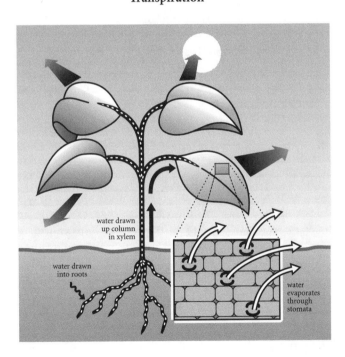

water drawn up column in xylem

water drawn into roots

water evaporates through stomata

Phloem cells, on the other hand, are thin-walled and are found on the outside of the vascular bundle. They usually transport nutrients, especially carbohydrates produced in the leaves down the stem. These living cells include the sieve tube cells and companion cells.

Growth of Tissues in Plants

Growth in higher plants is restricted to embryonic or undifferentiated cells called **meristem**. Meristem cells can undergo cell division to produce new organs through the life of the plant. Meristem cells elongate and differentiate into cell types characteristic of the tissues of the plant. Different types of meristem include the **apical meristem**, which is found in the tips of roots and stems. These cells provide for growth in length, which occurs only at the root and stem tips. Lateral meristem or **cambium** is located in stems between the xylem and phloem. This tissue permits growth in diameter. It is not a predominant tissue in monocots (grasses) or herbaceous dicots (alfalfa). Instead, cambium is found mostly in woody dicots such as oak.

Cambium cells (two layers thick) are the actively dividing, undifferentiated cells that upon differentiation give rise to new xylem and phloem as the plant grows. The cambium lies between the xylem and phloem cell layers. As it divides, the cells near the phloem differentiate into phloem cells, and those near the xylem differentiate into xylem cells.

Primary growth of plants occurs at the apical meristem and elongates the roots and stems. Primary growth causes a plant to grow taller. Secondary growth occurs in the cambium and increases the diameter of stems. **Secondary growth** results from xylem cells in the interior of the stem that die over time and are replaced by new xylem further from the center of the stem. An additional layer of cambium beneath the bark replaces the surface of the tree as secondary growth increases the diameter of the stems.

The gross structure of a nonwoody stem is as follows (proceeding from the outside inwards): epidermis (outer bark) cortex, phloem, cambium, xylem, and finally pith (tissue concerned with storage of nutrients and plant support). The phloem, cambium, and xylem layers are known as the fibrovascular bundle.

Control of Growth in Plants

The regulation of growth patterns is largely accomplished by plant hormones, which are almost exclusively devoted to this function. These hormones are produced by actively growing parts of the plant, such as the meristematic tissues in the apical region (apical meristem) of shoots and roots. They are also produced in young, growing leaves and developing seeds. Some of these hormones and their specific functions are discussed below, including auxins, gibberelins, cytokinins, ethylene and abscisic acid.

Auxins

This is an important class of plant hormone associated with several growth patterns, including the following types.

Auxins are responsible for **phototropism**, the tendency of the shoots of plants to bend toward light sources, particularly the sun. When light strikes the tip of a plant from one side, the auxin supply on that side is reduced and the illuminated side of the plant grows more slowly than the shaded side. This asymmetrical growth in the cells of the stem causes the plant to bend toward the slower growing light side; thus the plant turns toward the light.

Geotropism is the term given to the growth of portions of plants towards or away from gravity. With negative geotropism, shoots tend to grow upward, away from the force of gravity. If the plant is turned on its side (horizontally), the shoots will eventually turn upward again. Gravity itself increases the concentration of auxin on the lower side of the horizontally placed plant, while the concentration on the upper side decreases. This unequal distribution of auxins stimulates cells on the lower side to elongate faster than cells on the upper side. Thus the shoots turn upward until they grow vertically once again.

With positive geotropism, roots, unlike shoots, grow toward the pull of gravity. In a horizontally placed stem, however, the effect on the root cells is the opposite. Those exposed to a higher concentration of auxin (the lower side) are inhibited from growing, while the cells on the upper side continue to grow. In consequence, the root turns downward. Auxins produced in the terminal bud of a plant's growing tip move downward in the shoot and inhibit development of lateral buds. Auxins also initiate the formation of lateral roots, while they inhibit root elongation.

Gibberellins

Gibberellins are plant hormones that stimulate rapid stem elongation, particularly in plants that normally do not grow tall. Gibberellins also inhibit the formation of new roots, and stimulate the production of new phloem cells by the cambium (where the auxins stimulate the production of new xylem cells). Finally, these hormones terminate the dormancy of seeds and buds, and induce some biennial plants to flower during their first year of growth.

Cytokinins

Cytokinins are a class of hormones that promote cell division. Kinetin is an important type of cytokinin, and is involved in general plant growth, breaking seed dormancy, and expanding leaves. The ratio of cytokinin to auxin is of particular importance in the differentiation of buds and roots. If auxins are dominant then roots form but if cytokinins are more abundant then buds form from a stem.

Ethylene

Ethylene stimulates the ripening of fruit and the loss of leaves during seasonal changes. Fruit is often harvested before it is ripe and then sprayed with ethylene when it reaches the market to stimulate ripening. Ethylene inhibitors are important to the maintenance of dormancy in the lateral buds and seeds of plants during autumn and winter. They break down gradually with time and, in some cases, are destroyed by the cold, so that buds and seeds can become active in the next growing season.

Abscisic acid

Abscisic acid is a hormone that blocks the growth of stems and is commonly produced by the plant to promote dormancy and reduce damage that would be caused otherwise by harsh weather during cold periods in temperate climates.

Asexual Reproduction in Plants

Many plants utilize asexual reproduction, such as spore formation and vegetative propagation, to increase their numbers.

Spore Formation

Some plants produce specialized cells with hard coverings known as spores. The covering prevents loss of water from the spore contents so that they can remain viable for long periods. The spores are scattered by the air. Under favorable conditions (warmth, food, and moisture), these spores will germinate to create an organism.

Vegetative Propagation

Undifferentiated tissue (meristem) in plants provides a source of cells from which new plants can develop. Vegetative propagation offers a number of advantages to plants, including speed of reproduction. Disadvantages of vegetative propagation are lack of genetic variation and the ability to produce seedless fruit. This process can occur either naturally or artificially.

See chapter 4 for a detailed discussion of asexual reproduction in plants.

Sexual Reproduction in Plants

Most plants are able to reproduce both sexually and asexually; some do both in the course of their life cycles, while others do one or the other. In the life cycles of mosses, ferns, and vascular plants, there are two kinds of individuals associated with different stages of the life cycles: the diploid and the haploid.

Diploid and Haploid Generations

In the diploid or sporophyte generation, the asexual stage of a plant's life cycle, diploid nuclei divide meiotically to form haploid spores (not gametes) and the spores germinate to produce the haploid or gametophyte generation. The haploid or gametophyte generation is a separate haploid form of the plant concerned with the production of male and female gametes. Union of the gametes at fertilization restores the diploid sporophyte generation. Since there are two distinct generations, one haploid and the other diploid, this cycle is sometimes referred to as the alternation of generations. The relative lengths of the two stages vary with the plant type. In general, the evolutionary trend has been toward a reduction of the gametophyte generation, and increasing importance of the sporophyte generation.

How do these generations express themselves in common plants? In moss, the gametophyte is the green plant that you see growing on the north side of trees. The sporophyte variety is smaller, nongreen (nonphotosynthetic), and short-lived. It is attached to the top of the gametophyte, and is dependent upon it for its food supply. Spores from the sporophyte germinate directly into gametophytes. In ferns, on the other hand, the reverse pattern may be observed, with the sporophyte of the species dominant. The gametophyte is a heart-shaped leaf the size of a dime. Fertilization produces a zygote from which the commonly seen green fern sporophyte develops. The sporophyte fern's leaves (the fronds) develop spores on the underside of the leaf. These spores germinate to form the next generation of gametophyte.

Sexual Reproduction in Flowering Plants

In flowering plants or angiosperms, the gametophyte consists of only a few cells and survives for a very short time. The woody plant that is seen (for example, a rose) is the sporophyte stage of the species.

The Flower

The flower (see diagram on p. 180) is the organ for sexual reproduction of angiosperms and consists of male and female organs. The flower's male organ is known as the stamen. It consists of a thin, stalklike filament with a sac at the top. This structure is called the anther, and produces haploid spores. The haploid spores develop into pollen grains. The haploid nuclei within the spores will become the sperm nuclei, which fertilize the ovum. The flower's female organ is termed the pistil. It consists of three parts: the stigma, the style, and the ovary. The stigma is the sticky top part of the flower, protruding beyond the flower, which catches the pollen. The tubelike structure connecting the stigma to the ovary at the base of the pistil is known as the style; this organ permits the sperm to reach the ovules. The ovary, the enlarged base of the pistil, contains one or more ovules. Each ovule contains the monoploid egg nucleus.

Petals are specialized leaves that surround and protect the pistil. They attract insects with their characteristic colors and odors. This attraction is essential for cross-pollination—the transfer of pollen from the anther of one flower to the stigma of another. Note that some species of plants have flowers that contain only stamens ("male plants") and other flowers that contain only pistils ("female plants").

The pollen grain develops from the spores made by the sporophyte. Pollen grains are transferred from the anther to the stigma. Agents of cross-pollination include insects, wind, and water. The flower's reproductive organ is brightly colored and fragrant in order to attract insects and birds, which help to spread these male gametophytes. Carrying pollen directly from plant to plant is more efficient than relying on wind-borne pollen and helps to prevent self-pollination, which does not create diversity. When the pollen grain reaches the stigma (pollination), it releases enzymes that enable it to absorb and utilize food and water from the stigma and to generate a pollen tube. The pollen tube is the remains of the evolutionary gametophyte. The pollen's enzymes proceed to digest a path down the pistil to the ovary. Contained within the pollen tube are the tube nucleus and two sperm nuclei; all are haploid.

Fertilization in Angiosperms

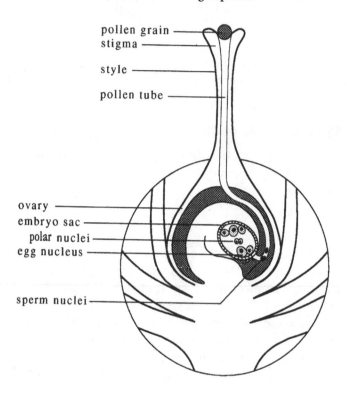

The female gametophyte in angiosperms develops in the ovule from one of four spores. This embryo sac contains nuclei, including the two polar (endosperm) nuclei and an egg nucleus. During fertilization (see figure) the sperm nuclei of the male gametophyte (pollen tube) enters the female gametophyte (embryo sac), and a double fertilization occurs. One sperm nucleus fuses with the egg nucleus to form the diploid zygote, which develops into the embryo. The other sperm nucleus fuses with the two polar bodies to form the endosperm (triploid or 3n). The endosperm provides food for the embryonic plant. In dicot plants, the endosperm is absorbed by the seed leaves (cotyledons).

1 sperm nucleus (n) + 1 egg nucleus (n) = zygote (2n) = embryo (2n)
1 sperm nucleus (n) + 2 polar nuclei (n) = endosperm (3n)

The zygote produced in the sequence above divides mitotically to form the cells of the embryo. This embryo consists of the following parts, each with its own function:

- The **epicotyl** develops into leaves and the upper part of the stem.
- The **cotyledons** or seed leaves store food for the developing embryo.
- The **hypocotyl** develops into the lower stem and root.
- The **endosperm** grows and feeds the embryo. In dicots, the cotyledon absorbs the endosperm.
- The **seed coat** develops from the outer covering of the ovule. The embryo and its seed coat together make up the seed. Thus, the seed is a ripened ovule.

The fruit, in which most seeds develop, is formed from the ovary walls, the base of the flower, and other consolidated flower pistil components. Thus, the fruit represents the ripened ovary. The fruit may be fleshy (as in the tomato) or dry (as in a nut). It serves as a means of seed dispersal; it enables the seed to be carried more frequently or effectively by air, water, or animals, through ingestion and subsequent elimination. Eventually, the seed is released from the ovary, and will germinate under proper conditions of temperature, moisture, and oxygen.

Chapter 8: Review Questions

1. In oogenesis, meiosis is completed

   (A) at fertilization
   (B) at birth
   (C) at ovulation
   (D) at menstruation
   (E) none of the above

2. Cells which have the potential to develop into any tissue are called

   (A) Leydig cells
   (B) determined
   (C) totipotent
   (D) mesoderm
   (E) ectoderm

3. Which hormone is secreted by the embryo and functions to maintain the corpus luteum?

   (A) estrogen
   (B) progesterone
   (C) androgen
   (D) insulin
   (E) hCG

4. Which statement regarding insulin and glucagon is true?

   (A) Glucagon is produced by beta-cells in the pancreas.
   (B) Insulin raises blood sugar.
   (C) Insulin inhibits glycogen formation in the liver.
   (D) Glucagon stimulates glycogen breakdown in the liver.
   (E) None of the above.

5. The circulatory system of which organism involves five aortic arches or "hearts" which pump blood through two main vessels, one dorsal and one ventral?

   (A) human
   (B) earthworm
   (C) grasshopper
   (D) fish
   (E) none of the above

6. Which cells produce antibodies (immunoglobulins)?

   (A) erythrocytes
   (B) B cells
   (C) helper T cells
   (D) killer T cells
   (E) neutrophils

Questions 7–9:

   (A) urea
   (B) uric acid
   (C) ammonia
   (D) nephridia
   (E) Malphigian tubules

7. nitrogenous waste secreted by humans

8. part of the excretory system of annelids

9. nitrogenous waste secreted by fish

Questions 10–11 refer to the diagram below.

Questions 12–13 refer to the diagram below.

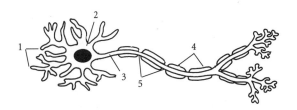

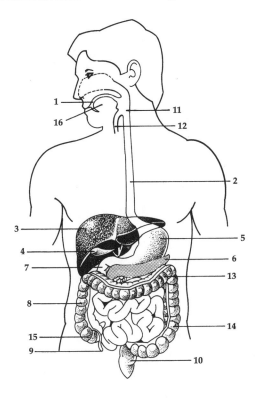

12. These insulated areas are important in saltatory conduction.

    (A) 1
    (B) 2
    (C) 3
    (D) 4
    (E) 5

13. These areas are known as Nodes of Ranvier.

    (A) 1
    (B) 2
    (C) 3
    (D) 4
    (E) 5

10. Mechanical digestion of food occurs here.

    (A) 1
    (B) 3
    (C) 5
    (D) 6
    (E) 10

11. This organ produces bile, which is involved in the digestion of fats.

    (A) 1
    (B) 3
    (C) 5
    (D) 6
    (E) 10

Questions 14–15:

    (A) root hairs
    (B) stomata
    (C) phloem
    (D) xylem
    (E) auxin

14. Plant hormone which is involved in phototropism

15. These structures increase the surface area of roots

Please see the answers and explanations to these review questions in Section V of this book.

Ecology

Biologists often study individual organisms, analyzing their cells, tissues and organs to learn about their structure and behavior. Organisms do not live on Earth in an isolated state, however. All organisms, including humans, live by interacting with other organisms and with the nonliving (abiotic) environment. Life on Earth is a network of interacting organisms that depend on each other for survival. Ecology is the study of the interactions between organisms and their environment and how these shape both the organisms and the environments they live in.

9.1   POPULATIONS IN ECOLOGY

Since ecology seeks to understand life at a broader level than the organism, it is the population rather than the individual that is the basic unit of study in this discipline. A **population** is a group of individuals in a species that interbreed and share the same gene pool, the same definition used in population genetics. Every environment will include many different interacting populations from different species. There are properties of populations, such as population growth and maximal population size, that are not applicable to individuals. Also, populations are concerned with the maintenance of the population, and are not as concerned with individual members of the population. These distinct properties of populations are important for ecosystems.

Patterns of Population Growth

One of the key characteristics of a population is the rate of population growth. At any given time a population can grow, stay the same, or shrink in size. The birth rate, the death rate, and the population size determine the rate of growth and are affected by the environment. If the birth rate is high and the death rate is low, as in an environment where resources are unlimited, a population will grow rapidly. If a population of mice breeds once every 3 months, producing 12 mice in each generation, a single pair of mice could produce almost 280,000 mice in two years (6 breeding pairs every 3 months, for 7 generations = 67 mice). A single bacteria reproducing by binary fission every thirty minutes can produce 8 million bacteria in 12 hours. This form of population growth produces a curve with rapidly increasing slope and is termed exponential growth, since every generation increases the population size in an exponential manner (see figure).

Exponential Population Growth

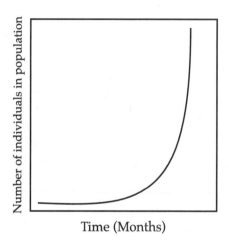

In nature, this rapid population growth may be observed when a population first encounters a favorable new environment, such as a rich growth medium inoculated with a small number of bacteria, or a fertile empty field invaded by a weed. Dispersal of organisms to colonize new territory is a common part of the life cycle of both plants and animals.

Exponential growth cannot be maintained forever, though. A population of mice growing exponentially in a field of wheat will soon eat so much of the available food that starvation will occur; growth will slow and then halt. Bacteria reproducing without check would in a few days weigh more than the mass of the earth. Limitations of the environment prevent exponential growth from proceeding indefinitely. Reasons for a slowdown in the growth rate include a lack of food, competition for other resources, predation, disease, accumulation of waste, or lack of space. All of these factors act more strongly to slow growth as the population becomes denser. Under these conditions, the growth curve may appear sigmoidal in shape (see figure), with rapid exponential growth at first, followed by a slowing and leveling off of growth. In this curve, the population size at the point where the growth curve is flat is the maximum sustainable number of individuals, called the **carrying capacity**, and is observed when the birth rate and death rate are equal.

Sigmoidal Population Growth

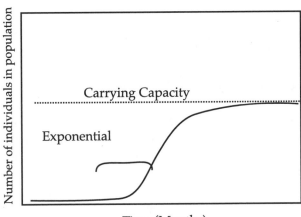

In natural populations, the environment is constantly changing, and the carrying capacity varies with it. For example, the carrying capacity for rabbits in a grassy plain will be greater in a year of plentiful rain and lush vegetation growth than in a year of drought. Populations often have regular fluctuations in size, suggesting that the carrying capacity changes in a periodic manner. If a population of rabbits consumes all available vegetation, the carrying capacity will be reduced and the population size will fall until the vegetation regrows and the carrying capacity for rabbits is increased once again. An example that is often cited is the size of hare and lynx populations in Canada. In this environment, the primary food of the lynx is the hare. The hare population size regularly cycles up and then crashes, perhaps due to the rapid spread of disease in crowded conditions. The lynx population size cycles along with that of the hare, crashing in size after the hare population crashes, then building in size once again after the hare population rebounds.

Population Density

In addition to the size of a population and its rate of growth, another characteristic of a population is its density in individuals per unit of territory. Density affects the behavior of a population in many ways, called **density dependent factors**. Density dependent factors that affect populations can include the spread of disease, competition for food, competition for living space, predation, and success in mating and reproduction. When a population is denser, disease spreads more easily from individual to individual. Denser populations consume food more rapidly than less dense populations. It can also be easier to find a mate in a denser population than if individuals are spread across very wide territories. If both the birth rate and the death rate are density dependent for a population, the density of the population will tend to be balanced at a constant level.

Some factors that affect populations are **density independent**. A volcanic explosion will tend to kill organisms without regard to their density. Natural disasters often cause disturbances in populations that are density independent. A density independent factor such as a flood can lead to density dependent factors such as food shortage as well.

Reproductive Strategies and Population Growth

Different strategies of reproduction can produce different patterns of population growth. An example is the choice between sexual and asexual reproduction. Sexual reproduction occurs in most complex multicellular organisms, and helps to create and maintain diversity in the gene pool of a population. Asexual reproduction can allow a population to grow very rapidly, as occurs in some plant species that reproduce from shoots rather than seeds. The benefit of asexual reproduction is a reduced cost to produce new organisms. For example, it takes less energy for a plant to produce a shoot and reproduce asexually than for it to flower or to produce fruit and seeds. By reproducing through parthenogenesis (in which an egg divides in the absence of fertilization by a male), only female offspring are produced, all of which produce their own young, increasing by twofold the rate of population growth. The cost of asexual reproduction is a reduction in genetic variability. Some species will reproduce asexually in times of abundant resources to maximize the opportunity for rapid growth, and reproduce sexually when resources are limited, perhaps generating the genetic diversity required to survive in a changing environment.

Species also use two different strategies that affect the number of offspring produced in each generation and the amount of care they receive. In the first strategy, offspring reach sexual maturity very rapidly, and produce a large number of young in each generation that receive little or no parental care. Insects and plants often reproduce in this way, producing large numbers of eggs or seeds that are left in the environment to fend for themselves. In colonizing a new environment, a species with this reproduction pattern will initially grow exponentially. The lack of parental care for a species with this strategy can cause a high death rate early in life when these species are in a competitive environment. For example, some marine species release a large number of eggs that develop without care. As a result, many are consumed early in life. The large number of young produced ensures survival despite the high mortality rate. Species with this reproductive strategy are also prone to rapid crashes in population size as the environment rapidly becomes depleted of resources after a period of exponential growth.

The other reproductive strategy is to delay sexual maturity, have few young, and put a great deal of parental care into offspring. Large mammals often fall into this category. These species have long life spans and are highly adapted to compete for resources in a competitive environment that is at, or near, the carrying capacity. Since these organisms invest so much into their young, they have a lower mortality rate. For example, birds invest a great deal of maternal care in young, increasing their survival rate, but decreasing the numbers of young that they are able to produce. Such species also have difficulty recovering from catastrophic decreases in their population size; examples include California condors and whooping cranes.

9.2 THE ROLE OF THE ABIOTIC ENVIRONMENT IN POPULATIONS AND ECOSYSTEMS

The size and growth of a population are affected both by the biotic (living) and abiotic (nonliving) portions of the environment. The abiotic portions include the air, water, soil, light and temperature that living organisms require. Not only are organisms dependent on the abiotic environment, but they in turn modify it. Plants create shade that alters the light environment for other plants and preserves water in the soil; they also consume carbon dioxide, and produce oxygen. The modifications of the environment by a population affect the species that live in that environment.

Water Cycle

Water is essential to all life and is a major component of all living things. Our bodies are made mostly of water. Animals must regulate their water content to ensure proper volume and salt concentrations inside and outside of the cell. The need for water is one of the factors that can affect a population. One of the key characteristics of terrestrial environments is the availability of water, ranging from tropical rainforests to deserts. Most of the water vapor in the atmosphere comes from evaporation from the oceans. Much of this water is returned to the oceans directly through rain that falls on the oceans, but some is carried over land as precipitation where it is available to terrestrial plants and animals.

Sunlight

Sunlight serves as the ultimate source of energy for almost all organisms. Green plants must compete for sunlight in forests. To this end, they develop adaptations to capture as much sunlight as possible, including broad leaves, branching, greater height, and vine growth. In water, the photic zone, the top layer through which light can penetrate, is where all photosynthetic activity takes place. In the aphotic zone, only heterotrophic life forms exist. The amount of sunlight is affected also by the latitude of an ecosystem, since locations further from the equator receive less sunlight; and the season, with a significantly decreased length of day during the winter in temperate and polar environments.

Oxygen Supply

Oxygen supply poses no problem for most terrestrial life, since air is composed of approximately 21 percent oxygen. At higher altitudes, atmospheric pressure and the partial pressure of oxygen decrease, forcing adaptations in organisms that live there. Aquatic plants and animals utilize oxygen dissolved in water. Fertilizer waste and other pollution can lead to the consumption of aquatic oxygen, lowering the oxygen content in water and threatening aquatic life.

Substratum (Soil or Rock)

Soil is a complex mixture of organic and inorganic materials and water.

Substratum determines the nature of plant and animal life in the soil. Some soil factors include:

- **Acidity** (pH). Acid rain may make soil pH too low for most plant growth.

- **Texture of soil and clay content.** These determine the quantity of water the soil can hold.

- **Minerals.** Nitrates, phosphates, and other minerals determine the type of vegetation soil will support.

- **Humus quantity.** Humus is a rich organic mixture in soils of decaying material that supports plants.

Chemical Cycles

Also included in the abiotic environment are inorganic chemicals required for life such as carbon and nitrogen. The movement of these essential elements between the biotic and abiotic environment form cycles that are central to all life on earth. Some organisms take the simple inorganic starting chemicals up from the soil and air and convert them into a biologically useful form. After passing through the biological community, respiration and decay organisms return these chemicals to their inorganic state to begin the cycle again.

Carbon Cycle

Carbon is held in the atmosphere as relatively small quantities of carbon dioxide, in rocks as carbonate minerals, and in the oceans as carbonate ions. Carbon cycles between these stores and the living world. Carbon is taken into living systems mostly from CO_2 in the atmosphere when plants and other photosynthetic organisms take it in and use it to produce glucose in photosynthesis. Plants use energy stored in glucose to make starch, proteins, and fats. Next, animals eat plants and use the digested nutrients to form carbohydrates, fats, and proteins characteristic of the species. Part of these organic compounds is used as fuel in respiration in plants and animals. The metabolically produced CO_2 is then released to the air (see figure). Aside from expelled wastes, the rest of the organic carbon remains locked within an organism until its death, at which time decay processes return the CO_2 to the air. Consumption of fossil fuels such as oil, gas and coal by humans has increased the amount of carbon dioxide in the atmosphere and is likely to affect the biosphere in many ways.

Carbon Cycle

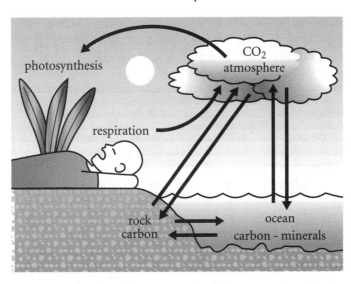

Nitrogen Cycle

Nitrogen is an essential component of amino acids and nucleic acids, which are the building blocks of all living things. Nitrogen makes up most of the atmosphere, but the N_2 in the atmosphere cannot be used by most organisms. Nitrogen fixation is the process by which atmospheric nitrogen is transformed into forms that are more usable in biological processes. Although all life needs nitrogen, only a few organisms are actually able to fix nitrogen out of the atmosphere: a few types of bacteria that convert nitrogen into nitrates and other forms of nitrogen. All other life is dependent on these organisms for the continuation of the essential nitrogen cycle.

Nitrogen Cycle

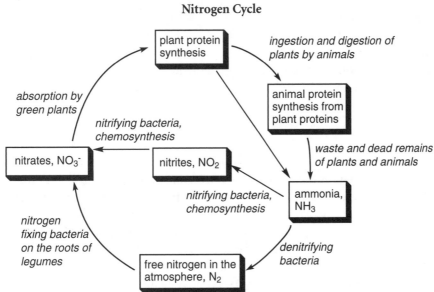

The following refer to the figure of the nitrogen cycle on the preceding page.

- Elemental (free) nitrogen, at the bottom of the figure, is chemically inert and cannot be used by most organisms. Lightning and nitrogen-fixing bacteria in the roots of legumes change the nitrogen to usable, soluble nitrates.

- The nitrates are absorbed by plants and are used to synthesize nucleic acids and plant proteins.

- Animals eat the plants and synthesize specific animal proteins from the plant proteins. Both plants and animals give off wastes and, eventually, die.

- The nitrogen locked up in the wastes and dead tissue is released by decay, which converts the proteins into ammonia.

- Two fates await the ammonia (NH_3): Part of it is nitrified to nitrites by chemosynthetic bacteria and then to usable nitrates by nitrifying bacteria. The rest of the ammonia is denitrified. This means that the ammonia is broken down to release free nitrogen, which returns us to the beginning of the cycle.

- Note that there are four kinds of bacteria: decaying, nitrifying, denitrifying, and nitrogen-fixing. The bacteria have no use for the excretory ammonia, nitrites, nitrates, and nitrogen they produce. These materials are essential, however, for the existence of other living organisms.

9.3 POPULATIONS IN COMMUNITIES

The next level of biological organization beyond a population is a **community**, which is all the interacting populations living together in an environment. The populations within a community interact with each other in a variety of ways, including predation, competition, or symbiosis. These interactions affect the number of individuals in each population in the community and the number of different species in the community. The living community combined with the abiotic environment, the interactions between populations, and the flow of energy and molecules within the system define an ecosystem.

Predation

Predation is the consumption of one organism by another, usually resulting in the death of the organism that is eaten. Both carnivores that consume meat and herbivores that consume only plants are types of predators. Examples of predation include a zebra eating grass, a lion eating a zebra, a whale eating plankton, a paramecium eating yeast, and a Venus flytrap eating a housefly. Predators often select weak or sick members of the prey population, removing alleles with poor fitness, and placing selective pressure on the prey to adapt effective means of escaping predation. Predator and prey often **coevolve**, with the predator evolving to become more effective as the prey evolves to escape predation. As the cheetah evolves to run more rapidly while hunting, so does the impala prey evolve to escape more rapidly as well, maintaining a balance between predator and prey in the ecosystem.

Prey species have evolved many different means of escaping predators. Rather than running more quickly, some prey species such as toads and caterpillars have evolved toxic chemicals that prevent them from being eaten. Brightly colored species often advertise by their coloration that they are toxic. The adaptation of toxic defenses by prey must prevent predation to be effective—this adaptation would be ineffective if the predators only learned the problem after consumption. Plants often use chemical defenses against either insect or vertebrate predators with toxic or bad tasting components to discourage consumption. Hiding from predators is another effective means of escaping predation, often involving camouflaged coloration that blends with the surroundings. **Mimicry** is a defense in which an edible species evolves to resemble a species that cannot be eaten, sometimes due to chemical defenses such as toxins. For example, many species of butterflies that are edible have evolved to resemble toxic species to deter predators.

Predator-prey relationships between populations in a community can influence the carrying capacity of prey populations involved and tend to achieve a balance in which the predator is effective enough to maintain its own population without decimating the prey it is dependent upon. Predation can cause a community to maintain a greater diversity of species. Without predation, one prey species will often come to predominate in the ecosystem.

Parasitism

Parasitism might be considered a form of predation since one organism gains nutrition at the expense of another. It is distinct from most predator-prey relationships since the prey is not usually killed. Also, the parasite is an unusual predator since it is smaller than the prey it feeds on. A parasite takes from the host but gives nothing in return; thus, the parasite benefits at the expense of the host. Examples of parasites include leeches, ticks, and sea lampreys. Few autotrophs (green plants) exist as parasites (mistletoe is an exception).

Ectoparasites cling to the exterior surface of the host with suckers or clamps, bore through the skin, and suck out juices. Endoparasites, on the other hand, live within the host. In order to gain entry into the host, they must break down formidable defenses, including skin, digestive juices, antibodies, and white blood cells. Parasites possess special adaptations to overcome these defenses. Parasitism is advantageous and efficient, since the parasite lives with a minimum expenditure of energy. Parasites may even be parasitic on other parasites. Thus, a mammal may have parasitic worms, which in turn are parasitized by bacteria, which in turn are targets of bacteriophages.

A prominent example of a parasitic relationship is that between the virus and its host cell. All viruses are parasites. As viral nucleic acid enters the host, the virus takes over the host cell functions and redirects them into replication of the virus. The tapeworm-human relationship is a particularly good example of parasitism. Tapeworms can live inside their hosts' intestines for many years, growing longer and longer. It is interesting to note that successful parasites do not kill their hosts, as this would, counterproductively, lead to the death of the parasite itself. The more dangerous the parasite is to its host, the less chance it has of ultimate survival. In some cases death of the host does not prevent transmission of the pathogenic parasite, and the organisms may evolve to more virulent and deadly forms.

Competition and the Niche

A competitive relationship between populations in a community exists when different populations in the same location use a limiting resource. Competition can be interspecific (between species) or intraspecific (between organisms of the same species). Integral to understanding interspecific competition is the idea of the **ecological niche**. If the habitat is the physical environment in which the population lives, the niche is the way it lives within the habitat, including what it eats, where it lives, how it reproduces, and all other aspects of the species that define the role it plays in the ecosystem. The niche occupied by each species is unique to that species and can in part define that species. There cannot be two species in the same niche for an extended period of time. Another way to understand the niche is to say that if the habitat is the address of a population, the niche is its profession.

Interspecific Competition

When two populations have overlap in their niches, such as by eating the same insects or occupying the same nesting sites, there is competition between the populations. The more the niches overlap, the greater the competition. Generally, when two populations compete, one will compete more effectively than the other and grow more rapidly. Competition can drive the less efficient population out of the community, with the 'winner' occupying the niche on its own. Another result of competition for a niche can be that evolution drives the two populations to occupy niches that overlap less, reducing the competition. For example, if two species of related birds compete for the same nesting site, then they may evolve to reduce competition by using different nesting sites (see the figure below). Even in an environment with several different herbivores, their niches are unique since they evolve to have different heights, different sizes, different teeth and digestive tracts to avoid competition for the same plants. Several closely related species of birds can live in the same tree and eat similar food, and yet occupy distinct niches by living in different part of the tree, with some near the crown, others in the middle, and still others close to the ground.

Evolution Drives Reduced Niche Overlap

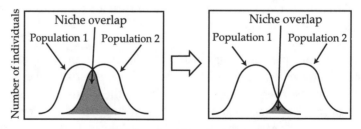

Height of nesting site in apple trees

Beneficial Interactions

Commensalism

In **commensalism,** one organism is benefited by the association and the other is not affected (this is symbolized as +/0). The host neither discourages nor fosters the relationship. Barnacles attach to the underside of a whale. Through this association, the barnacles obtain more food, wide geographic dispersal, and protection. The whale is totally indifferent to the association.

Mutualism

Mutualism is a symbiotic relationship from which both organisms derive some benefit (+/+). In the example of the tick bird and rhinoceros, the rhinoceros aids the bird through the provision of food in the form of parasites on its skin. The bird in its turn aids the rhinoceros by removing the parasites and by warning the rhinoceros of danger when it suddenly flies away. Small fish that eat parasites are often found in mutualistic relationships with large fish they clean—the small fish gets food and the large fish gets parasites removed.

A more intimate mutualistic association exists between a fungus and an algae in the form of the lichen. Lichens are found on rocks and tree barks. The green algae produces food for itself and the fungus by photosynthesis. Meshes of fungal threads support the algae and conserve rain water. Thus, the fungus provides water, respiratory carbon dioxide, and nitrogenous wastes for the algae, all of which are needed for photosynthesis and protein synthesis. Lichens are pioneer organisms in the order of ecological succession on bare rock.

Nitrogen-fixing bacteria and legumes also engage in mutualism. Nitrogen-fixing bacteria invade the roots of legumes and infected cells grow to form root nodules. In the nodule, the legume provides nutrients for the bacteria, and the bacteria fixes nitrogen (by changing it to soluble nitrate, a mineral essential for protein synthesis by the plant). These bacteria are a major source of usable nitrogen, which is needed by all plants and animals.

Protozoa and termites work together in a similar fashion. Termites chew and ingest wood, but are unable to digest cellulose. Protozoa in the digestive tract of the termite secrete an enzyme that is capable of digesting cellulose, and both organisms share the carbohydrates. In this manner, the protozoa are guaranteed protection and a steady food supply, while the termite obtains nourishment from the ingested wood. Likewise, in the case of intestinal bacteria and humans, bacteria utilize some of the food material not fully digested by humans and, in turn, manufacture vitamin B_{12}.

Angiosperms have mutualistic relationships with animals as well. The elaborate flowers of angiosperms attract pollinators, providing food for the animal and providing efficient transfer of pollen for the plant. The fruit produced by angiosperms is often encased in fruits that are eaten by the animals. Animals receive food and the plant gets broader dispersal of seeds that are eaten and later eliminated.

Intraspecific Competition

Competition is not restricted to interspecific interactions. Individuals belonging to the same species utilize the same resources; if a particular resource is limited, these organisms must compete with one another. Members of the same species compete, but they must also cooperate. Intraspecific cooperation may be extensive (as in the formation of societies in animal species) or may be nearly nonexistent. Hence, within a species, relationships between individuals are influenced by both disruptive and cohesive forces. Competition (for food or a mate, for example) is the chief disruptive force, while cohesive forces include reproduction, protection from predators, and destructive weather.

Community Structure

Producers

The populations within a community are organized in many different ways. Within the community, each population plays a different role depending on the source of energy for that population. **Producers** are autotrophs, organisms that get energy from the environment (the sun or inorganic molecules) and use this energy along with simple molecules (carbon dioxide, water and minerals) to drive the biosynthesis of their own proteins, carbohydrates, and lipids. The energy a producer such as a plant gets from the sun is stored in chemical bonds in the biological molecules it produces. Producers use some of the energy derived from photosynthesis to drive their own respiration and biosynthesis and store the rest of the energy. Producers form the foundation of any community, passing on their stored energy to other organisms. In a terrestrial environment, green plants, photosynthetic bacteria, or mosses are producers, using the energy of sunlight to produce biosynthetic energy through photosynthesis. In marine environments, green plants or algae are the main producers. There are also marine ecosystems at deep, dark ocean geothermal vents based not on photosynthetic producers but on chemosynthetic bacteria that use the energy of inorganic molecules released from the volcanic vent to drive biosynthesis.

Consumers

Consumers get the energy to drive their own biosynthesis and to maintain life by ingesting and oxidizing the complex molecules synthesized by other organisms. Since they get their energy by consuming other organisms, they are called heterotrophs. Herbivores (plant eaters), carnivores (meat eaters), and omnivores (eating both plants and animals) are all consumers. The adaptations of each consumer depend on the type of food it eats. Herbivores tend to have teeth for grinding and long digestive tracts that allow for the growth of symbiotic bacteria to digest cellulose found in plants. Carnivores are more likely to have pointed, fanglike teeth for catching and tearing prey and shorter digestive tracts than herbivores. Primary consumers such as the cow, the grasshopper, and the elephant are animals that consume producers such as green plants. Secondary consumers, for example, frogs, tigers, and dragonflies, are carnivorous and consume primary consumers. Finally, tertiary consumers, for example, snakes that eat frogs, feed on secondary consumers.

Decay Organisms (Saprophytes)

Decay organisms, also called **saprophytes** or decomposers, are heterotrophs, since they derive their energy from oxidizing complex biological molecules, but they do not consume living organisms. Decay organisms get energy from the biological organic molecules they encounter left as waste by producers and consumers, or the debris of dead organisms. They perform respiration to derive energy, and return carbon dioxide, nitrogen, phosphorous, and other inorganic compounds to the environment to renew the cycles of these materials between the biotic and physical environments. Bacteria and fungi are the primary examples of decay organisms. Scavengers such as hyenas or vultures play a similar role, living on the stored chemical energy found in dead organisms.

9.4   ECOSYSTEMS

Food Chains and Food Webs

The term **food chain** is often used to describe a community, depicting a simple linear relationship between a series of species, with one eating the other. For example, a food chain might contain shrubs as the producers, mice as the primary consumers, snakes as the secondary consumers, and hawks as the tertiary consumers. The different levels in the food chain, such as producers and primary consumers, are sometimes called **trophic levels**. A more realistic depiction of the relationships with in the community is a **food web** (see figure), in which every population interacts not with one other population, but several other populations. An animal in an ecosystem is often preyed upon by several different predators, and predators commonly have a diet of several different prey. The greater the number of potential interactions in a community food web, the more stable the system will be, and the better able it will be to withstand and rebound from external pressures such as disease or weather.

Food Web—Desert

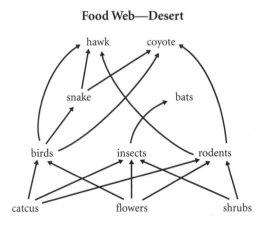

Energy Flow in Communities

Each trophic level in a food web contains a different quantity of stored chemical energy. When consumers eat producers and secondary consumers eat primary consumers, some energy is lost in each transfer from one level to another. As producers get energy from the sun, not all of the energy is converted into stored energy in chemical bonds. Some of the energy is lost at that level to the metabolic energy an organism requires to maintain life. Plants consume some of the energy they produce in respiration to support their own metabolic activities. The total chemical energy generated by producers is **gross primary productivity**, and the total with losses to respiration by plants subtracted is the **net primary productivity**.

At the next level of the energy pyramid, herbivores consume primary producers, incorporating about 10 percent of the energy consumed into their own stored chemical energy. The remainder is lost through respiration. Only about 10 percent of the stored chemical energy is present in the next higher trophic level at every stage (see the figure below). The energy contained in a community can be visualized as a pyramid, with the most energy in the producers and less energy at successive levels of consumers (see figure). The efficiency of energy transfer between levels can differ greatly from 10 percent, but the pyramid will always have the most energy in the producer level, with less in each level of consumers. A similar pyramid is observed if one compares the biomass or numbers of individuals in a community, with each successive level about 10 percent the size of the level beneath it. In terms of numbers, the shape of the pyramid can often vary, with a single large producer such as a tree supporting a large number of primary consumers such as birds or insects.

Energy Structure of a Community

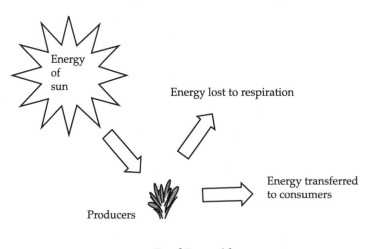

Food Pyramid

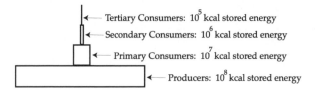

Tertiary Consumers: 10^5 kcal stored energy
Secondary Consumers: 10^6 kcal stored energy
Primary Consumers: 10^7 kcal stored energy
Producers: 10^8 kcal stored energy

Community Diversity

The number of species within a community defines community diversity. The types of interactions between populations within a community affect the number of species in the community, as well as the physical environment. Predation has been noted as one factor that increases the diversity of species in a community, and competition may do the same by driving populations into distinct niches. Warm environments with very high productivity have the greatest diversity, while colder environments have less diversity. Topographic diversity increases community diversity, perhaps by creating a greater number of niches in the environment. Larger land masses or ecosystems also have a greater community diversity.

Often a single species will play a particularly important role in determining the composition of species in a community. Species that play this role are called **keystone species**. Keystone species in terrestrial environments are often plants. For example, grasses are the dominant keystone species in grasslands and conifers are dominant in northern forests.

Changes in Ecosystems Over Time: Ecological Succession

Communities and ecosystems can change over time, either as a result of changes in the physical environment such as the climate, or as the result of changes created by the populations that live in the community. When a population changes the environment it lives in, it may make the environment more favorable for some populations and less favorable for others, including itself. When a community changes as a result of organisms that live in the community, this is termed **succession**. For example, grassland may provide abundant sunlight and rich soil that lead to colonization by trees, followed by other trees that grow best in the shade of the pioneer trees. Successive communities are composed of populations best able to exist in each new set of conditions, both biotic and abiotic.

The community will continue to change until it arrives at a combination of populations that do not change the environment further; this is called a **climax community**. The climax community is stable over time, with each generation leaving the environment unchanged, and it will remain unchanged unless disturbed by climate change, fire, humans, or catastrophes. If the climax community is disturbed, the series of community successions will begin again until a climax community is achieved once again. The type of climax community that is present in an environment depends on the abiotic factors of the ecosystem, including rainfall, temperature, soil, and sunlight.

Consider, for example, a rocky area in the northeastern United States that is barren as a result of a severe forest fire. Lichen would be a good candidate to be the first or pioneer organism to resettle this area. Recall that a lichen is an association between an algae and a fungus that can live on a rocky surface. Acids produced by the lichen attack rock, breaking it down to form the first layers of soil. Since lichens thrive only on a solid surface, conditions at this stage are worse for the lichen but better for mosses. Airborne spores of mosses land on the soil and germinate. The result is a new sere, or series of ecological communities formed in ecological succession, with moss supplanting lichen as the dominant species in the community.

As the remains of the moss build up the soil still more, annual grasses and then perennial grasses with deeper roots become the dominant species. With time, we find shrubs and then trees. The first trees will be the sun-loving gray birch and poplar. As more and more trees compete for sun, the birch and poplar will be replaced by white pine and, finally, maples and beeches, which grow in shade.

The growth of maples and beeches produces the same conditions that originally favored their appearance. And so this community remains for a thousand years. In the final maple-beech community, you would also find foxes, deer, chipmunks, and plant-eating insects. These are all animals that would not have been found in the original barren rock terrain.

To summarize this example of ecological succession:

Lichen → mosses → annual grasses → perennial grasses
→ shrubs → sun-loving trees (poplar) → thick shade trees (hemlock, beech, maple)
Time elapsed: About 1,000 years

Here's an example of the progression of a climax community in an aquatic environment. This community starts with a pond:

1. **Pond.** This pond contains plants such as algae and pondweed and animals such as protozoa, water insects, and small fish.
2. **Shallow water.** The pond begins to fill in with reeds, cattails, and water lilies.
3. **Moist land.** The former pond area is now filled with grass, herbs, shrubs, willow trees, frogs, and snakes.
4. **Woodland.** Pine or oak becomes the dominant tree of the climax community.

It is important to remember that the dominant species of the climax community is determined by such physical factors as temperature, nature of the soil, and rainfall. Thus the climax community at higher elevations in New York state is hemlock-beech-maple, while at lower elevations, it is more often oak-hickory. In cold Maine, the climax community is dominated by the pine; in the wet areas of Wisconsin, by cypress; in sandy New Jersey, by pine; in Georgia, by oak, hickory, and pine; and on a cold, windy mountain top, by scrub oak.

9.5   BIOMES

The conditions in a particular terrestrial and climatic region select plants and animals possessing suitable adaptations for that particular region. Each geographic region is inhabited by a distinct community called a biome.

Terrestrial Biomes

Land biomes are characterized and named according to the climax vegetation of the region in which they are found. The climax vegetation is the vegetation that becomes dominant and stable after years of evolutionary development. Since plants are important as food producers, they determine the nature of the inhabiting animal population; hence the climax vegetation determines the climax animal population. There are eight types of terrestrial biomes that can be formed as a result of all these factors: tropical forest, savanna/grasslands, desert, temperate deciduous forest, Northern coniferous forest, taiga, tundra, and polar regions.

Tropical Forest

Tropical forests are characterized by high temperatures and, in tropical rain forests, by high levels of rainfall. The climax community includes a dense growth of vegetation that does not shed its leaves. Vegetation such as vines and epiphytes (plants growing on other plants) and animals such as monkeys, lizards, snakes, and birds inhabit the typical tropical forest, or rain forest. Trees grow closely together in a dense canopy high above the ground; sunlight barely reaches the forest floor. The floor is inhabited by saprophytes living off dead organic matter. Tropical rain forests are found in central Africa, Central America, the Amazon basin, and southeast Asia, and are one of the most productive and diverse communities.

Savanna/Grasslands

The savanna (grassland) is characterized by low rainfall (usually 10–30 inches per year), although it gets considerably more rain than the desert biomes do. Grassland has few trees and provides little protection for herbivorous mammals (such as bison, antelopes, cattle, and zebras) from carnivorous predators. That is why animals that do inhabit the savanna have generally developed long legs and hoofs, enabling them to run fast. The grasses of the grasslands are subjected to grazing and fires but many grasses regrow quickly afterwards. Examples of savanna include the prairies east of the Rockies, the Steppes of the Ukraine, and the Pampas of Argentina.

Desert

The desert receives less than ten inches of rain per year, and this rain is concentrated within a few heavy cloudbursts. The growing season in the desert is restricted to those days after rain falls. Generally, small plants and animals inhabit the desert. Most desert plants (for example, cactus, sagebrush, and mesquite) conserve water actively and avoid extreme heat, often by being nocturnal. Desert animals such as the lizard, meanwhile, live in burrows. Birds and mammals found in the deserts also have developed adaptations for maintaining constant body temperatures. Examples of desert biomes include the Sahara in Africa, the Mojave in the United States, and the Gobi in Asia.

Temperate Deciduous Forest

Temperate deciduous forests have cold winters, warm summers, and a moderate rainfall. Trees such as beech, maple, oaks, and willows are angiosperms that shed their leaves during the cold winter months. There is greater productivity and greater diversity than in northern coniferous forests. Animals found in temperate deciduous forests include the deer, fox, woodchuck, and squirrel. The forest floor is a rich soil of decaying matter, inhabited by worms and fungi. Temperate deciduous forests are located in the northeastern and central eastern United States and in central Europe.

Northern Coniferous Forest

Northern coniferous forests are cold, dry, and inhabited by fir, pine, and spruce trees, all conifers. As gymnosperms, these plants rely on wind pollination during the brief spring and summer. Much of the vegetation here has evolved needle-shaped leaves. Northern coniferous forests are found in the extreme northern part of the United States and in Canada. Although the growing season is short, the long lives of these conifers create very large biomass of wood in these forests. The forest floor is dry and contains a layer of needles with fungi, moss, and lichens. Common animals include (in North America) moose, deer, black bears, hares, wolves, and porcupines.

Taiga

The taiga receives less rainfall than the temperate forests, has long, cold winters, and is inhabited by a single type of coniferous tree, the spruce. The forest floors in the taiga contain moss and lichens. Birds are the most common animal; however, the black bear, the wolf, and the moose are also found here. Taiga exists in the northern parts of Canada and Russia.

Tundra

Tundra is a treeless, frozen plain located between the taiga and the northern icesheets. Although the ground is always frozen in a state called **permafrost**, the surface can melt during the summer and trap water above the permafrost. It has a very short summer and a very short growing season, during which time the ground becomes wet and marshy. Lichens, moss, polar bears, musk oxen, and arctic hens make their homes here.

Polar Region

The polar region is a frozen area with very few types of vegetation or terrestrial animals. Animals that do inhabit polar regions, such as seals, walruses, and penguins, generally live near the polar oceans, surviving by preying on marine life.

Terrestrial Biomes and Altitude

The sequence of biomes between the equator and the pole is comparable to the sequence of regions on mountains. The nature of those regions is determined by the same decisive factors —temperature and rainfall. The base of the mountain, for example, would resemble the biome of a temperate deciduous area. As one ascends the mountain, one would pass through a coniferous-like biome, then taigalike, tundralike, and polarlike biomes.

Aquatic Biomes

In addition to the eight terrestrial biomes, there are aquatic biomes, each with its own characteristic plants and animals. More than 70 percent of the earth's surface is covered by water, and most of the earth's plant and animal life is found there. As much as 90 percent of the earth's food and oxygen production (photosynthesis) takes place in the water. Aquatic biomes are classified according to criteria quite different from the criteria used to classify terrestrial biomes. Plants have little controlling influence in communities of aquatic biomes, as compared to their role in terrestrial biomes.

Aquatic areas are also the most stable ecosystems on Earth. The conditions affecting temperature, amount of available oxygen and carbon dioxide, and amount of suspended or dissolved materials are stable over very large areas, and show little tendency to change. For these reasons, aquatic food webs and aquatic communities tend to be balanced. There are two types of major aquatic biomes: marine and freshwater.

Marine Biomes

The oceans connect to form one continuous body of water that controls the earth's temperature by absorbing solar heat. Water has the ability to absorb large amounts of heat without undergoing a great temperature change. Marine biomes contain a relatively constant amount of nutrient materials and dissolved salts. Although ocean conditions are more uniform than those on land, distinct zones in the marine biomes do exist, including the intertidal zone, littoral zone, and pelagic zone.

The intertidal zone is a region exposed at low tide that undergoes variations in temperature and periods of dryness. Populations in the intertidal zone include algae, sponges, clams, snails, sea urchins, sea stars (starfish), and crabs. The littoral zone is a region on the continental shelf that contains ocean area with depths of up to 600 feet, and extends several hundred miles from the shores. Populations in littoral zone regions include algae, crabs, crustacea, and many different species of fish. The pelagic zone is typical of the open seas and can be divided into photic and aphotic zones. The photic zone is the sunlit layer of the open sea extending to a depth of 250–600 feet. It contains plankton—passively drifting masses of microscopic photosynthetic and heterotrophic organisms. Active swimmers such as fish, sharks, and whales feed on plankton and smaller fish. The chief autotroph is the diatom, an algae. The open ocean is low in nutrients and productivity is low, but the vast size of the oceans contributes greatly to global productivity.

Meanwhile, the aphotic zone may be defined as the region beneath the photic zone with no sunlight and no photosynthesis; only heterotrophs and chemosynthetic bacteria can survive here. Deep-sea organisms in this zone have adaptations that enable them to survive in very cold water, under high pressure, and in complete darkness. The zone contains nekton and benthos—crawling and sessile organisms. Some are scavengers and some are predators. The habitat of the aphotic zone is fiercely competitive.

Freshwater Biomes

Rivers, lakes, ponds, and marshes are the links between the oceans and land and contain freshwater. Rivers are the routes by which ancient marine organisms reached land and evolved terrestrial adaptations. Many forms failed to adapt to land and developed adaptations for freshwater. Others developed special adaptations suitable for both land and freshwater. As in marine biomes, factors affecting life in freshwater include temperature, transparency (illumination due to suspended mud particles), depth of water, available CO_2 and O_2, and most importantly, salt concentration.

Freshwater biomes differ from salt water biomes in three basic ways:

- Freshwater has a lower concentration of salt (greater concentration of water) than the cell, creating a diffusion gradient that results in the passage of water into the cell. Freshwater organisms have homeostatic mechanisms to maintain water balance by the regular removal of excess water, such as the contractile vacuole of protozoa and excretory systems of fish.

- In rivers and streams, strong, swift currents have caused fish to develop strong muscles and plants to develop rootlike holdfasts.

- Freshwater biomes (except for very large lakes) are affected by variations in climate and weather. They might freeze, dry up, or have mud from their floors stirred up by storms. Temperatures of freshwater bodies vary considerably.

9.6 HUMAN INFLUENCES ON ECOSYSTEMS

At the end of the last ice age as humans were spreading across much of the globe, a series of extinctions of large mammals occurred that coincided with the spread of humans. It is likely that humans were responsible for these extinctions through over-hunting of species such as the wooly mammoth. There are over six billion humans on the earth today. Agriculture, industry, hygiene and medicine have increased the carrying capacity of the earth for humans, but man is having wide-spread impact on the ecosystems of the world. These impacts include loss of habitats, loss of species, increases in carbon dioxide, soil and water pollution, acid rain, ozone loss and climate change. Conservation efforts in many countries work to minimize the negative impacts of man on Earth's ecosystems.

Impact of Man on Species Diversity

Some species are threatened or have been lost due to direct overexploitation. Whales, rhinoceroses, and many other species have been threatened with extinction by hunting. The loss of large carnivores such as wolves and large cats in some North American habitats has resulted in explosions in deer populations. Humans have also disturbed ecosystems through introduction of foreign species into ecosystems. Rabbits were not found in Australia until humans introduced them and they have now become a major pest, killing native marsupials.

As a result of the rapid increase in the numbers of humans on the planet, land has been converted from wild habitat to agricultural, industrial, and urban living areas. Forests and other areas are exploited for their short term economic resources. Tropical forests are currently being cleared in many parts of the world at a rapid rate. These forests contain the greatest biological diversity on earth, and as the habitat is lost, species are lost as well. The species diversity that a habitat can support is related to the size of the habitat. Often, remaining habitats are highly fragmented. Fragmented habitats are more prone to losing diversity than larger intact habitats. Species with large ranges are often unable to support themselves in small fragmented habitats. Some species do not migrate between habitat fragments, limiting the gene pool of the populations that interbreed. With reduced variability in the gene pool, fragmented populations are more vulnerable to decline. Some scientists estimate that the current rate of habitat loss caused by man will result in the loss of half of the world's species in the next few decades.

The loss of species could carry a significant economic cost for humans. The genetic information in the lost species may hold value that we will never realize, such as potential medicines that extinct forest species may have held. Careful management of land can create economically viable long term uses that preserve habitats as alternatives to the clearing and burning of forest for short term gains with the complete loss of habitats. Careful creation of parks and preserves can avoid habitat fragmentation and allow for long term preservation of species. Dealing with the size of the human population may be another aspect of preventing habitat loss.

Changes in Chemical Cycles Caused by Humans

The cycling of inorganic minerals in the environment occurs naturally but has been altered by man. Phosphorous and nitrates are often limiting factors in terrestrial and aquatic productivity, due to their extensive use as fertilizers. The runoff water from agricultural land carries phosphorous and nitrates into aquatic environments. Algae and bacteria in lakes and rivers will proliferate under these conditions; this increased productivity will consume oxygen, changing the overall ecosystem and killing many fishes that cannot get sufficient oxygen.

Acid rain is a well known example of the effect of man on the chemistry of environments. Industrial air pollution often contains acidic subtances that precipitate in rain and harm the vegetation and bodies of water that the acid rain contacts. Wind currents can carry acidic pollutants far from their origin, often across national boundaries, making cooperation between nations important to solve this ecological problem.

The changes in the world's carbon cycle are global and are likely to affect the world's climate. The consumption of fossil fuels as an energy source emits carbon dioxide as a byproduct. In the past, the rate of carbon dioxide production through respiration was equal to the rate of carbon dioxide removal through photosynthesis and inorganic processes. The amount of carbon dioxide in the atmosphere has been increasing steadily since the 1800s. Carbon dioxide in the atmosphere tends to trap infrared heat that is radiated from the earth's surface. The climate of the earth has probably already warmed about half a degree, and projections for the next century estimate that the climate will warm at least a few degrees.

This change in global climate could have profound consequences for humans as well as the other species of the planet. The warming is likely to change prevailing weather patterns, shift regions of rain and desert, and alter agricultural areas. Polar ice likely would melt, raising the sea level significantly. A large percentage of people around the world live near the coast in areas that would be flooded by elevated sea level. This issue is difficult, but with cooperation between nations, through energy conservation or use of alternatives to fossil fuels, the severity of the future consequences of today's actions could be reduced.

Chapter 9: Review Questions

1. Which of the following is a non–density-dependent factor that affects a population?

   (A) spread of disease
   (B) space
   (C) earthquake
   (D) food
   (E) mating and reproduction

2. Which of the following is/are true regarding asexual reproduction?

   (A) Asexual reproduction increases genetic variability.
   (B) Asexual reproduction decreases genetic variability.
   (C) Less energy is required to reproduce asexually.
   (D) More energy is required to reproduce asexually.
   (E) More than one of the above.

3. A species which releases eggs which develop without parental care is likely to

   (A) have a low death rate among young offspring
   (B) have a long life span
   (C) delay sexual maturity
   (D) produce large numbers of offspring
   (E) none of the above

4. Which organism would be dominant in a climax community?

   (A) grass
   (B) moss
   (C) pine
   (D) lichen
   (E) shrub

5. Which of the following statements is false?

   (A) Grass would be a producer within an ecological community.
   (B) Humans would be consumers within an ecological community.
   (C) Omnivores consume both plants and animals.
   (D) Herbivores consume animals.
   (E) Bacteria and fungi would be decomposers or saprophytes within an ecological community.

6. Gross primary productivity differs from net primary productivity in that

   (A) Gross primary productivity is the total chemical energy generated by roducers while the net primary productivity subtracts out the loss of energy to respiration by plants.
   (B) Net primary productivity is the total chemical energy generated by plants while the gross primary productivity subtracts out the loss of energy to respiration by plants.
   (C) Gross primary productivity is the total chemical energy generated by producers while the net primary productivity subtracts out the loss of energy to respiration by animals.
   (D) Net primary productivity is the total chemical energy generated by plants while the gross primary productivity subtracts out the loss of energy to respiration by animals.
   (E) None of the above.

7. If a toxin is present in producers in a food chain
    (A) the toxin will not be found in organisms at the top of the food chain
    (B) the toxin will be more concentrated in organisms at the top of the food chain
    (C) the toxin will be less concentrated in organisms at the top of the food chain
    (D) the toxin will be found in the same concentration in organisms at the top of the food chain
    (E) none of the above

Questions 8–10:

    (A) tropical rain forest
    (B) desert
    (C) taiga
    (D) tundra
    (E) temperate deciduous forest

8. treeless, frozen plain; ground always frozen

9. high temperatures; dense vegetation; high productivity

10. cold winters, warm summers, moderate rainfall

Please see the answers and explanations to these review questions in Section V of this book.

ANSWERS AND EXPLANATIONS TO CHAPTER REVIEW QUESTIONS

In this Section, you will find answers and detailed explanations to the Review Questions at the end of chapters 1 through 9. Use these review exercises to determine your strengths and weaknesses, and help you prepare to take the full-length Practice Tests that follow in the next section.

Answer and Explanations to Chapter Review Questions

CHAPTER 1

1. **(D)** The quaternary level of protein structure refers to the interaction between two or more polypeptide subunits. The primary level of protein structure refers to the linear sequence of amino acids joined by peptide bonds. The secondary level of protein structure refers to the structure of a polypeptide which is generated by hydrogen bonding between amino acids at regular intervals. Some examples of the secondary level of protein structure are the alpha helix and the beta pleated sheet. The tertiary level of protein structure refers to the three dimensional folding of a polypeptide chain brought about by the bonding of two amino acid side chains which may be far apart on the chain of amino acids. An example of the tertiary level of protein structure is the formation of disulfide bonds between two cysteines.

2. **(D)** SDS is used to denature proteins, giving them a globular shape. The other statements are true: SDS adds charge to the denatured proteins. SDS-polyacrylamide gel electrophoresis is used to separate proteins by molecular weight: smaller proteins move more quickly through the gel while larger proteins move more slowly through the gel.

3. **(C)** Adenine, thymine, cytosine, and guanine are the nitrogenous bases found in DNA while uracil is found in RNA.

4. **(A)** Many of the unique physical properties of water may be explained by its polar nature. The other statements are false; here are the corrected statements: water has a high surface tension, is able to easily dissolve polar substances, has a neutral pH, and expands when it freezes.

5. **(D)** Enzymes increase the rate of a reaction by decreasing the activation energy. The other statements are true: enzymes are almost always proteins, enzyme activity is affected by changes in pH, enzymes increase the rate of reaction, and enzymes do not change the free energy of products.

6. **(A)** In feedback inhibition, the presence of the end product of a metabolic pathway inhibits an enzyme which functions in an early step in the pathway. In competitive inhibition, a molecule which resembles the normal substrate of an enzyme competes for the active site. In non-competitive inhibition, a molecule binds to a part of the enzyme other than the active site, causing a conformational change in the enzyme. This conformational change alters the effectiveness of the enzyme. In irreversible inhibition, a molecule covalently bonds to the active site of the enzyme, preventing substrate from accessing the active site.

7. **(C)** In irreversible inhibition, a molecule covalently bonds to the active site of the enzyme, preventing substrate from accessing the active site. In feedback inhibition, the presence of the end product of a metabolic pathway inhibits an enzyme which functions in an early step in the pathway. In

noncompetitive inhibition, a molecule binds to a part of the enzyme other than the active site, causing a conformational change in the enzyme. This conformational change alters the effectiveness of the enzyme.

8. **(B)** If the free energy change (ΔG) of a reaction is greater than zero, then the reaction is nonspontaneous. If the free energy change of a reaction is less than zero, then the reaction is spontaneous. If the free energy change of a reaction is equal to zero, then the reaction is at equilibrium. The terms exothermic and endothermic do not refer to the free energy change of a reaction, but to the enthalpy of a reaction. If the change in enthalpy (ΔH) is positive, then heat is absorbed and the reaction is endothermic. If the change in enthalpy is negative, then heat is released and the reaction is exothermic.

9. **(E)** Proteins do not contain nitrogenous bases (nucleotides do contain nitrogenous bases). The other statements are true: proteins can function as enzymes, consist of amino acids joined by peptide bonds, are important in cell signaling (hormones, for example), and may be used as an energy source (amino acids are converted by enzymes into intermediates of glycolysis and the Krebs cycle).

10. **(D)** All of the statements are true: altering the three dimensional structure of an enzyme disrupts its activity; each enzyme is specific for a given reaction; and hydrophobic, ionic, and hydrogen bonds play a role in binding substrate.

CHAPTER 2

1. **(C)** Animal cells do not have a cell wall. Plant cells have cell walls which provide structural support and protection to the cell. Both plant and animal cells have a nucleus (an exception being the mature red blood cell in humans) and ribosomes. Both plant and animal cells contain DNA which codes genetic information.

2. **(E)** Statements A, B, and C are true about the protein products of tumor suppressor genes. Tumor suppressor genes code for proteins which are present in noncancerous cells, may signal cell death if a cell begins dividing when it should not, and function to regulate the cell cycle. Choice D is incorrect because protein products of tumor suppressor genes prevent (not cause) unregulated cell growth.

3. **(C)** When a cell undergoes mitosis, the nuclear membrane disintegrates during prophase.

4. **(D)** During mitosis, the chromosomes line up along the midline of the cell (sometimes referred to as the metaphase plate) during metaphase.

5. **(C)** In exocytosis, a vesicle from the interior of the cell fuses with the cell membrane; the contents of the vesicle are released outside of the cell.

6. **(E)** Active transport is the net movement of dissolved particles against their concentration gradient via transport proteins within the cell membrane. Energy is required to move particles against their concentration gradient.

7. **(B)** Facilitated diffusion is the net movement of dissolved particles down their concentration gradient via carrier proteins within the cell membrane. Energy is not required to move particles down their concentration gradient.

8. **(D)** The plasma membrane encloses the cell and regulates what substances pass across it; the plasma membrane is a semi-permeable cell barrier.

9. **(D)** Prokaryotes have a plasma membrane, but lack a nucleus and membrane bound organelles, such as Golgi apparatus, lysosomes, and endoplasmic reticulum.

10. **(B)** The Golgi apparatus receives proteins from the endoplasmic reticulum, modifies them, and packages them for transport out of the cell.

CHAPTER 3

1. **(C)** The products of the light reactions of photosynthesis are ATP and NADPH. In glycolysis, a molecule of glucose is split into two molecules of pyruvate. Glucose is the product of the Calvin cycle, the second stage of photosynthesis. CO_2 and H_2O are the starting materials for photosynthesis. Ribulose biphosphate (RuBP) is a 5 carbon sugar to which carbon is fixed in the Calvin cycle.

2. **(C)** 3 ATP are produced via the chemiosmotic principle for every molecule of NADH that transfers high-energy electrons to the electron transport chain.

3. **(B)** 2 ATP are produced via the chemiosmotic principle for every molecule of $FADH_2$ that transfers high-energy electrons to the electron transport chain.

4. **(E)** Most of the ATP produced in aerobic respiration results from oxidative phosphorylation. Oxidative phosphorylation yields a net 32 ATP. Glycolysis yields a net 2 ATP (via substrate-level phosphorylation); substrate-level phosphorylation yields a net 2 ATP in glycolysis and 2 ATP in the Krebs cycle; fermentation yields a net 2 ATP (via substrate-level phosphorylation). Active transport is a form of molecular transport across membranes which moves molecules against their concentration gradient and requires ATP.

5. **(B)** Pyruvate is a product of glycolysis.

6. **(C)** Oxygen is the final electron acceptor in the electron transport chain. At the end of the electron transport chain, oxygen accepts electrons, and, combining with hydrogen ions, is reduced to water.

7. **(D)** The light reactions occur within the thylakoid.

8. **(A)** Stomata regulate the entrance of CO_2 into the leaves of a plant.

9. **(E)** The Calvin cycle takes place in the stroma.

10. **(A)** Water is split to form hydrogen ions, oxygen, and electrons in photosystem II. The splitting of water replaces electrons which exit photosystem II when activated by sunlight. The electrons pass through an electron transport chain, producing ATP, and enter photosystem I. There the electrons enter another electron transport chain, producing more ATP. Cyclic photophosphorylation takes place in photosystem I and does not require water, since photosystem I receives electrons from photosystem II. The Calvin cycle fixes CO_2 into glucose, and fermentation splits glucose into pyruvate.

CHAPTER 4

1. **(E)** Meiosis is the form of cell division used to produce haploid gametes. Binary fission, budding, parthenogenesis, and regeneration are all forms of asexual reproduction.

2. **(D)** Let's represent the dominant allele by A, and the recessive allele by a. An individual who is homozygous dominant has the genotype AA, and an individual who is homozygous recessive has the genotype aa. Their offspring will all receive an A allele from the AA parent, and an a allele from the aa parent, so all the offspring are heterozygotes.

3. **(B)** The AB blood type in humans is best described as codominance. A person with blood type AB expresses alleles for both the A antigen and the B antigen. Incomplete dominance results in a blending of the effects of different alleles. Sex-linkage occurs when an allele for a trait is on the X chromosome or the Y chromosome rather than an autosome. In polygenic inheritance, a trait is produced by an interaction between many different genes. An example of cytoplasmic inheritance is the inheritance of mitochondria. All of a person's mitochondria are descendants of their mother's mitohondria.

4. **(B)** Mitochondria follow a pattern of inheritance known as cytoplasmic inheritance. At fertilization, the zygote receives all of its cytoplasm from its mother; thus all of the mitochondria will be descendants of the mother's mitochondria.

5. **(A)** An allele that has 0% penetrance is heavily influenced by the environment. The penetrance of an allele is the degree to which its expression is influenced by the environment. The more penetrant an allele, the less it is influenced by the environment.

6. **(D)** Interaction between two or more genes is termed epistasis.

7. **(B)** The observable appearance of an organism that reflects the expression of genes is called the organism's phenotype.

8. **(A)** The genetic makeup of an individual is the individual's genotype.

9. **(E)** An organism that has two different forms of a gene is heterozygous.

10. **(C)** Nondisjunction can occur in anaphase and may result in extra copies of chromosomes or missing chromosomes.

CHAPTER 5

1. **(E)** DNA is composed of nucleotides. Each nucleotide consists of deoxyribose, which is a five-carbon sugar, a phosphate group, and a nitrogenous base.

2. **(B)** In forming the double helix, one strand of DNA is joined to another strand via hydrogen bonds.

3. **(D)** Oswald Avery identified DNA as the genetic material of inheritance. Mendel worked out the basic principles of heredity. Watson and Crick deduced the structure of DNA. Charles Darwin is known for his theory of Natural Selection. Okazaki is the name of the Japanese scientist who discovered the fragments of replicated DNA on the lagging strand of DNA during DNA replication.

4. **(C)** Mutagens are substances that cause mutations; carcinogens are substances that cause cancer. Not all mutations cause cancer; for example, if a mutation is found by DNA repair enzymes and corrected, the cell will not multiply abnormally and become cancerous. Also, if a mutation occurs in a gene that is not expressed, the mutation may not result in cancer. The other statements are true: Mutations are important in evolution, since without mutations, there would be no new alleles that could be selected for; errors in DNA replication can cause mutations; products of cellular metabolism, such as superoxide, can cause mutations; and DNA polymerase proofreads newly synthesized DNA, correcting incorrectly placed nucleotides.

5. **(E)** Splicing is the removal of introns, or sections of DNA that are not expressed. Splicing takes place only in eukaryotes.

6. **(B)** A guanine nucleotide forms three hydrogen bonds when linked with a cytosine nucleotide.

7. **(D)** Transcription is the process of transferring genetic information in DNA to an RNA message that is decoded to produce protein.

8. **(A)** Uracil nucleotides are found only in RNA.

9. **(E)** Viruses have DNA or RNA as their genetic material, and a protein coat. Viruses do not have cytoplasm or organelles. Viruses do not have a nucleus, mitochondria, or ribosomes.

10. **(C)** DNA fragments are negatively charged, so in gel electrophoresis, DNA fragments migrate toward the positive electrode. The gel acts as a sieve, sorting out the fragments by size; the smaller the fragment, the faster it moves through the gel, since larger fragments will encounter more resistance.

CHAPTER 6

1. **(E)** Oxygen was not present in the prebiotic environment. Carbon dioxide, ammonia, water, and methane were present in the prebiotic environment of the earth.

2. **(B)** The earliest enzymes are thought to have consisted of RNA.

3. **(E)** Modern life forms use L-amino acids and D-sugars. It is thought that the fact that all life forms use the same configuration of amino acids and sugars is an indication of common ancestry.

4. **(C)** Mammals evolved during the Jurassic period.

5. **(B)** Gymnosperms developed during the Devonian period.

6. **(A)** Chordates developed during the Cambrian period.

7. **(D)** Finding the ratio of ^{14}C to ^{12}C is one method used to date organic material.

8. **(D)** Analogous structures have similar functions/features and the same evolutionary origin while homologous structures have similar functions/features but different evolutionary origins.

9. **(E)** Random mating is a condition of Hardy-Weinberg equilibrium. Other conditions of Hardy-Weinberg equilibrium are no natural selection, no mutation, no migration into or out of a population, and large population size.

10. **(C)** "Fitness" in an evolutionary sense means that an organism survives long enough to reproduce and pass along genetic information to the next generation of organisms.

CHAPTER 7

1. **(C)** The correct order of classification categories, from largest to smallest is: **k**ingdom, **p**hylum, **c**lass, **o**rder, **f**amily, **g**enus, **s**pecies. A mnemonic for remembering the order of the categories is: **K**ings **p**lay **c**hess **o**n **f**unny **g**reen **s**eats. The first letter of each word corresponds to a classification category. Choice A is incorrect because family and genus have been switched. Choice B is incorrect because phylum and class have been switched. Choice D is incorrect because class and order have been switched. Choice E is incorrect because genus and species have been switched.

2. **(E)** The evolutionary relationships between organisms are called phylogeny. Taxonomy is a system of naming and classifying organisms. Embryology is the study of embryos and embryological development. Physiology is the study of the functioning of organisms. Genetics is the study of heredity and heritable traits.

3. **(A)** Slime molds are heterotrophic and can exist in a cellular form or a multinucleated acellular form that is capable of amoebalike motion. In unfavorable conditions, slime molds form fruiting bodies which produce spores. Sponges are a unique phylum within the kingdom Animalia. Sponges have no true tissues and are not capable of moving. Ferns are vascular plants which bear spores in sporangia on the underside of leaves. Ferns usually grow near water, since moisture is required for fertilization. Ferns, algae, and violets are capable of photosynthesis and are therefore photoautotrophs, not heterotrophs as are slime molds.

4. **(B)** The sea urchin has radial symmetry, while the frog, rabbit, kangaroo, and crab have bilateral symmetry. In radial symmetry, lines of symmetry meet at a point at the center of the organism. In bilateral symmetry, an organism has two equal parts; a plane divides the body into two halves which are mirror images of each other.

5. **(C)** Gymnosperms is the group of plants which includes the pine.

6. **(A)** Endosperm is the tissue which nurtures the developing plant zygote.

7. **(E)** Tracheid cells in plants produce tubes and fibers which function in fluid transport.

5–7: Let's look at the other choices: Endoderm is the tissue in animals which develops into the liver, lungs, and digestive tract. Angiosperms are flowering plants.

8. **(D)** Organisms in phylum *Platyhelminthes* have bilateral symmetry, in which the body has two halves which are mirror images of each other.

9. **(B)** Organisms classified as deuterostomes show radial cleavage during early development and have a dorsal nerve cord.

10. **(E)** The coelom is a fluid-filled body cavity which is lined with muscle.

8–10: Let's look at the other choices: Protostomes show spiral cleavage during early development and have bilateral symmetry. An organism with radial symmetry has several lines of symmetry which meet at a point at the center of the organism.

CHAPTER 8

1. **(A)** In oogenesis, which is the formation of an egg cell, meiosis is not completed until fertilization. The unfertilized egg remains in meiosis II until fertilization by a sperm cell, when a polar body is released.

2. **(C)** Cells which have the potential to develop into any tissue are called totipotent cells. Leydig cells produce testosterone in males. Cells which are determined have become committed to a specific developmental fate. Gastrulation results in three germ layers: the endoderm, mesoderm, and ectoderm. The endoderm develops into the digestive and respiratory tracts, parts of the liver and pancreas, and the bladder lining. The mesoderm develops into the muscles and skeletal system, circulatory system, excretory system, gonads, and dermis, or inner layer of skin. The ectoderm develops into the epidermis (outer layer of skin), nervous system, and sweat glands.

3. **(E)** Once implanted, embryonic tissues secrete human chorionic gonadotropin (hCG), which takes over the function of maintaining the corpus luteum from luteinizing hormone (LH). Estrogen, produced by the ovary in females, initiates and maintains female sex characteristics, while both estrogen and progesterone function in the development of the uterine lining in preparation for pregnancy. Androgens are produced by the testis in males, and function to initiate and maintain male sex characteristics and support spermatogenesis. Insulin is produced by the pancreas, and lowers blood sugar levels.

4. **(D)** Glucagon stimulates glycogen breakdown in the liver while insulin stimulates glycogen formation. Choice A is incorrect because insulin (not glucagon) is produced by beta-cells in the pancreas. Choice B is incorrect because insulin lowers blood sugar. Choice C is incorrect because insulin stimulates (not inhibits) glycogen formation in the liver.

5. **(B)** The circulatory system of the earthworm involves five aortic arches or "hearts" which pump blood through a dorsal vessel and a ventral vessel. The circulatory system of the human involves one four-chambered heart which pumps blood through arteries, capillaries, and veins. The circulatory system of the grasshopper is an open one that bathes body tissues directly in a fluid called hemolymph. The circulatory system of the fish involves one two-chambered heart and one circuit of blood flow.

6. **(B)** Antibodies, or immunoglobulins, are produced by B cells. Erythrocytes (red blood cells) do not produce antibodies. Helper T cells

coordinate the immune response of other immune cells. Killer T cells kill cells which have been infected by pathogens. Neutrophils consume bacteria and foreign particles which have not entered body cells.

7. **(A)** Mammals excrete their nitrogenous waste in the form of urea. Urea is many times less toxic than ammonia, and therefore may be stored in greater concentrations than ammonia. The use of urea as a means of disposing nitrogenous waste is also an important adaptation to land. Because urea is less toxic than ammonia, less water is required to dispose of urea.

8. **(D)** Annelids have nephridia or metanephridia which function in excretion.

9. **(C)** Fish dispose of their nitrogenous wastes in the form of ammonia, a small molecule which is easily dissolved in water. Ammonia is a very toxic substance; thus it must be transported through the body in low concentrations. The ammonia diffuses across the epithelium of the gills of the fish in saltwater fish. In freshwater fish, ammonium ion, NH_4^+, is exchanged for Na^+ in the epithelium of the gills.

10. **(A)** Mechanical digestion of food occurs in the mouth (structure 1). Structure 3 is the liver, which secretes bile, among other functions. Bile is involved in the digestion of fats. Structure 5 is the stomach, where the chemical digestion of proteins begins. Structure 6 is the pancreas, which secretes insulin and amylase. Insulin regulates blood sugar levels, while amylase breaks down carbohydrates. Structure 10 is the anus, through which solid wastes are eliminated.

11. **(B)** Structure 3 is the liver, which secretes bile, among other functions. Bile is involved in the digestion of fats. Mechanical digestion of food occurs in the mouth (structure 1). Structure 5 is the stomach, where the chemical digestion of proteins begins. Structure 6 is the pancreas, which

secretes insulin and amylase. Insulin regulates blood sugar levels, while amylase breaks down carbohydrates. Structure 10 is the anus, through which solid wastes are eliminated.

12. **(D)** Structure 4 indicates areas of the neuron covered in myelin, an insulating material. In saltatory conduction, an action potential jumps from one Node of Ranvier to another Node of Ranvier, skipping areas of the neuron covered by myelin. Structures labeled 1 are the dendrites, which receive input signals. Structure 2 is the cell body, which contains the nucleus. Structure 3 is the axon. Structures labeled 5 are Nodes of Ranvier, uninsulated areas of the neuron.

13. **(E)** Structures labeled 5 are Nodes of Ranvier, uninsulated areas of the neuron. Structures labeled 1 are the dendrites, which receive input signals. Structure 2 is the cell body, which contains the nucleus. Structure 3 is the axon. Structure 4 indicates areas of the neuron covered in myelin, an insulating material. In saltatory conduction, an action potential jumps from one Node of Ranvier to another Node of Ranvier, skipping areas of the neuron covered by myelin.

14. **(E)** Auxin is the plant hormone which is involved in phototropism. Phototropism is the tendency of plant shoots to grow toward light. Root hairs increase the surface area of roots and enhance water absorption. Stomata are openings in the leaves through which plants take in carbon dioxide. Phloem are vascular tissues in plants which transport nutrients. Xylem are vascular tissues in plants which transport water.

15. **(A)** Root hairs increase the surface area of roots and enhance water absorption. Stomata are openings in the leaves through which plants take in carbon dioxide. Phloem are vascular tissues in plants which transport nutrients. Xylem are vascular tissues in plants which transport water. Auxin is the plant hormone which is involved in phototropism.

CHAPTER 9

1. **(C)** Density dependent factors which affect a population include the spread of disease, the availability of space and food, and success in mating and reproduction. If the number of individuals increases, the population density increases, and there is greater competition for available space and food. In addition, the more dense a population, the more easily diseases are spread. However, an advantage to a dense population is that the greater number of individuals may make it easier to find a mate and reproduce.

2. **(E)** Asexual reproduction requires less energy than sexual reproduction and asexual reproduction does not increase genetic variability. Sexual reproduction increases genetic variability, but requires more energy (to develop and maintain reproductive structures such as flowers and fruit, for example).

3. **(D)** A species that releases eggs which develop without parental care is likely to produce large numbers of offspring, to compensate for the fact that many offspring may be eaten before they reach maturity. This type of species usually has a high death rate among young offspring, has a short life span, and produces large numbers of offspring which reach sexual maturity quickly.

4. **(C)** Shade trees (such as pine, hemlock, beech, maple, and oak) dominate in a climax community. The order of succession would be: lichen → moss → grass → shrub → pine. Lichen can grow on rocks; as the lichen transforms the rock to soil, moss appears; as the moss improve the soil, grass appears; finally shrubs and shade trees complete the ecological succession.

5. **(D)** Choice D is false because herbivores consume plants (not animals). Carnivores consume animals. The other choices are true: grass would be a producer within an ecological community; humans would be consumers within an ecological community; omnivores consume both plants and animals; and bacteria and fungi would be decomposers or saprophytes within an ecological community.

6. **(A)** Gross primary productivity is the total chemical energy generated by producers while the net primary productivity subtracts out the loss of energy to respiration by plants. Choice B is incorrect because net primary productivity and gross primary productivity have been switched. Choices C and D are incorrect because primary productivity does not take into account losses to respiration by animals.

7. **(B)** If a toxin is ingested by producers in a food chain, the toxin will be more concentrated in organisms at the top of the food chain. Primary consumers consume producers which contain the toxin; secondary consumers consume primary consumers that contain the toxin from consumed producers; thus, the concentration of toxin is greater the higher up on the food chain an organism is.

8. **(D)** Tundra is the biome characterized by treeless, frozen plain, and permanently frozen ground called permafrost.

9. **(A)** Tropical rain forest is the biome characterized by high temperatures, dense vegetation, and high productivity.

10. **(E)** Temperate deciduous forest is the biome characterized by cold winters, warm summers, and moderate rainfall.

PRACTICE TESTS

In this section are two full-length Practice Tests. Also included are the official test directions for the AP Biology Exam. Familiarize yourself with these directions, then go on to take the Practice Tests.

Answer keys and detailed explanations for each question follow each test.

The format of the Practice Tests in this book varies slightly from that of the actual AP Biology exam since you will not be using separate booklets or answering 15 multiple-choice questions regarding exam preparation.

AP BIOLOGY

Three hours are allotted for this examination: 1 hour and 30 minutes for Section I, which consists of multiple-choice questions; and 1 hour and 30 minutes for Section II, which consists of essay questions. Section I is printed in this examination booklet; Section II, in a separate booklet.

SECTION I

Time—1 hour and 30 minutes

Number of Questions: 120

Percent of total grade: 60

Section I of this examination contains 120 multiple-choice questions, followed by 15 multiple-choice questions regarding your preparation for this exam. Please be careful to fill in only the ovals that are preceded by numbers 1 through 135 on your answer sheet.

General Instructions

DO NOT OPEN THIS BOOKLET UNTIL YOU ARE INSTRUCTED TO DO SO.

INDICATE ALL YOUR ANSWERS TO QUESTIONS IN SECTION I ON THE SEPARATE ANSWER SHEET ENCLOSED. No credit will be given for anything written in this examination booklet, but you may use the booklet for notes or scratchwork. After you have decided which of the suggested answers is best, COMPLETELY fill in the corresponding oval on the answer sheet. Give only one answer to each question. If you change an answer, be sure that the previous mark is erased completely.

Example: Sample Answer

Chicago is a

(A) state
(B) city
(C) country
(D) continent
(E) village

Many candidates wonder whether or not to guess the answers to questions about which they are not certain. In this section of the examination, as a correction for haphazard guessing, one-fourth of the number of questions you answer incorrectly will be subtracted from the number of questions you answer correctly. It is improbable, therefore, that mere guessing will improve your score significantly; it may even lower your score, and it does take time. If, however, you are not sure of the correct answer but have some knowledge of the question and are able to eliminate one or more of the answer choices as wrong, your chance of getting the right answer is improved, and it may be to your advantage to answer such a question.

Use your time effectively, working as rapidly as you can without losing accuracy. Do not spend too much time on questions that are too difficult. Go on to other questions and come back to the difficult ones later if you have time. It is not expected that everyone will be able to answer all the multiple-choice questions.

Copyright © 1994 by Educational Testing Service. All rights reserved. Princeton, NJ 08541.

PRACTICE TEST I
ANSWER SHEET

1 Ⓐ Ⓑ Ⓒ Ⓓ Ⓔ   25 Ⓐ Ⓑ Ⓒ Ⓓ Ⓔ   49 Ⓐ Ⓑ Ⓒ Ⓓ Ⓔ   73 Ⓐ Ⓑ Ⓒ Ⓓ Ⓔ   97 Ⓐ Ⓑ Ⓒ Ⓓ Ⓔ

2 Ⓐ Ⓑ Ⓒ Ⓓ Ⓔ   26 Ⓐ Ⓑ Ⓒ Ⓓ Ⓔ   50 Ⓐ Ⓑ Ⓒ Ⓓ Ⓔ   74 Ⓐ Ⓑ Ⓒ Ⓓ Ⓔ   98 Ⓐ Ⓑ Ⓒ Ⓓ Ⓔ

3 Ⓐ Ⓑ Ⓒ Ⓓ Ⓔ   27 Ⓐ Ⓑ Ⓒ Ⓓ Ⓔ   51 Ⓐ Ⓑ Ⓒ Ⓓ Ⓔ   75 Ⓐ Ⓑ Ⓒ Ⓓ Ⓔ   99 Ⓐ Ⓑ Ⓒ Ⓓ Ⓔ

4 Ⓐ Ⓑ Ⓒ Ⓓ Ⓔ   28 Ⓐ Ⓑ Ⓒ Ⓓ Ⓔ   52 Ⓐ Ⓑ Ⓒ Ⓓ Ⓔ   76 Ⓐ Ⓑ Ⓒ Ⓓ Ⓔ   100 Ⓐ Ⓑ Ⓒ Ⓓ Ⓔ

5 Ⓐ Ⓑ Ⓒ Ⓓ Ⓔ   29 Ⓐ Ⓑ Ⓒ Ⓓ Ⓔ   53 Ⓐ Ⓑ Ⓒ Ⓓ Ⓔ   77 Ⓐ Ⓑ Ⓒ Ⓓ Ⓔ   101 Ⓐ Ⓑ Ⓒ Ⓓ Ⓔ

6 Ⓐ Ⓑ Ⓒ Ⓓ Ⓔ   30 Ⓐ Ⓑ Ⓒ Ⓓ Ⓔ   54 Ⓐ Ⓑ Ⓒ Ⓓ Ⓔ   78 Ⓐ Ⓑ Ⓒ Ⓓ Ⓔ   102 Ⓐ Ⓑ Ⓒ Ⓓ Ⓔ

7 Ⓐ Ⓑ Ⓒ Ⓓ Ⓔ   31 Ⓐ Ⓑ Ⓒ Ⓓ Ⓔ   55 Ⓐ Ⓑ Ⓒ Ⓓ Ⓔ   79 Ⓐ Ⓑ Ⓒ Ⓓ Ⓔ   103 Ⓐ Ⓑ Ⓒ Ⓓ Ⓔ

8 Ⓐ Ⓑ Ⓒ Ⓓ Ⓔ   32 Ⓐ Ⓑ Ⓒ Ⓓ Ⓔ   56 Ⓐ Ⓑ Ⓒ Ⓓ Ⓔ   80 Ⓐ Ⓑ Ⓒ Ⓓ Ⓔ   104 Ⓐ Ⓑ Ⓒ Ⓓ Ⓔ

9 Ⓐ Ⓑ Ⓒ Ⓓ Ⓔ   33 Ⓐ Ⓑ Ⓒ Ⓓ Ⓔ   57 Ⓐ Ⓑ Ⓒ Ⓓ Ⓔ   81 Ⓐ Ⓑ Ⓒ Ⓓ Ⓔ   105 Ⓐ Ⓑ Ⓒ Ⓓ Ⓔ

10 Ⓐ Ⓑ Ⓒ Ⓓ Ⓔ   34 Ⓐ Ⓑ Ⓒ Ⓓ Ⓔ   58 Ⓐ Ⓑ Ⓒ Ⓓ Ⓔ   82 Ⓐ Ⓑ Ⓒ Ⓓ Ⓔ   106 Ⓐ Ⓑ Ⓒ Ⓓ Ⓔ

11 Ⓐ Ⓑ Ⓒ Ⓓ Ⓔ   35 Ⓐ Ⓑ Ⓒ Ⓓ Ⓔ   59 Ⓐ Ⓑ Ⓒ Ⓓ Ⓔ   83 Ⓐ Ⓑ Ⓒ Ⓓ Ⓔ   107 Ⓐ Ⓑ Ⓒ Ⓓ Ⓔ

12 Ⓐ Ⓑ Ⓒ Ⓓ Ⓔ   36 Ⓐ Ⓑ Ⓒ Ⓓ Ⓔ   60 Ⓐ Ⓑ Ⓒ Ⓓ Ⓔ   84 Ⓐ Ⓑ Ⓒ Ⓓ Ⓔ   108 Ⓐ Ⓑ Ⓒ Ⓓ Ⓔ

13 Ⓐ Ⓑ Ⓒ Ⓓ Ⓔ   37 Ⓐ Ⓑ Ⓒ Ⓓ Ⓔ   61 Ⓐ Ⓑ Ⓒ Ⓓ Ⓔ   85 Ⓐ Ⓑ Ⓒ Ⓓ Ⓔ   109 Ⓐ Ⓑ Ⓒ Ⓓ Ⓔ

14 Ⓐ Ⓑ Ⓒ Ⓓ Ⓔ   38 Ⓐ Ⓑ Ⓒ Ⓓ Ⓔ   62 Ⓐ Ⓑ Ⓒ Ⓓ Ⓔ   86 Ⓐ Ⓑ Ⓒ Ⓓ Ⓔ   110 Ⓐ Ⓑ Ⓒ Ⓓ Ⓔ

15 Ⓐ Ⓑ Ⓒ Ⓓ Ⓔ   39 Ⓐ Ⓑ Ⓒ Ⓓ Ⓔ   63 Ⓐ Ⓑ Ⓒ Ⓓ Ⓔ   87 Ⓐ Ⓑ Ⓒ Ⓓ Ⓔ   111 Ⓐ Ⓑ Ⓒ Ⓓ Ⓔ

16 Ⓐ Ⓑ Ⓒ Ⓓ Ⓔ   40 Ⓐ Ⓑ Ⓒ Ⓓ Ⓔ   64 Ⓐ Ⓑ Ⓒ Ⓓ Ⓔ   88 Ⓐ Ⓑ Ⓒ Ⓓ Ⓔ   112 Ⓐ Ⓑ Ⓒ Ⓓ Ⓔ

17 Ⓐ Ⓑ Ⓒ Ⓓ Ⓔ   41 Ⓐ Ⓑ Ⓒ Ⓓ Ⓔ   65 Ⓐ Ⓑ Ⓒ Ⓓ Ⓔ   89 Ⓐ Ⓑ Ⓒ Ⓓ Ⓔ   113 Ⓐ Ⓑ Ⓒ Ⓓ Ⓔ

18 Ⓐ Ⓑ Ⓒ Ⓓ Ⓔ   42 Ⓐ Ⓑ Ⓒ Ⓓ Ⓔ   66 Ⓐ Ⓑ Ⓒ Ⓓ Ⓔ   90 Ⓐ Ⓑ Ⓒ Ⓓ Ⓔ   114 Ⓐ Ⓑ Ⓒ Ⓓ Ⓔ

19 Ⓐ Ⓑ Ⓒ Ⓓ Ⓔ   43 Ⓐ Ⓑ Ⓒ Ⓓ Ⓔ   67 Ⓐ Ⓑ Ⓒ Ⓓ Ⓔ   91 Ⓐ Ⓑ Ⓒ Ⓓ Ⓔ   115 Ⓐ Ⓑ Ⓒ Ⓓ Ⓔ

20 Ⓐ Ⓑ Ⓒ Ⓓ Ⓔ   44 Ⓐ Ⓑ Ⓒ Ⓓ Ⓔ   68 Ⓐ Ⓑ Ⓒ Ⓓ Ⓔ   92 Ⓐ Ⓑ Ⓒ Ⓓ Ⓔ   116 Ⓐ Ⓑ Ⓒ Ⓓ Ⓔ

21 Ⓐ Ⓑ Ⓒ Ⓓ Ⓔ   45 Ⓐ Ⓑ Ⓒ Ⓓ Ⓔ   69 Ⓐ Ⓑ Ⓒ Ⓓ Ⓔ   93 Ⓐ Ⓑ Ⓒ Ⓓ Ⓔ   117 Ⓐ Ⓑ Ⓒ Ⓓ Ⓔ

22 Ⓐ Ⓑ Ⓒ Ⓓ Ⓔ   46 Ⓐ Ⓑ Ⓒ Ⓓ Ⓔ   70 Ⓐ Ⓑ Ⓒ Ⓓ Ⓔ   94 Ⓐ Ⓑ Ⓒ Ⓓ Ⓔ   118 Ⓐ Ⓑ Ⓒ Ⓓ Ⓔ

23 Ⓐ Ⓑ Ⓒ Ⓓ Ⓔ   47 Ⓐ Ⓑ Ⓒ Ⓓ Ⓔ   71 Ⓐ Ⓑ Ⓒ Ⓓ Ⓔ   95 Ⓐ Ⓑ Ⓒ Ⓓ Ⓔ   119 Ⓐ Ⓑ Ⓒ Ⓓ Ⓔ

24 Ⓐ Ⓑ Ⓒ Ⓓ Ⓔ   48 Ⓐ Ⓑ Ⓒ Ⓓ Ⓔ   72 Ⓐ Ⓑ Ⓒ Ⓓ Ⓔ   96 Ⓐ Ⓑ Ⓒ Ⓓ Ⓔ   120 Ⓐ Ⓑ Ⓒ Ⓓ Ⓔ

AP BIOLOGY
PRACTICE TEST I

Section I

Time—1 hour and 30 minutes

<u>Directions:</u> Each of the questions or incomplete statements below is followed by five suggested answers or completions. Select the one that is best in each case and then fill in the corresponding oval on the answer sheet.

1. If the free energy change of a reaction is less than zero, then the reaction

   (A) is spontaneous
   (B) is nonspontaneous
   (C) is at equilibrium
   (D) is endothermic
   (E) is exothermic

2. If the diploid number of a female organism is 36, then the number of chromosomes in each egg cell is

   (A) 9
   (B) 18
   (C) 36
   (D) 48
   (E) 72

3. If a plant is grown in the presence of CO_2 containing ^{18}O, the ^{18}O would be found in which product of photosynthesis?

   (A) glucose
   (B) oxygen
   (C) glucose and oxygen
   (D) the ^{18}O would be not be found in any of the photosynthetic products since it exits the cells as $^{18}O_2$
   (E) the ^{18}O would be not be found in any of the photosynthetic products since CO_2 is not required in photosynthesis

4. Which of the following pairings is correct?

   (A) *Euglena*—cilia
   (B) *Paramecium*—flagella
   (C) human sperm cell—cilia
   (D) human oviduct—flagella
   (E) human respiratory tract—cilia

5. Which of the following will not cause a frameshift mutation?

   (A) insertion of one nucleotide pair
   (B) insertion of two nucleotide pairs
   (C) insertion of three nucleotide pairs
   (D) deletion of one nucleotide pair
   (E) deletion of two nucleotide pairs

6. Viruses contain

   (A) DNA or RNA, ribosomes, mitochondria
   (B) DNA polymerase, protein coat
   (C) mitochondria, ribosomes, DNA polymerase
   (D) ribosomes, DNA polymerase, protein coat
   (E) DNA or RNA, protein coat

7. Two closely related species breed during different seasons. This is an example of

   (A) behavioral isolation
   (B) temporal isolation
   (C) habitat isolation
   (D) mechanical isolation
   (E) gametic isolation

GO ON TO THE NEXT PAGE. ➡

8. Glasslike shells consisting of silica are a characteristic of members of which phylum?

   (A) Bacillariophyta (diatoms)
   (B) Chrysophyta (golden algae)
   (C) Phaeophyta (brown algae)
   (D) Euglenophyta (*Euglena*)
   (E) Dinoflagellata (dinoflagellates)

9. In the human kidney

   (A) water is actively transported from of the loop of Henle to the medulla
   (B) water is actively transported from the medulla to the loop of Henle
   (C) salt is passively transported from the loop of Henle to the medulla
   (D) salt is actively transported from the loop of Henle to the medulla
   (E) salt is actively transported from the medulla to the loop of Henle

10. Nitrogen fixing bacteria

    (A) convert CO to CO_2
    (B) convert CO_2 to CO
    (C) convert NO_3^- to N_2
    (D) convert N_2 to NO_3^-
    (E) convert N_2 to NH_3

11. The heavy and light chain subunits of an antibody are held together by

    (A) hydrogen bonds
    (B) ionic bonds
    (C) disulfide bonds
    (D) polar covalent bonds
    (E) nonpolar covalent bonds

12. Ethanol is a product of

    (A) the electron transport chain
    (B) fermentation
    (C) Krebs cycle
    (D) the Calvin cycle
    (E) photosystem I

13. In active transport, unlike facilitated diffusion

    (A) molecules are moved with their concentration gradient
    (B) membrane proteins are not involved
    (C) molecules cross the cell membrane
    (D) energy is required
    (E) energy is not required

14. DNA synthesis requires a primer which consists of

    (A) primase
    (B) DNA
    (C) RNA
    (D) an Okazaki fragment
    (E) helicase

15. Which of the following is in the correct order regarding blood flow from the heart through the body?

    (A) left ventricle to aorta
    (B) arteries to venules
    (C) pulmonary vein to right atrium
    (D) lung to pulmonary artery
    (E) vena cava to left atrium

16. Which of the following is not a steroid hormone?

    (A) estrogen
    (B) cortisol
    (C) androgen
    (D) progesterone
    (E) insulin

17. A chemical reaction which is endergonic

    (A) cannot occur within a cell
    (B) can occur within a cell at very high temperatures
    (C) can occur because enzymes lower the energy of activation
    (D) can be coupled to a more favorable reaction
    (E) cannot occur at normal body temperature

GO ON TO THE NEXT PAGE. ➡

18. Which of the following is found only in eukaryotic cells?

    (A) mRNA
    (B) snRNA
    (C) rRNA
    (D) tRNA
    (E) DNA

19. Imperfect flowers

    (A) are missing petals
    (B) are missing either stamens or carpels
    (C) are missing sepals
    (D) are missing ovaries
    (E) are missing both stamens and carpels

20. The earliest forms of life were most likely

    (A) unicellular autotrophs
    (B) multicellular autotrophs
    (C) unicellular heterotrophs
    (D) multicellular heterotrophs
    (E) photoautotrophs

21. Which sequence best represents embryonic development?

    (A) zygote → morula → blastula → gastrula
    (B) zygote → blastula → morula → gastrula
    (C) morula → zygote → blastula → gastrula
    (D) morula → zygote → gastrula → blastula
    (E) zygote → blastula → gastrula → morula

22. In a food chain which consists of grass → insects → mice → snakes → owls, which class of organism is missing?

    (A) producer
    (B) primary consumer
    (C) secondary consumer
    (D) tertiary consumer
    (E) decomposer

23. Unlike animal cells, cytokinesis in plants involves the formation of a

    (A) cleavage furrow
    (B) cell plate
    (C) nuclear envelope
    (D) mitotic spindle
    (E) contractile ring

24. Organic compounds, unlike inorganic compounds, contain

    (A) hydrogen
    (B) oxygen
    (C) nitrogen
    (D) sulfur
    (E) carbon

25. Members of this class undergo torsion during early embryonic development, resulting in a mantle cavity and anus which is above the head in the adult.

    (A) Bivalvia
    (B) Cephalopoda
    (C) Polychaeta
    (D) Gastropoda
    (E) Cestoda

26. Which organism has an open circulatory system?

    (A) grasshopper
    (B) earthworm
    (C) fish
    (D) amphibian
    (E) human

GO ON TO THE NEXT PAGE. ➡

27. Which of the following sequences best represents the flow of filtrate through a nephron?

    (A) Bowman's capsule → proximal convoluted tubule → loop of Henle → collecting duct → distal convoluted tubule
    (B) Bowman's capsule → distal convoluted tubule → loop of Henle → collecting duct → proximal convoluted tubule
    (C) Bowman's capsule → proximal convoluted tubule → loop of Henle → distal convoluted tubule → collecting duct
    (D) Bowman's capsule → proximal convoluted tubule → collecting duct → loop of Henle → distal convoluted tubule
    (E) Bowman's capsule → loop of Henle → proximal convoluted tubule → distal convoluted tubule → collecting duct

28. Which of the following best describes the function of oncogenes?

    (A) They kill cancerous cells.
    (B) They function in the normal regulation of the cell cycle.
    (C) They prevent cells from multiplying.
    (D) They proofread newly synthesized DNA for errors.
    (E) They may result in cancer.

29. The binding of a molecule other than the substrate to the active site of an enzyme is involved in

    (A) non-competitive inhibition
    (B) feedback inhibition
    (C) irreversible inhibition
    (D) increasing the energy of activation
    (E) modifying the free energy change of the reaction

30. When water evaporates, what bonds are broken?

    (A) polar covalent bonds
    (B) nonpolar covalent bonds
    (C) ionic bonds
    (D) hydrogen bonds
    (E) no bonds are broken when water evaporates

31. Which of the following is true regarding phage lambda, a virus which infects bacteria?

    (A) In the lytic cycle, the bacterial host replicates viral DNA, passing it on to daughter cells during binary fission.
    (B) In the lysogenic cycle, the bacterial host replicates viral DNA, passing it on to daughter cells during binary fission.
    (C) In the lytic cycle, viral DNA is integrated into the host genome.
    (D) In the lysogenic cycle, the host bacterial cell bursts, releasing phages.
    (E) Phage lambda is able to replicate without entering a bacterial host.

32. In humans, color blindness is a sex-linked trait. The gene for color blindness is a recessive allele on the X chromosome. Affected males inherit the recessive allele from their mother. If a female who is a carrier for color blindness mates with a male who is color blind, what percentage of their female children will be color blind?

    (A)   0%
    (B)   25%
    (C)   50%
    (D)   66%
    (E)   100%

33. Which of the following was not a major constituent of the atmosphere of the primitive earth, according to the Oparin-Haldane hypothesis?

    (A) O_2
    (B) H_2
    (C) H_2O
    (D) CH_4
    (E) NH_3

34. Which of the following ecosystems has the greatest primary productivity?

    (A) savanna
    (B) tropical rainforest
    (C) tundra
    (D) desert
    (E) temperate grassland

GO ON TO THE NEXT PAGE. ➡

35. Double fertilization occurs in

    (A) mostly gymnosperms
    (B) mostly angiosperms
    (C) both gymnosperms and angiosperms
    (D) neither gymnosperms nor angiosperms
    (E) mammals

36. Which of the following would not be a result of the release of epinephrine and norepinephrine?

    (A) increased heart rate
    (B) increased blood pressure
    (C) increased alertness
    (D) increased digestive system activity
    (E) decreased urine output

37. Which of the following is responsible for the ripening of fruit?

    (A) chlorophyll A
    (B) auxin
    (C) chlorophyll B
    (D) gibberellin
    (E) ethylene

38. Which of the following organisms uses an arrangement of blood vessels called a countercurrent heat exchanger to help maintain body temperature?

    (A) fish
    (B) amphibians
    (C) reptiles
    (D) birds
    (E) humans

39. In what stage of meiosis does crossing-over occur?

    (A) prophase I
    (B) prophase II
    (C) metaphase I
    (D) anaphase II
    (E) telophase I

40. Which of the following is not found in the cell membrane?

    (A) cholesterol
    (B) phospholipid
    (C) protein
    (D) carbohydrate
    (E) nucleic acid

41. Members of which of the following groups do not lay eggs?

    (A) reptiles, placental mammals
    (B) marsupials, monotremes
    (C) monotremes, aves (birds)
    (D) amphibians (frogs), marsupials
    (E) marsupials, placental mammals

42. Food storing tissue in the developing plant embryo is called

    (A) mesoderm
    (B) endoderm
    (C) ectoderm
    (D) endosperm
    (E) megaspore

43. A person feeds a cat at the same time every day. After a few days, the cat begins to salivate at the sound of the electric can opener. This is called

    (A) classical conditioning
    (B) operant conditioning
    (C) imprinting
    (D) instinct
    (E) insight

44. If 20000 Joules of energy are stored by producers in a food chain, how much energy is available to secondary consumers?

    (A) 0 Joules
    (B) 20 Joules
    (C) 200 Joules
    (D) 2000 Joules
    (E) 20000 Joules

GO ON TO THE NEXT PAGE. ➡

45. Which of the following hormones is incorrectly matched with its function?

    (A) glucocorticoids — promote release of glucose
    (B) mineralocorticoides — promote release of water
    (C) vasopressin (antidiuretic hormone) — promotes retention of water
    (D) aldosterone — promotes sodium retention
    (E) all are correctly matched

46. Nondisjunction during which phases of meiosis may result in an extra copy of a chromosome in a sperm cell?

    (A) prophase I, prophase II
    (B) metaphase I, anaphase II
    (C) anaphase I, anaphase II
    (D) anaphase I, telophase II
    (E) metaphase I, telophase II

47. In the Krebs cycle, electrons are passed to

    (A) pyruvate
    (B) ethanol
    (C) oxygen
    (D) photosystem I
    (E) NADH and FADH$_2$

48. Cells regulate membrane fluidity by varying the amount of what substance in the cell membrane?

    (A) carbohydrate
    (B) phospholipid
    (C) protein
    (D) cholesterol
    (E) nucleic acid

49. Which sequence represents a possible pathway in the production of a secretory protein?

    (A) rough ER → secretory vesicle → ribosomes → Golgi apparatus
    (B) ribosomes → rough ER → Golgi apparatus → secretory vesicle
    (C) secretory vesicle → Golgi apparatus → ribosomes → rough ER
    (D) ribosomes → Golgi apparatus → rough ER → secretory vesicle
    (E) rough ER → ribosomes → secretory vesicle → Golgi apparatus

50. Which of the following are characteristic of both animal and plant cells?

    (A) nucleus, mitochondria, cell wall
    (B) plasma membrane, central vacuole, ER
    (C) DNA, RNA, chloroplast
    (D) centrioles, ribosomes, nucleolus
    (E) nuclear membrane, Golgi apparatus, mitochondria

51. Water is the source of electrons for which of the following?

    (A) photosystem I
    (B) photosystem II
    (C) the Calvin cycle
    (D) glycolysis
    (E) anaerobic respiration

52. Both eukaryotic and prokaryotic cells contain

    (A) a nucleus
    (B) endoplasmic reticulum
    (C) DNA
    (D) mitochondria
    (E) chloroplasts

53. The pancreas produces

    (A) amylase, lipase, glycogen, insulin
    (B) trypsin, glycogen, insulin
    (C) amylase, glycogen, insulin
    (D) amylase, lipase, glycogen, insulin
    (E) amylase, lipase, glucagon, insulin

GO ON TO THE NEXT PAGE. ➡

KAPLAN

54. Which of the following is false regarding antibodies?

(A) They are produced by cells descended from T-cells
(B) They are proteins
(C) They bind antigen
(D) There are five different classes of antibodies
(E) They consist of heavy and light chain subunits

55. Which of the following correctly represents DNA organization, from least organized to most organized?

(A) DNA → nucleosomes → chromosome → looped domain
(B) nucleosomes → DNA → chromosome → looped domain
(C) DNA → looped domain → nucleosome → chromosome
(D) nucleosome → DNA → looped domain → chromosome
(E) DNA → nucleosomes → looped domain → chromosome

56. How do the membranes of eukaryotic cells differ?

(A) Certain membranes consist of a single layer of phospholipids.
(B) In some cells, the hydrophobic tails of the phospholipids face the cytoplasm.
(C) Different proteins are found in different membranes.
(D) Some membranes do not contain phospholipids.
(E) Some membranes are not selectively permeable.

57. In a certain type of pea plant, the gene for yellow color (Y) is dominant to the gene for green color (y), and the gene for round shape (R) is dominant to the gene for wrinkled shape (r). If a homozygous dominant plant is mated with a homozygous recessive plant, what proportion of the plants in the F_2 generation will be homozygous recessive?

(A) 3/4
(B) 1/16
(C) 1/4
(D) 3/8
(E) 5/8

58. Oogenesis differs from spermatogenesis in that

(A) oogenesis involves unequal cytoplasmic division
(B) oogenesis involves one mitotic division
(C) oogonia are produced throughout a woman's lifetime
(D) spermatogenesis produces only one viable sperm cell per spermatogonium
(E) spermatogenesis involves two meiotic divisions

59. In a food chain which consists of grass → spiders → mice → snakes → hawks, the organism with the most biomass is the

(A) grass
(B) spiders
(C) mice
(D) snakes
(E) hawks

60. Which of the following is not produced by the pituitary gland?

(A) antidiuretic hormone (ADH)
(B) prolactin
(C) human chorionic gonadotropin (hCG)
(D) follicle-stimulating hormone (FSH)
(E) luteinizing hormone (LH)

61. Which of the following does not apply to HIV, the virus which causes AIDS?

    (A) is a retrovirus
    (B) is a DNA virus
    (C) attaches to helper T cells
    (D) incorporates viral DNA into host genome
    (E) produces reverse transcriptase

62. Which of the following contains thymine?

    (A) mRNA
    (B) rRNA
    (C) tRNA
    (D) snRNA
    (E) DNA

63. Which of the following is not a condition of Hardy-Weinberg equilibrium?

    (A) no net mutations
    (B) isolation from other populations
    (C) no natural selection
    (D) small population size
    (E) random mating

64. Restriction enzymes

    (A) are only found in eukaryotes
    (B) cut DNA at every third base
    (C) recognize a specific sequence of DNA and cut at a specific location within this sequence
    (D) methylate DNA
    (E) prevent DNA from being cut

65. Which of the following correctly represents the pathway from gene to trait?

    (A) DNA → transcription → RNA → translation → protein → phenotype
    (B) DNA → translation → RNA → transcription → protein → phenotype
    (C) RNA → translation → DNA → transcription → phenotype → protein
    (D) RNA → transcription → DNA → translation → protein → phenotype
    (E) DNA → transcription → RNA → translation → phenotype → protein

66. Avascular plants which alternate between haploid and diploid forms and have flagellated sperm are members of which group?

    (A) Rhizopoda (amoebas)
    (B) Bryophyta (mosses)
    (C) Pterophyta (ferns)
    (D) Conifers (pines)
    (E) Gymnosperms (flowering plants)

67. The sea floor is in which zones?

    (A) intertidal, aphotic
    (B) intertidal, photic
    (C) neritic, aphotic
    (D) benthic, aphotic
    (E) oceanic, photic

68. If a toxin is introduced into a food chain

    (A) organisms at the top of the food chain will have the greatest concentration of toxin
    (B) organisms in the middle of the food chain will have the greatest concentration of toxin
    (C) organisms at the bottom of the food chain will have the greatest concentration of toxin
    (D) all organisms will have the same concentration of toxin
    (E) no organisms will ingest the toxin

69. Myrcorrhizae are formed on plant roots which have developed a

    (A) parasitic relationship with fungi in the soil
    (B) mutualistic relationship with fungi in the soil
    (C) commensal relationship with fungi in the soil
    (D) mutualistic relationship with bacteria in the soil
    (E) commensal relationship with bacteria in the soil

GO ON TO THE NEXT PAGE. ➡

70. Humoral immunity differs from cell mediated immunity in that

    (A) humoral immunity involves antibodies secreted by descendants of T-cells
    (B) cell mediated immunity involves antibodies secreted by descendants of B-cells
    (C) humoral immunity is involved in recognizing and destroying pathogens which have not yet entered cells
    (D) cell mediated immunity is involved in recognizing and destroying cells which have been infected by pathogens
    (E) humoral immunity is involved in recognizing and destroying cells which have been infected by pathogens

71. A type of mutation in which one nitrogenous base is exchanged for another is called a

    (A) nondisjunction
    (B) insertion
    (C) deletion
    (D) base substitution
    (E) translocation

72. Which of the following are both diploid?

    (A) sporophyte, gametophyte
    (B) sporophyte, zygote
    (C) spore, zygote
    (D) gamete, zygote
    (E) sporophyte, spore

73. Which of the following is NOT a characteristic of monocots?

    (A) one cotyledon
    (B) parallel leaf veins
    (C) fibrous root system
    (D) vascular bundles complexly arranged
    (E) floral parts in multiples of four or five

74. CAM photosynthesis differs from C_3 photosynthesis in that

    (A) in CAM photosynthesis, CO_2 is first incorporated into organic acids
    (B) CO_2 is not required in CAM photosynthesis
    (C) CO_2 is not required in C_3 photosynthesis
    (D) in C_3 photosynthesis, CO_2 is first incorporated into organic acids
    (E) CAM photosynthesis does not involve the Calvin cycle

Directions: Each group of questions below consists of five lettered headings followed by a list of numbered phrases or sentences. For each numbered phrase or sentence select the one heading that is most closely related to it and fill in the corresponding oval on the answer sheet. Each heading may be used once, more than once, or not at all in each group.

Questions 75–77

(A) bottleneck effect
(B) gradualism
(C) stabilizing selection
(D) punctuated equilibrium
(E) disruptive selection

75. Extreme phenotypes are favored

76. A small group of individuals is separated from a larger population

77. Evolutionary change occurs in spurts

Questions 78–80

(A) comparative embryology
(B) taxonomy
(C) comparative anatomy
(D) paleontology
(E) comparative physiology

78. The wing of a bat and the forelimb of a cat have similar bone structure

79. Cytochrome c has the same basic structure in all aerobic species

80. System of naming and classifying organisms

Questions 81–84

(A) Platyhelminthes
(B) Cnidaria
(C) Porifera
(D) Echinodermata
(E) Annelida

81. water vascular system

82. gastrovascular cavity

83. spongocoel

84. flame-cell excretory system

Questions 85–87

(A) oviduct
(B) ovary
(C) uterus
(D) cervix
(E) vagina

85. Contains egg cells

86. In mammals, estrogen and progesterone act on the lining of this organ

87. Location where fertilization normally occurs in humans

GO ON TO THE NEXT PAGE. ➡

Questions 88–91

    (A) amylase
    (B) pepsin
    (C) dipeptidase
    (D) lipase
    (E) bile salts

88. Hydrolyzes fat into glycerol, fatty acids, and glycerides

89. Breaks proteins into smaller polypeptides

90. Functions in the emulsification of fat

91. Hydrolyzes carbohydrates

Questions 92–95 refer to the following diagram.

The graph below shows the change in membrane potential of a neuron as it carries an impulse.

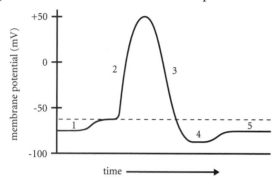

92. In what phase of the action potential are only Na$^+$ channels open?

(A) 1
(B) 2
(C) 3
(D) 4
(E) 5

93. In what phase of the action potential are K$^+$ channels open?

(A) 1
(B) 2
(C) 3
(D) 4
(E) 5

94. What phase of the action potential represents the refractory period?

(A) 1
(B) 2
(C) 3
(D) 4
(E) 5

95. The level of depolarization which must be reached before an action potential is triggered is called the

(A) membrane potential
(B) water potential
(C) graded potential
(D) equilibrium potential
(E) threshold potential

GO ON TO THE NEXT PAGE. ➡

Questions 96–99 refer to the diagrams of organic molecules below.

(A)

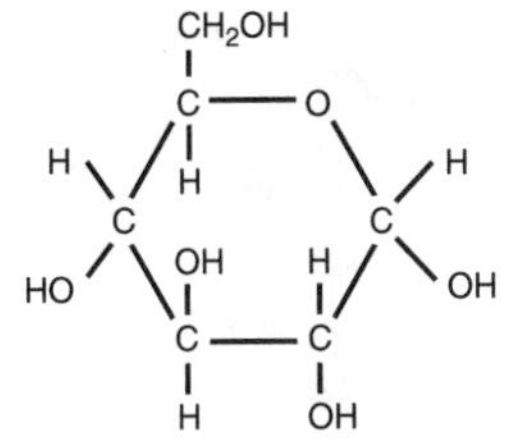

(B)

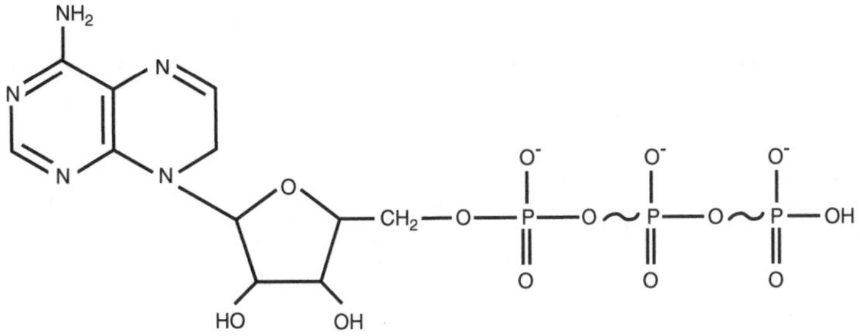

(C)

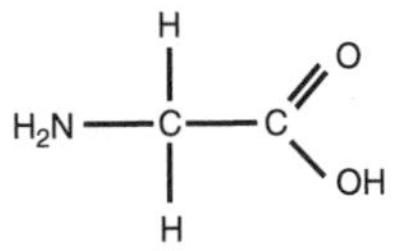

(D)

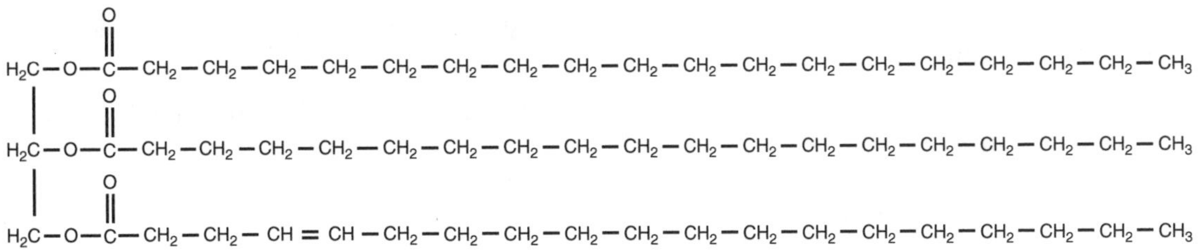

(E)

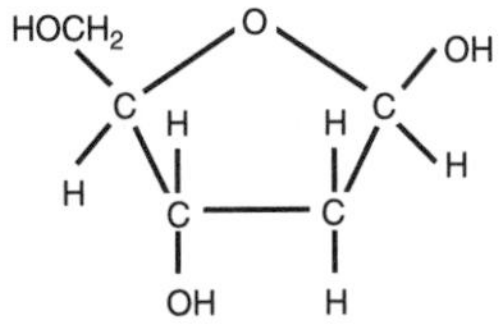

96. This molecule contains ester bonds

97. Hydrolysis of this molecule is coupled to energetically unfavorable reactions in cells

98. This molecule may be found in protein

99. Glycogen is a polymer of this molecule

Questions 100–103 refer to the diagrams of organelles below.

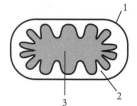 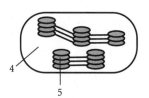

mitochondria   **chloroplast**

100. Site of the Krebs cycle

101. Hydrogen ions accumulate here, creating a proton gradient used to drive ATP synthesis in cellular respiration

102. Site of the light reactions

103. Site of the Calvin cycle

Questions 104–106 refer to the following diagram.

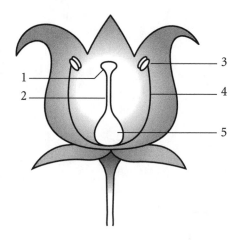

104. Structure which generates pollen grains

105. Structure where fertilization occurs

106. Structure which captures pollen grains

GO ON TO THE NEXT PAGE. ➡

Questions 107–109 refer to the information and table below.

Second Base

		U	C	A	G	
First Base (5′)	U	UUU } Phe UUC UUA } Leu UUG	UCU UCC } Ser UCA UCG	UAU } Tyr UAC UAA } Stop UAG } Stop	UGU } Cys UGC UGA } Stop UGG } Trp	U C A G
	C	CUU CUC } Leu CUA CUG	CCU CCC } Pro CCA CCG	CAU } His CAC CAA } Gln CAG	CGU CGC } Arg CGA CGG	U C A G
	A	AUU AUC } Ile AUA AUG } Start or Met	ACU ACC } Thr ACA ACG	AAU } Asn AAC AAA } Lys AAG	AGU } Ser AGC AGA } Arg AGG	U C A G
	G	GUU GUC } Val GUA GUG	GCU GCC } Ala GCA GCG	GAU } Asp GAC GAA } Glu GAG	GGU GGC } Gly GGA GGG	U C A G

Third Base (3′)

mRNA   AUG CCU A̍GU GGA UUA

protein   Met Pro Ser Gly Leu

107. If a mutation occurs and the adenine marked by an asterisk is deleted, what is the resulting amino acid sequence?

   (A) Met-Pro-Ser-Gly-Leu
   (B) Met-Pro-Ser-Leu
   (C) Met-Pro-Gly
   (D) Met-Pro-Ser-Gly
   (E) Met-Pro-Leu

GO ON TO THE NEXT PAGE. ➡

108. If a mutation occurs and the adenine marked by a dot is changed to a cytosine, what is the resulting amino acid sequence?

(A) Met-Pro-Ser-Gly-Leu
(B) Met-Pro-Gly-Leu
(C) Met-Pro
(D) Met-Pro-Gly-Gly-Leu
(E) Met-Pro-Arg-Gly-Leu

109. The type of mutation in Question 108 is known as a

(A) base substitution
(B) insertion
(C) deletion
(D) nondisjunction
(E) translocation

GO ON TO THE NEXT PAGE.

Questions 110–112 refer to the following information and graph.

In an experiment, five test tubes were prepared, each containing the same amount of starch and salivary amylase. Each of the five test tubes was maintained at a different pH. After ten minutes, the amount of disaccharide in each test tube was measured and recorded. The results are summarized in the graph below.

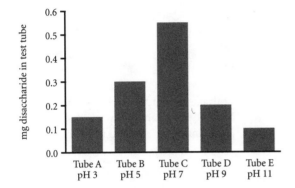

110. What is the product of the breakdown of starch by salivary amylase?

(A) maltose
(B) fructose
(C) glucose
(D) glycogen
(E) cellulose

111. The best explanation for the fact that the least amount of disaccharide was observed at pH 11 is that

(A) competitive inhibition reduced the number of enzyme molecules free to react with substrate
(B) substrate was not available to react with enzyme at pH 11
(C) the number of collisions between substrate and enzyme decreased
(D) the enzyme was denatured at pH 11
(E) the activation energy increases at higher pH

112. What is the optimal pH for salivary amylase based on this experiment?

(A) 3
(B) 5
(C) 7
(D) 9
(E) 11

Questions 113–115 refer to the information and graphs below.

The graphs below show carbon dioxide intake by two different plant species over a 24-hour period.

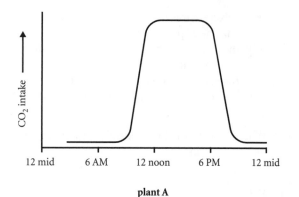

plant A

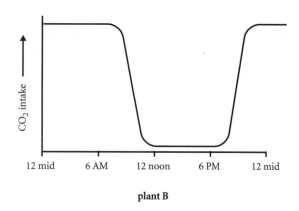

plant B

113. Which plant is most likely a CAM plant?

(A) Plant A
(B) Plant B
(C) Both plant A and plant B.
(D) Neither plant A nor plant B.
(E) It cannot be determined from the information given.

114. The turgor pressure is greater in the guard cells of which plant at 12 noon?

(A) Plant A
(B) Plant B
(C) There is no turgor pressure in the guard cells.
(D) The turgor pressure is the same in the guard cells of both plants.
(E) It cannot be determined from the information given.

115. Which plant uses the Calvin cycle to produce glucose?

(A) Plant A
(B) Plant B
(C) Both plant A and plant B.
(D) Neither plant A nor plant B.
(E) It cannot be determined from the information given.

GO ON TO THE NEXT PAGE. ➡

KAPLAN

Questions 116–118 refer to the following diagram.

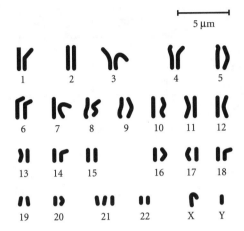

116. This picture of chromosomes which has been arranged in homologous pairs is known as a

(A) genotype
(B) phenotype
(C) karyotype
(D) blood type
(E) Punnett square

117. This individual most likely has

(A) trisomy 18
(B) cri du chat syndrome
(C) Klinefelter syndrome
(D) Turner syndrome
(E) Down syndrome

118. This individual is

(A) a male
(B) a female
(C) a metafemale
(D) a hermaphrodite
(E) It cannot be determined from the information given

Questions 119–120 refer to the graph below.

	radial symmetry	bilateral symmetry	coelomate		pseudo-coelomate	acoelomate
			protostome	deuterostome		
organism I		+				+
organism II		+		+		

119. Organism I could be classified into which phylum?

   (A) Annelida
   (B) Chordata
   (C) Nematoda
   (D) Platyhelminthes
   (E) Porifera

120. Organism II could be classified into which phylum?

   (A) Annelida
   (B) Chordata
   (C) Nematoda
   (D) Platyhelminthes
   (E) Porifera

GO ON TO THE NEXT PAGE. ➡

AP BIOLOGY
PRACTICE TEST I

Section II

Time—1 hour and 30 minutes

Answer all questions. Number your answer as the question is numbered below.

Answers must be in essay form. Outline form is NOT acceptable. Labeled diagrams may be used to supplement discussion, but in no case will a diagram alone suffice. It is important that you read each question completely before you begin to write.

1. Living organisms require a means of obtaining energy. State and describe the major reactions of the following catabolic pathways. Give an example of an organism that uses each pathway. Compare and contrast the pathways: how efficient is each pathway relative to the other?

   (A) anaerobic respiration
   (B) aerobic respiration

2. Accurate DNA replication is essential to cell division.

   (A) The details of DNA structure and function were pieced together by many scientists. Discuss the experiments of Hershey and Chase. How did the results of their experiments give evidence that the genetic material is DNA, not protein?
   (B) Draw a short segment of DNA. Show at least one example of each type of nucleotide pairing. Note the polarity of each strand. How does DNA replication proceed on the "lagging" strand?

3. Different organisms have evolved different mechanisms to deal with disposal of nitrogenous waste materials. Compare and contrast the methods of excretion employed by the following organisms. What specific waste product is produced by each organism? How does the organism deal with the toxicity of nitrogenous waste compounds?

   (A) fish
   (B) birds
   (C) mammals

4. The human digestive tract is composed of specialized structures which perform specific tasks.

   (A) Name in sequence the structures of the digestive tract. Name at least one structural modification which allows for more effective functioning of the digestive tract.
   (B) In what specific area of the digestive tract is each of the following first chemically digested? What specific enzyme performs this initial metabolism? What organ produces each enzyme, and where is it secreted?
   (I) fat
   (II) sugar
   (III) carbohydrate

PRACTICE TEST I
ANSWER KEY

Section I—Multiple-Choice

1. A	25. D	49. B	73. E	97. B
2. B	26. A	50. E	74. A	98. C
3. A	27. C	51. A	75. E	99. A
4. E	28. E	52. C	76. A	100. C
5. C	29. C	53. E	77. D	101. B
6. E	30. D	54. A	78. C	102. E
7. B	31. B	55. E	79. E	103. D
8. A	32. C	56. C	80. B	104. C
9. D	33. A	57. B	81. D	105. E
10. E	34. B	58. A	82. B	106. A
11. C	35. B	59. A	83. C	107. D
12. B	36. D	60. C	84. A	108. E
13. D	37. E	61. B	85. B	109. A
14. C	38. D	62. E	86. C	110. A
15. A	39. A	63. D	87. A	111. D
16. E	40. E	64. C	88. D	112. C
17. D	41. E	65. A	89. B	113. B
18. B	42. D	66. B	90. E	114. A
19. B	43. A	67. D	91. A	115. C
20. C	44. C	68. A	92. B	116. C
21. A	45. B	69. B	93. C	117. E
22. E	46. C	70. E	94. D	118. A
23. B	47. E	71. D	95. E	119. D
24. E	48. D	72. B	96. D	120. B

Answers and Explanations to Practice Test I

SECTION I: MULTIPLE-CHOICE

1. **(A)** If the free energy change of a reaction is less than zero, then the reaction is spontaneous. If the free energy change (ΔG) of a reaction is greater than zero, then the reaction is nonspontaneous. If the free energy change of a reaction is equal to zero, then the reaction is at equilibrium. The terms exothermic and endothermic do not refer to the free energy change of a reaction, but to the enthalpy of a reaction. If the change in enthalpy (ΔH) is positive, then heat is absorbed and the reaction is endothermic. If the change in enthalpy is negative, then heat is released and the reaction is exothermic.

2. **(B)** If the diploid number of an organism is 36, then $2n = 36$. The number of chromosomes in an egg cell would be half the diploid number, or 18.

3. **(A)** If a plant is grown in the presence of CO_2 containing ^{18}O, the ^{18}O would be found in the glucose produced through photosynthesis. The CO_2 containing ^{18}O is incorporated into glucose in the Calvin cycle. The molecular oxygen produced in photosynthesis results from the hydrolysis of water in photosystem I. Choice D is false, since the CO_2 containing ^{18}O is incorporated into glucose. Choice E is false because CO_2 is required in photosynthesis.

4. **(E)** Cilia line the human respiratory tract, helping to clear dust and debris. Euglena has a flagella, while paramecium has cilia. A human sperm cell has a flagella, and cilia line the human oviduct.

5. **(C)** If the number of nucleotide pairs added or deleted from a strand of DNA is not a multiple of three, a frameshift mutation occurs. Each group of three nucleotides constitutes one codon, and if the number of nucleotides which is added or deleted is not a multiple of three, the triplet grouping of the nucleotides changes. For example, if the original string of DNA has the sequence AATGCGCTAC, the RNA has the following code: UUACGCGAUG. The corresponding codons would be AAU GCG and CUA, which code for the following amino acids: asparagine, alanine, and leucine. If the first adenine were deleted, the new codons would be AUG CGC UAC, which code for the following amino acids: methionine, arginine, and tyrosine.

6. **(E)** Viruses consist of DNA or RNA, and a protein coat. Viruses do not have mitochondria, ribosomes, or DNA polymerase.

7. **(B)** If two species breed during different seasons, even if they are closely related, their gametes will not meet. This is a form of temporal isolation. An example of behavioral isolation is a courtship ritual which is unique to a species: members of other species do not respond to the courtship ritual of a given species. In habitat isolation, two species cannot mate because they live in different habitats. An example of habitat isolation would be two closely related species: one which lives near mountains, another which lives near rivers; the two species would rarely encounter each other. An example of mechanical isolation would be two species which could not mate because they are anatomically incompatible. An example of gametic isolation would be two species which could not mate because the sperm of one species does not recognize the egg of another species.

8. **(A)** Members of phylum Bacillariophyta (diatoms) have glasslike shells consisting of silica.

9. **(D)** In the human kidney, salt is actively transported from the loop of Henle to the medulla. Water follows passively along its osmotic gradient, from the loop of Henle (which has a low salt concentration) to the medulla (which has a high salt concentration).

10. **(E)** Nitrogen fixing bacteria convert N_2 to NH_3.

11. **(C)** The heavy and light chain subunits of an antibody are held together by disulfide bonds.

12. **(B)** Ethanol is the product of fermentation. Anaerobic organisms such as yeast break down glucose into pyruvate via glycolysis, then convert the pyruvate to ethanol. The products of the electron transport chain are ATP and H_2O. The products of the Krebs cycle are ATP, NADH and $FADH_2$. The product of the Calvin cycle is glucose. The products of photosystem I are ATP and NADPH.

13. **(D)** In active transport, molecules are moved through the cell membrane against their concentration gradient, with the help of membrane proteins. Since the molecules are moving against their concentration gradient, energy is required. In facilitated diffusion, molecules are moved through the cell membrane with their concentration gradient, with the help of membrane proteins. Since the molecules are moving with their concentration gradient, no energy is required. Facilitated diffusion transports molecules which are too large or polar to cross the cell membrane by themselves.

14. **(C)** The enzyme which synthesizes DNA is DNA polymerase. DNA polymerase requires an RNA primer. Primase synthesizes the short fragment of RNA which serves as a primer during DNA synthesis. Okazaki fragments are formed during the replication of DNA on the lagging strand. Helicase is the enzyme which unwinds the double helix of DNA at the base of the replication fork.

15. **(A)** Blood flows from the left ventricle to the aorta. The other statements have been corrected here: blood flows from arteries to arterioles, from the pulmonary vein to the left atrium, from the lung to the pulmonary vein, and from the vena cava to the right atrium.

16. **(E)** Insulin is a polypeptide hormone, while estrogen, cortisol, androgen, and progesterone are steroid hormones.

17. **(D)** An endergonic chemical reaction requires an input of energy, but may occur within the cell at normal body temperature because the reaction may be coupled to a more favorable reaction, such as the breakdown of ATP to ADP and inorganic phosphate. Very high temperatures cause denaturing of proteins and cell death. Enzymes do not lower the energy of activation of a reaction, but only increase the rate of reaction.

18. **(B)** Small nuclear RNA (snRNA), is found in the nucleus of eukaryotic cells only. Messenger RNA (mRNA), ribosomal RNA (rRNA), transfer RNA (tRNA), and DNA are all found in both prokaryotic and eukaryotic cells.

19. **(B)** Imperfect flowers are missing either stamens or carpels.

20. **(C)** According to the heterotroph hypothesis, the earliest forms of life were probably unicellular organisms which used organic molecules as their source of food.

21. **(A)** The union of male and female gametes produces a zygote. The zygote undergoes many cleavages, or divisions, forming a morula, a solid ball of cells. The next stage in embryonic development is the blastula, which is a hollow ball of cells. The blastula invaginates to form a gastrula, with three germ layers.

22. **(E)** In a food chain, decomposers break down waste products left by producers and composers. Decomposers are not represented in this food chain. Producers use energy from the sun or inorganic molecules to drive synthesis of organic molecules. In this food chain, grass is the producer. Primary consumers consume producers; in this food chain, insects eat grass so the insects are primary consumers. Secondary consumers eat primary consumers; in this food chain, mice are secondary consumers since they feed on insects. Tertiary consumers feed on secondary consumers; in this food chain, snakes are tertiary consumers since they feed on mice.

23. **(B)** Since plant cells have cell walls, plant cells cannot pinch along the middle to divide the cytoplasm, as animal cells do. Instead, cytokinesis in plants involves membrane vesicles which fuse together, forming a double-membraned cell plate. The new cell wall is formed between the two membranes of the cell plate. Cytokinesis in animal cells involves a cleavage furrow, which is formed by a contractile ring. This contractile ring is composed of microfilaments which pinch the parent cell into two cells. Formation of a mitotic spindle and regeneration of the nuclear envelope are events which occur in mitosis in both plant and animal cells.

24. **(E)** Organic compounds contain carbon atoms. Each carbon atom is able to form four covalent bonds, accounting for the wide variety of organic molecules.

25. **(D)** Members of the class Gastropoda (snails, for example) undergo torsion during early embryonic development, resulting in a mantle cavity and anus which is above the head in the adult.

26. **(A)** Grasshoppers have an open circulatory system in which fluid called hemolymph leaves vessels and directly bathes body tissues. In a closed circulatory system, blood does not exit vessels. Earthworms, fish, amphibians, and humans all have a closed circulatory system.

27. **(C)** Filtrate first enters Bowman's capsule, then the proximal convoluted tubule, the loop of Henle, the distal convoluted tubule, and then the collecting duct.

28. **(E)** Oncogenes cause normal cells to become cancerous. Tumor suppressor genes function in the normal regulation of the cell cycle by preventing cells from multiplying inappropriately. Tumor suppressor genes may signal cell death if a cell is dividing inappropriately, preventing the cells from becoming cancerous. DNA polymerase is the enzyme which proofreads newly synthesized DNA for errors.

29. **(C)** In irreversible inhibition, a molecule other than the substrate covalently bonds to the active site of the enzyme, preventing substrate from accessing the active site. In non-competitive inhibition, a molecule binds to a part of the enzyme other than the active site, causing a conformational change in the enzyme. This conformational change alters the effectiveness of the enzyme. In feedback inhibition, the presence of the end product of a metabolic pathway inhibits an enzyme which functions in an early step in the pathway. Enzymes do not modify the overall free energy change of a reaction (the difference in free energy between products and reactants). They increase the rate of a reaction by decreasing the activation energy of a reaction.

30. **(D)** Liquid water consists of H_2O molecules joined by hydrogen bonds. Within each molecule of H_2O, two hydrogen atoms are connected to one oxygen atoms via polar covalent bonds. Each hydrogen carries a slight positive charge, and the oxygen carries a slight negative charge. Liquid water is made up of many H_2O molecules; a hydrogen atom of one molecule of H_2O forms a hydrogen bond with an oxygen atom of another molecule of H_2O. When water evaporates, the hydrogen bonds between the molecules are broken.

31. **(B)** Phage lambda must enter a bacterial host to replicate. Phage lambda has two methods of reproduction, the lytic cycle and the lysogenic cycle. In the lytic cycle, the phage attaches to a host bacterial cell and injects its DNA into the bacterium. The virus utilizes the nucleotides, enzymes, and ribosomes of the host bacterium to replicate, and organizes the DNA and coat proteins into new phages. The host cell bursts, releasing the phages. In the lysogenic cycle, the phage attaches to a host bacterial cell and injects its DNA into the bacterium. The phage DNA is integrated into the genome of the bacterial host, and when the bacterium divides by binary fission, the viral DNA is replicated and passed on to daughter cells.

32. **(C)** Let us call the normal X chromosome X^+, and we'll call the X chromosome with the gene for color blindness Xc. A female who is a carrier for color blindness has a genotype of X^+X^c, while a male who is color blind has a genotype of X^cY. If we work through the Punnett square, the genotypes of their offspring are 1 X^+X^c, 1 X^cX^c, 1 X^+Y, 1 X^cY. Thus, of the female children, 1/2 are carriers of the gene for color blindness, and 1/2 are color blind. 50% of their female children will be color blind.

33. **(A)** According to the Oparin-Haldane hypothesis, the early atmosphere was an oxidizing one which consisted mainly of H_2, H_2O, CH_4, and NH_3, with very little free O_2.

34. **(B)** Of the savanna, tropical rainforest, tundra, desert, and temperate grassland ecosystems, tropical rainforest has the greatest primary productivity.

35. **(B)** Double fertilization is the union of two sperm cells with two cells of the embryo sac. It occurs in angiosperms, and one species of gymnosperm.

36. **(D)** Epinephrine and norepinephrine are released by the adrenal medulla in response to perceived danger or stress (the "fight-or-flight" response). The hormones decrease digestive system activity and urine output, while increasing heart rate, blood pressure, and alertness.

37. **(E)** Ethylene is the plant hormone responsible for the ripening of fruit. Chlorophyll A and Chlorophyll B are pigments in the chloroplasts which are involved in photosynthesis. Auxin and gibberellin are plant hormones which are involved in growth.

38. **(D)** Birds have a unique arrangement of arteries and veins called a countercurrent heat exchanger which helps reduce heat loss and maintain body temperature.

39. **(A)** In meiosis, crossing over occurs in prophase I.

40. **(E)** Nucleic acids make up DNA and RNA, which are contained in the interior of the cell. Cholesterol, phospholipid, protein, and carbohydrate are all found in the cell membrane.

41. **(E)** Marsupials and placental mammals do not lay eggs. Monotremes are the only mammals that lay eggs. Reptiles, aves (birds), and amphibians (frogs) all lay eggs.

42. **(D)** Food storing tissue in the developing plant embryo is called endosperm. Mesoderm, endoderm, and ectoderm are the three germ layers in developing animals. The megaspore is a precursor of the female gametophyte in plants.

43. **(A)** In classical conditioning, an organism learns an association between two events after repeated pairing. Since the person feeds the cat at the same time every day, the cat always hears the sound of a can opener before being fed. After a few days, the cat associated the sound of the can opener with food.

44. **(C)** Only about 10 percent of the stored energy at a given trophic level is present at the next higher trophic level. If producers store 20,000 Joules of energy, primary consumers who feed on producers store 10 percent of 20,000 Joules, or 2,000 Joules. Secondary consumers who feed on primary consumers store 10 percent of 2,000 Joules, or 200 Joules.

45. **(B)** Mineralocorticoids promote the retention of water. Aldosterone is an example of a mineralocorticoid. Aldosterone promotes the retention of sodium in the kidney, which increases water reabsorption.

46. **(C)** Nondisjunction may occur during anaphase I and anaphase II of meiosis. In anaphase I, homologous chromosomes may fail to separate, while in anaphase II, sister chromatids may fail to separate. Nondisjunction results in gametes which are missing chromosomes or have extra copies of chromosomes.

47. **(E)** In the Krebs cycle, electrons are passed to NADH and $FADH_2$. Electrons are not passed to pyruvate in either cellular respiration or photosynthesis. Electrons are passed to ethanol in alcohol fermentation. Electrons are passed to oxygen in the final step of aerobic respiration, the electron transport chain. In photosynthesis, electrons are passed to photosystem I from photosystem II.

48. **(D)** The amount of cholesterol found in cell membranes is related to membrane fluidity. Cholesterol molecules fill in the spaces between phospholipid molecules, stiffening the cell membrane. Fewer cholesterol molecules in a membrane results in a more fluid cell membrane. Cholesterol also helps maintain the fluidity of cell membranes in freezing temperatures.

49. **(B)** A protein that is to be secreted is synthesized by ribosomes, which are located on the rough ER. Next the protein enters the Golgi apparatus where it is processed and packaged. From the Golgi apparatus, the protein is packaged into a secretory vesicle. When the secretory vesicle fuses with the cell membrane, the protein is released outside the cell.

50. **(E)** Both animal and plant cells have nuclear membranes, Golgi apparatus, and mitochondria. In choice A, both animal and plant cells have a nucleus and mitochondria, but only plant cells have a cell wall. In choice B, both animal and plant cells have a plasma membrane and ER, but only plant cells have a central vacuole. In choice C, both animal and plant cells have DNA and RNA, but only plant cells have chloroplasts. In choice D, both plant and animal cells have ribosomes and a nucleolus, but only animal cells have centrioles.

51. **(A)** Water is the source of electrons for photosystem II. In photosystem II, water is split to form hydrogen ions, oxygen, and electrons. The splitting of water replaces electrons which exit photosystem II when activated by sunlight. The electrons pass through an electron transport chain, producing ATP, and enter photosystem I. Electrons which enter photosystem I enter another electron transport chain, producing ATP, then these electrons are passed to $NADP^+$, reducing it to NADPH. This NADPH is the source of electrons for the Calvin cycle. Oxygen is not required for glycolysis (the breakdown of glucose into pyruvate), or anaerobic respiration (which takes place in the absence of oxygen).

52. **(C)** Both prokaryotic and eukaryotic cells contain DNA as their genetic material. Eukaryotic cells contain membrane-bound organelles, so they contain a nucleus, endoplasmic reticulum, and mitochondria. Plant cells, which are eukaryotic, may contain chloroplasts. Prokaryotic cells do not contain membrane-bound organelles, so they do not have a nucleus, endoplasmic reticulum, mitochondria, or chloroplasts.

53. **(E)** The pancreas produces amylase, lipase, trypsin, glucagon, and insulin. Glycogen is a polymer of glucose. The liver and muscles store glucose in the form of glycogen.

54. **(A)** Antibodies are produced by plasma cells which are descendants of B-cells. The other statements are true: antibodies are proteins which consist of heavy and light chain subunits. There are five different classes of antibodies: IgG, IgM, IgA, IgD, and IgE. Antibodies bind antigen at the antigen binding site.

55. **(E)** DNA is first wrapped around histone proteins, forming nucleosomes. Nucleosomes are organized into chromatin fibers which form looped domains. Following the formation of looped domains, DNA is folded even more, forming the densely packed chromosome.

56. **(C)** Different membrane proteins are found in the various membranes of eukaryotic cells. Membranes consist of a double layer of phospholipids. Phospholipids are polar molecules, with a hydrophilic head and a hydrophobic tail. The hydrophobic tails of each layer face each other, reducing their interaction with water. The hydrophilic heads of the phospholipid molecules face the cytoplasm of the cell. Membranes are selectively permeable, regulating the molecules which pass though them.

57. **(B)** In the F_1 generation, a homozygous dominant plant (YYRR) is mated with a homozygous recessive plant (yyrr). Their offspring will all be heterozygous for both traits (YyRr). In the F_2 generation, a plant which is heterozygous for both traits (YyRr) is mated with another plant which is heterozygous for both traits (YyRr). If we work through the Punnett square, the genotypes of the offspring are 1 YYRR, 2 YYRr, 2 YyRR, 1 YYrr, 4 YyRr, 2 Yyrr, 1 yyRR, 2 yyRr, 1 yyrr. Thus 1/16 of the plants in the F_2 generation will be yyrr or homozygous recessive.

58. **(A)** Oogenesis involves unequal cytoplasmic division, producing one large ovum and three polar bodies which degenerate. Both oogenesis and spermatogenesis involve one meiotic division. Oogonia are not produced throughout a woman's lifetime; at birth, a female has all the oogonia she will ever have. In contrast, spermatogenesis continues throughout the reproductive life of a male. Spermatogenesis results in four sperm cells per spermatogonium.

59. **(A)** Organisms at the top of the food chain have the least biomass, while organisms at the bottom have the greatest biomass. In this food chain, grass is at the bottom and has the greatest biomass.

60. **(C)** Human chorionic gonadotropin (hCG) is secreted by an embryo and maintains the supportive environment of the ovary. The other hormones are all secreted by the pituitary gland: antidiuretic hormone (ADH) functions to regulate absorption of water in the kidney, prolactin stimulates production and secretion of milk, follicle-stimulating hormone (FSH) stimulates the ovarian follicle in females and spermatogenesis in males, while luteinizing hormone (LH) stimulates the corpus luteum and ovulation in females and interstitial cells in males.

61. **(B)** HIV, the virus which causes AIDS, is an RNA virus. The other statements about HIV are true: HIV is a retrovirus which produces reverse transcriptase, attaches to helper T cells, and incorporates viral DNA into the host genome (in provirus form).

62. **(E)** DNA contains thymine, while all RNA contains uracil instead of thymine.

63. **(D)** The conditions of Hardy-Weinberg equilibrium are: very large population size, no net mutations, no natural selection, isolation from other populations, and random mating.

64. **(C)** Restriction enzymes are found only in prokaryotic cells (bacteria). Restriction enzymes exist to protect bacteria from foreign DNA. Foreign DNA that enters the bacterial cell is cut, a process known as restriction. Restriction enzymes recognize a specific sequence of DNA and cut at a specific location within this sequence. Bacteria prevent their own DNA from being cut by methylating certain bases using other enzymes.

65. **(A)** Genes are sections of DNA that are transcribed to RNA. The RNA transcripts are then translated to protein, which results in an observable phenotype.

66. **(B)** Bryophytes (mosses) are avascular plants which alternate between haploid and diploid forms and have flagellated sperm.

67. **(D)** The sea floor is in the benthic zone and the aphotic zone. The oceanic zone consists of deep water past the continental shelves. The intertidal zone is a shallow area where land meets water. The neritic zone is the water past the intertidal zone over the continental shelves. The photic zone is the zone of water that light penetrates. The aphotic zone is the zone of water that light does not penetrate.

68. **(A)** If a toxin is introduced into a food chain, organisms at the top of the food chain will have the greatest concentration of toxin, since they are feeding on consumers that have ingested toxin from producers.

69. **(B)** Myrcorrhizae are formed on plant roots which have developed a mutualistic relationship with fungi in the soil. The plant provides the fungus with nutrition, and the fungus provides the plant with phosphates, so both organisms benefit. In a parasitic relationship, one organism benefits,

but another organism is harmed. In a commensal relationship, one organism benefits while the other organism neither benefits nor is harmed.

70. **(E)** Humoral immunity is involved in recognizing and destroying cells which have been infected by pathogens. Cell mediated immunity is involved in recognizing and destroying pathogens which have not yet entered cells. Humoral immunity involves antibodies secreted by plasma cells, which are descendants of B-cells. Cell mediated immunity involves cytotoxic T-cells which destroy cells which have been infected by pathogens.

71. **(D)** In base substitution, one nitrogenous base is substituted for another; for example, a guanine is substituted for an adenine in a sequence of DNA. Nondisjunction may occur during anaphase I and anaphase II of meiosis. In anaphase I, homologous chromosomes may fail to separate, while in anaphase II, sister chromatids may fail to separate. Nondisjunction results in gametes which are missing chromosomes or have extra copies of chromosomes. Insertion is a type of point mutation in which an extra nitrogenous base is inserted into a sequence of DNA. Deletion is a type of point mutation in which a nitrogenous base is removed from a sequence of DNA. In translocation, a part of one chromosome is moved to another chromosome.

72. **(B)** The sporophyte and the zygote are both diploid. The gametophyte, spore, and gamete are all haploid.

73. **(E)** Floral parts (petals) in multiples of four or five is a characteristic of dicots. Floral parts in multiples of three is a characteristic of monocots. The other characteristics belong to monocots: one cotyledon, parallel leaf veins, a fibrous root system, and vascular bundles complexly arranged.

74. **(A)** In CAM photosynthesis, CO_2 is first incorporated into organic acids. CAM photosynthesis is an adaptation to hot, dry climates. Rather than get their CO_2 directly from openings in their leaves called stomata, CAM plants incorporate CO_2 into organic acids at night, keeping their stomata closed during the day. In this way, CAM plants are able to conserve water, since water cannot exit the leaf through closed stomata. Both CAM and C_3 photosynthesis require CO_2. The final step in both C_3 and CAM photosynthesis is the Calvin cycle, in which CO_2 is incorporated into organic molecules.

75. **(E)** In disruptive selection, individuals with more extreme phenotypes are favored.

76. **(A)** The bottleneck effect occurs when a small group of individuals is separated from a larger population. The small sample of individuals may not accurately represent the gene pool of the entire population.

77. **(D)** Punctuated equilibrium is the theory which says that evolutionary change occurs in spurts.

78. **(C)** According to comparative anatomy, organisms with similar bone structure may have evolved from a common ancestor.

79. **(E)** According to comparative physiology, organisms whose proteins are very similar may have evolved from a common ancestor.

80. **(B)** Taxonomy is a system of naming and classifying organisms.

81. **(D)** Organisms in the phylum Echinodermata, which include starfish, possess a water vascular system, which functions in gas exchange, feeding, and locomotion.

82. **(B)** Organisms in the phylum Cnidaria, which include jellyfish, hydra, sea anemones, and coral, have a central digestive compartment called the gastrovascular cavity.

83. **(C)** Sponges, phylum Porifera, have a central body cavity called the spongocoel.

84. **(A)** Flatworms, phylum Platyhelminthes, have a simple excretory system consisting of flame-cells.

85. **(B)** The ovary contains egg cells and associated follicle cells.

86. **(C)** Estrogen and progesterone act on the lining of the uterus to develop the endometrial tissue.

87. **(A)** In humans, fertilization normally occurs in the oviduct. If the embryo does not move to the uterus but instead implants in the oviduct, a tubal pregnancy results.

88. **(D)** Lipase hydrolyzes fat into glycerol, fatty acids, and glycerides.

89. **(B)** Pepsin breaks proteins into smaller polypeptides.

90. **(E)** Bile salts function to emulsify fat.

91. **(A)** Amylase hydrolyzes carbohydrates.

92. **(B)** When a neuron is stimulated, gated sodium channels open. Gated potassium channels remain closed.

93. **(C)** During the repolarization phase of the action potential, sodium channels close and potassium channels open.

94. **(D)** During this phase of the action potential, potassium channels remain open. Potassium channels are slow to close and potassium ions exit the neuron. Because positively charged ions are leaving the neuron, the charge within the neuron is more negative, and the membrane potential is lower than the resting membrane potential. Sodium channel inactivation gates remain closed and the neuron is unable to be stimulated. This phase is called the refractory period.

95. **(E)** The threshold potential is the level of depolarization which must be reached before an action potential is triggered. Membrane potential is a measure of the electrical charge across a membrane caused by a difference in ion concentration on different sides of a membrane. Water potential predicts the direction of water flow based on solute concentration. Graded potential refers to the fact that stimuli of different intensity produce voltage changes of varying intensity. The equilibrium potential of an ion is the potential at which there will be no net movement of the ion across a membrane.

96. **(D)** Triglycerides are made of three fatty acids joined to glycerol via ester bonds.

97. **(B)** Hydrolysis of ATP is coupled to energetically unfavorable reactions in cells.

98. **(C)** Proteins are made of amino acids.

99. **(A)** Glycogen is a polymer of glucose.

100. **(C)** The Krebs cycle takes place in the mitochondrial matrix.

101. **(B)** Hydrogen ions accumulate in the intermembrane space. This proton gradient is used to drive ATP synthesis via oxidative phosphorylation in the mitochondria.

102. **(E)** The light reactions occur within the thylakoid membrane of the chloroplast.

103. **(D)** The Calvin cycle takes place in the stroma of the chloroplast.

104. **(C)** Pollen grains are generated in the anther.

105. **(E)** Fertilization occurs within the ovary.

106. **(A)** The stigma is sticky and captures pollen grains.

107. **(D)** If the adenine marked by the asterisk is deleted, the mRNA would have the following code: AUG CCU AGU GGU UA. The resulting amino acid sequence would be Met-Pro-Gly since GGU also codes for Gly, but the deletion of the adenine means that one base is missing from the mRNA, so GGU is the last complete codon.

108. **(E)** If the adenine marked by the dot is changed to a cytosine, the mRNA would have the following code: AUG CCU CGU GGA UUA. The resulting amino acid sequence would be Met-Pro-Arg-Gly-Leu.

109. **(A)** In Question 108, a cytosine is substituted for an adenine. This type of mutation is called a base substitution.

110. **(A)** Starch is a polymer of glucose. Salivary amylase breaks down starch into maltose, a disaccharide consisting of two glucose subunits joined by a glycosidic linkage. Glycogen and cellulose are other polymers of glucose; fructose is a monosaccharide which is not found in starch.

111. **(D)** The change in pH disrupted the enzyme's three-dimensional structure, altering the shape of the active site. This change in the shape of the active site makes an enzyme less effective at catalyzing a reaction. Therefore, less disaccharide was observed in the test tube maintained at pH 11. The experiment did not involve a competitive inhibitor, eliminating choice A. A change in pH does not affect the availability of substrate or the number of collisions between substrate and enzyme, eliminating choices B and C. The activation energy of a reaction is not changed by an increase in pH, therefore choice E is not correct.

112. **(C)** The greatest amount of disaccharide was produced at pH 7, so based on this experiment, the optimal pH for salivary amylase is pH 7.

113. **(B)** CAM plants live in hot, dry environments. To conserve water, they keep their stomata closed during the day, preventing CO_2 from entering the leaves. At night, CAM plants open their stomata and incorporate CO_2 into organic acids. During the day, CO_2 is released from the organic acids for use in the Calvin cycle.

114. **(A)** Plant A takes in CO_2 during the day, so its stomata are open at 12 noon. Increased turgor pressure in guard cells results in the opening of stomata, so the turgor pressure is greater in the guard cells of plant A.

115. **(C)** C_3 plants (plant A) and CAM plants (plant B) both use the Calvin cycle to fix CO_2 into organic molecules (glucose).

116. **(C)** A karyotype is a picture of a set of chromosomes which has been arranged in homologous pairs. An organism's genotype is its genetic makeup, while an organism's phenotype is its appearance. A person's blood type is determined by the presence or absence of antigens on red blood cells. A Punnett square is used to predict the results of the mating of two individuals.

117. **(E)** Down syndrome, or trisomy 21, can be diagnosed from a karyotype. An individual with Down syndrome has three copies of chromosome 21. An individual with trisomy 18 has three copies of chromosome 18. An individual with cri du chat syndrome has a deletion on part of chromosome 5. An individual with Klinefelter syndrome has two X chromosomes and one Y chromosome. An individual with Turner syndrome has one X chromosome and no Y chromosome.

118. **(A)** This individual is XY and genetically male.

119. **(D)** Let's make a table with the choices we are given:

	radial symmetry	bilateral symmetry	coelomate protostome	coelomate deuterostome	pseudo-coelomate	acoelomate
Annelida		+	+			
Chordata		+		+		
Nematoda		+			+	
Platyhelminthes		+				+
Porifera	*	*	*	*	*	*

* Organisms in phylum Porifera have no true tissues and are separate from all other animal phyla.

From our chart, Organism I could be classified in phylum Platyhelminthes.

120. **(B)** From our chart above, Organism II could be classified in phylum Chordata.

SECTION II

It's important to note that essays will differ, and these sample essay responses are not necessarily the only responses that would earn credit for a particular essay question.

1. **(A)** Living organisms require a way to release the energy within food to provide ATP for cellular reactions. One way of accomplishing this is through anaerobic respiration. In anaerobic respiration (fermentation), a molecule of glucose is first converted to two molecules of pyruvate via glycolysis, yielding 2 ATP and 2 NADH. The electrons removed from glucose do not enter an electron transport chain but instead are transferred directly to pyruvate, yielding ATP via substrate-level phosphorylation. Two examples of fermentation are alcohol fermentation, performed by yeast, and lactic acid fermentation, performed by bacteria and fungi. Lactic acid

fermentation also takes place in human muscle, during anaerobic conditions when oxygen supplied by the blood is insufficient. Lactic acid buildup in the muscle results in soreness.

(B) In aerobic respiration, a molecule of glucose is first converted to two molecules of pyruvate via glycolysis, yielding 2 NADH. The pyruvate molecules enter the Krebs cycle, breaking the pyruvate into carbon dioxide. The electrons removed are transferred through an electron transport chain, with the final electron acceptor being oxygen. ATP is generated through substrate-level phosphorylation and oxidative phosphorylation. *Homo sapiens* is an example of an organism which utilizes aerobic respiration.

Aerobic respiration is the most efficient pathway, with a theoretical yield of 36 ATP per molecule of glucose, while fermentation yields just 2 ATP per molecule of glucose.

2. **(A)** Hershey and Chase's experiments involved bacteria and phages, viruses which infect bacteria. Hershey and Chase used radioactive sulfur to label phage protein. They found that viral protein did not enter the bacterial cells during infection. When they used radioactive phosphorus to label phage DNA, they found that radioactive DNA had entered the bacterial cells.

(B)

Structure of DNA

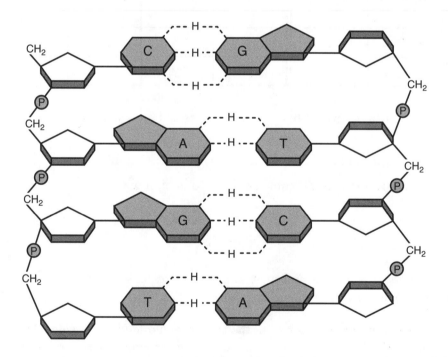

DNA polymerase is capable of synthesizing DNA in the 3′ to 5′ direction. Since the two strands of DNA run antiparallel, the strand which runs in the 3′ to 5′ direction from the origin of replication is synthesized continuously. This continuously synthesized strand is called the leading strand. The other strand runs in the 5′ to 3′ direction from the origin of replication; this strand is known as the lagging strand. The lagging strand is synthesized discontinuously, in short segments known as Okazaki fragments. Primase synthesizes a short primer sequence of RNA; DNA polymerase synthesizes short fragments of DNA, adding nucelotides to the RNA primer which are complementary to the lagging strand of DNA. After synthesizing a short fragment of DNA, the polymerase releases the lagging strand, and the process repeats itself: primase synthesizes a new RNA primer, and DNA polymerase synthesizes a fragment of DNA in the 3′ to 5′ direction. When the entire lagging strand of DNA has been duplicated, enzymes remove the RNA primers and replace them with DNA. DNA ligase connects the ends of the Okazaki fragments.

3.　(A)　Nitrogenous wastes are generated when cells use proteins for energy or convert protein to carbohydrate or fat. Fish dispose of their nitrogenous wastes in the form of ammonia, a small molecule which is easily dissolved in water. Ammonia is a very toxic substance; thus it must be transported through the body in low concentrations. The ammonia diffuses across the epithelium of the gills of the fish in saltwater fish. In freshwater fish, ammonium ion, $NH_4{}^+$, is exchanged for Na^+ in the epithelium of the gills.

　　(B)　Birds dispose of their nitrogenous waste in the form of uric acid, which is less soluble in water than ammonia and is excreted after most of the water has been removed. This is an important adaptation to land, since less water is required to dispose of uric acid.

　　(C)　Mammals excrete their nitrogenous waste in the form of urea. Urea is many times less toxic than ammonia, and therefore may be stored in greater concentrations than ammonia. The use of urea as a means of disposing nitrogenous waste is also an important adaptation to land. Because urea is less toxic than ammonia, less water is required to dispose of urea.

4.　(A)　The human digestive tract consists of the oral cavity, the pharynx, the esophagus, the stomach, the small intestine, and the large intestine. The mechanical digestion of food begins in the oral cavity. As food is ground up by the teeth, the surface area of the food is increased, allowing digestive enzymes to work more efficiently. The small intestine is the area of the digestive tract where most absorption takes place. The epithelial tissue which lines the small intestine is folded into fingerlike projections called villi which greatly increase the available surface area for digestion. The epithelium which form the villi also is folded into fingerlike projections, further increasing the available surface area for absorption.

　　(B)　The chemical digestion of fat begins in the small intestine. When chyme enters the small intestine, the release of cholecystokinin is triggered. Bile is secreted by the pancreas in response to cholecystokinin. Bile is composed of bile salts, cholesterol, and bile pigments. Bile salt molecules have one region which binds to fat, and another region which binds to water. This property allows bile salts to emulsify fats and contain them in packets known as micelles. Lipase then breaks down fat into fatty acids and glycerol.

　　The chemical digestion of carbohydrates begins in the mouth. Salivary amylase secreted by the salivary glands is present in the saliva. When starch is consumed, salivary amylase breaks down the starch into maltose, a disaccharide. Other enzymes break maltose into monosaccharides in the small intestine.

　　The chemical digestion of protein begins in the stomach. Pepsin breaks down protein into smaller polypeptide chains. Other enzymes then break down the polpeptide chains into amino acids; this takes place in the small intestine.

AP BIOLOGY

Three hours are allotted for this examination: 1 hour and 30 minutes for Section I, which consists of multiple-choice questions; and 1 hour and 30 minutes for Section II, which consists of essay questions. Section I is printed in this examination booklet; Section II, in a separate booklet.

SECTION I

Time—1 hour and 30 minutes

Number of Questions: 120

Percent of total grade: 60

Section I of this examination contains 120 multiple-choice questions, followed by 15 multiple-choice questions regarding your preparation for this exam. Please be careful to fill in only the ovals that are preceded by numbers 1 through 135 on your answer sheet

General Instructions

DO NOT OPEN THIS BOOKLET UNTIL YOU ARE INSTRUCTED TO DO SO.

INDICATE ALL YOUR ANSWERS TO QUESTIONS IN SECTION I ON THE SEPARATE ANSWER SHEET ENCLOSED. No credit will be given for anything written in this examination booklet, but you may use the booklet for notes or scratchwork. After you have decided which of the suggested answers is best, COMPLETELY fill in the corresponding oval on the answer sheet. Give only one answer to each question. If you change an answer, be sure that the previous mark is erased completely.

<u>Example:</u>   <u>Sample Answer</u>

Chicago is a

(A) state
(B) city
(C) country
(D) continent
(E) village

Many candidates wonder whether or not to guess the answers to questions about which they are not certain. In this section of the examination, as a correction for haphazard guessing, one-fourth of the number of questions you answer incorrectly will be subtracted from the number of questions you answer correctly. It is improbable, therefore, that mere guessing will improve your score significantly; it may even lower your score, and it does take time. If, however, you are not sure of the correct answer but have some knowledge of the question and are able to eliminate one or more of the answer choices as wrong, your chance of getting the right answer is improved, and it may be to your advantage to answer such a question.

Use your time effectively, working as rapidly as you can without losing accuracy. Do not spend too much time on questions that are too difficult. Go on to other questions and come back to the difficult ones later if you have time. It is not expected that everyone will be able to answer all the multiple-choice questions.

Copyright © 1994 by Educational Testing Service. All rights reserved. Princeton, NJ 08541.

PRACTICE TEST II
ANSWER SHEET

1 Ⓐ Ⓑ Ⓒ Ⓓ Ⓔ   25 Ⓐ Ⓑ Ⓒ Ⓓ Ⓔ   49 Ⓐ Ⓑ Ⓒ Ⓓ Ⓔ   73 Ⓐ Ⓑ Ⓒ Ⓓ Ⓔ   97 Ⓐ Ⓑ Ⓒ Ⓓ Ⓔ

2 Ⓐ Ⓑ Ⓒ Ⓓ Ⓔ   26 Ⓐ Ⓑ Ⓒ Ⓓ Ⓔ   50 Ⓐ Ⓑ Ⓒ Ⓓ Ⓔ   74 Ⓐ Ⓑ Ⓒ Ⓓ Ⓔ   98 Ⓐ Ⓑ Ⓒ Ⓓ Ⓔ

3 Ⓐ Ⓑ Ⓒ Ⓓ Ⓔ   27 Ⓐ Ⓑ Ⓒ Ⓓ Ⓔ   51 Ⓐ Ⓑ Ⓒ Ⓓ Ⓔ   75 Ⓐ Ⓑ Ⓒ Ⓓ Ⓔ   99 Ⓐ Ⓑ Ⓒ Ⓓ Ⓔ

4 Ⓐ Ⓑ Ⓒ Ⓓ Ⓔ   28 Ⓐ Ⓑ Ⓒ Ⓓ Ⓔ   52 Ⓐ Ⓑ Ⓒ Ⓓ Ⓔ   76 Ⓐ Ⓑ Ⓒ Ⓓ Ⓔ   100 Ⓐ Ⓑ Ⓒ Ⓓ Ⓔ

5 Ⓐ Ⓑ Ⓒ Ⓓ Ⓔ   29 Ⓐ Ⓑ Ⓒ Ⓓ Ⓔ   53 Ⓐ Ⓑ Ⓒ Ⓓ Ⓔ   77 Ⓐ Ⓑ Ⓒ Ⓓ Ⓔ   101 Ⓐ Ⓑ Ⓒ Ⓓ Ⓔ

6 Ⓐ Ⓑ Ⓒ Ⓓ Ⓔ   30 Ⓐ Ⓑ Ⓒ Ⓓ Ⓔ   54 Ⓐ Ⓑ Ⓒ Ⓓ Ⓔ   78 Ⓐ Ⓑ Ⓒ Ⓓ Ⓔ   102 Ⓐ Ⓑ Ⓒ Ⓓ Ⓔ

7 Ⓐ Ⓑ Ⓒ Ⓓ Ⓔ   31 Ⓐ Ⓑ Ⓒ Ⓓ Ⓔ   55 Ⓐ Ⓑ Ⓒ Ⓓ Ⓔ   79 Ⓐ Ⓑ Ⓒ Ⓓ Ⓔ   103 Ⓐ Ⓑ Ⓒ Ⓓ Ⓔ

8 Ⓐ Ⓑ Ⓒ Ⓓ Ⓔ   32 Ⓐ Ⓑ Ⓒ Ⓓ Ⓔ   56 Ⓐ Ⓑ Ⓒ Ⓓ Ⓔ   80 Ⓐ Ⓑ Ⓒ Ⓓ Ⓔ   104 Ⓐ Ⓑ Ⓒ Ⓓ Ⓔ

9 Ⓐ Ⓑ Ⓒ Ⓓ Ⓔ   33 Ⓐ Ⓑ Ⓒ Ⓓ Ⓔ   57 Ⓐ Ⓑ Ⓒ Ⓓ Ⓔ   81 Ⓐ Ⓑ Ⓒ Ⓓ Ⓔ   105 Ⓐ Ⓑ Ⓒ Ⓓ Ⓔ

10 Ⓐ Ⓑ Ⓒ Ⓓ Ⓔ   34 Ⓐ Ⓑ Ⓒ Ⓓ Ⓔ   58 Ⓐ Ⓑ Ⓒ Ⓓ Ⓔ   82 Ⓐ Ⓑ Ⓒ Ⓓ Ⓔ   106 Ⓐ Ⓑ Ⓒ Ⓓ Ⓔ

11 Ⓐ Ⓑ Ⓒ Ⓓ Ⓔ   35 Ⓐ Ⓑ Ⓒ Ⓓ Ⓔ   59 Ⓐ Ⓑ Ⓒ Ⓓ Ⓔ   83 Ⓐ Ⓑ Ⓒ Ⓓ Ⓔ   107 Ⓐ Ⓑ Ⓒ Ⓓ Ⓔ

12 Ⓐ Ⓑ Ⓒ Ⓓ Ⓔ   36 Ⓐ Ⓑ Ⓒ Ⓓ Ⓔ   60 Ⓐ Ⓑ Ⓒ Ⓓ Ⓔ   84 Ⓐ Ⓑ Ⓒ Ⓓ Ⓔ   108 Ⓐ Ⓑ Ⓒ Ⓓ Ⓔ

13 Ⓐ Ⓑ Ⓒ Ⓓ Ⓔ   37 Ⓐ Ⓑ Ⓒ Ⓓ Ⓔ   61 Ⓐ Ⓑ Ⓒ Ⓓ Ⓔ   85 Ⓐ Ⓑ Ⓒ Ⓓ Ⓔ   109 Ⓐ Ⓑ Ⓒ Ⓓ Ⓔ

14 Ⓐ Ⓑ Ⓒ Ⓓ Ⓔ   38 Ⓐ Ⓑ Ⓒ Ⓓ Ⓔ   62 Ⓐ Ⓑ Ⓒ Ⓓ Ⓔ   86 Ⓐ Ⓑ Ⓒ Ⓓ Ⓔ   110 Ⓐ Ⓑ Ⓒ Ⓓ Ⓔ

15 Ⓐ Ⓑ Ⓒ Ⓓ Ⓔ   39 Ⓐ Ⓑ Ⓒ Ⓓ Ⓔ   63 Ⓐ Ⓑ Ⓒ Ⓓ Ⓔ   87 Ⓐ Ⓑ Ⓒ Ⓓ Ⓔ   111 Ⓐ Ⓑ Ⓒ Ⓓ Ⓔ

16 Ⓐ Ⓑ Ⓒ Ⓓ Ⓔ   40 Ⓐ Ⓑ Ⓒ Ⓓ Ⓔ   64 Ⓐ Ⓑ Ⓒ Ⓓ Ⓔ   88 Ⓐ Ⓑ Ⓒ Ⓓ Ⓔ   112 Ⓐ Ⓑ Ⓒ Ⓓ Ⓔ

17 Ⓐ Ⓑ Ⓒ Ⓓ Ⓔ   41 Ⓐ Ⓑ Ⓒ Ⓓ Ⓔ   65 Ⓐ Ⓑ Ⓒ Ⓓ Ⓔ   89 Ⓐ Ⓑ Ⓒ Ⓓ Ⓔ   113 Ⓐ Ⓑ Ⓒ Ⓓ Ⓔ

18 Ⓐ Ⓑ Ⓒ Ⓓ Ⓔ   42 Ⓐ Ⓑ Ⓒ Ⓓ Ⓔ   66 Ⓐ Ⓑ Ⓒ Ⓓ Ⓔ   90 Ⓐ Ⓑ Ⓒ Ⓓ Ⓔ   114 Ⓐ Ⓑ Ⓒ Ⓓ Ⓔ

19 Ⓐ Ⓑ Ⓒ Ⓓ Ⓔ   43 Ⓐ Ⓑ Ⓒ Ⓓ Ⓔ   67 Ⓐ Ⓑ Ⓒ Ⓓ Ⓔ   91 Ⓐ Ⓑ Ⓒ Ⓓ Ⓔ   115 Ⓐ Ⓑ Ⓒ Ⓓ Ⓔ

20 Ⓐ Ⓑ Ⓒ Ⓓ Ⓔ   44 Ⓐ Ⓑ Ⓒ Ⓓ Ⓔ   68 Ⓐ Ⓑ Ⓒ Ⓓ Ⓔ   92 Ⓐ Ⓑ Ⓒ Ⓓ Ⓔ   116 Ⓐ Ⓑ Ⓒ Ⓓ Ⓔ

21 Ⓐ Ⓑ Ⓒ Ⓓ Ⓔ   45 Ⓐ Ⓑ Ⓒ Ⓓ Ⓔ   69 Ⓐ Ⓑ Ⓒ Ⓓ Ⓔ   93 Ⓐ Ⓑ Ⓒ Ⓓ Ⓔ   117 Ⓐ Ⓑ Ⓒ Ⓓ Ⓔ

22 Ⓐ Ⓑ Ⓒ Ⓓ Ⓔ   46 Ⓐ Ⓑ Ⓒ Ⓓ Ⓔ   70 Ⓐ Ⓑ Ⓒ Ⓓ Ⓔ   94 Ⓐ Ⓑ Ⓒ Ⓓ Ⓔ   118 Ⓐ Ⓑ Ⓒ Ⓓ Ⓔ

23 Ⓐ Ⓑ Ⓒ Ⓓ Ⓔ   47 Ⓐ Ⓑ Ⓒ Ⓓ Ⓔ   71 Ⓐ Ⓑ Ⓒ Ⓓ Ⓔ   95 Ⓐ Ⓑ Ⓒ Ⓓ Ⓔ   119 Ⓐ Ⓑ Ⓒ Ⓓ Ⓔ

24 Ⓐ Ⓑ Ⓒ Ⓓ Ⓔ   48 Ⓐ Ⓑ Ⓒ Ⓓ Ⓔ   72 Ⓐ Ⓑ Ⓒ Ⓓ Ⓔ   96 Ⓐ Ⓑ Ⓒ Ⓓ Ⓔ   120 Ⓐ Ⓑ Ⓒ Ⓓ Ⓔ

AP BIOLOGY
PRACTICE TEST II

Section I

Time—1 hour and 30 minutes

<u>Directions:</u> Each of the questions or incomplete statements below is followed by five suggested answers or completions. Select the one that is best in each case and then fill in the corresponding oval on the answer sheet.

1. Cell plate formation is involved in

   (A) viral replication
   (B) bacterial replication
   (C) bacterial conjugation
   (D) plant cell mitosis
   (E) animal cell mitosis

2. The ectoderm gives rise to

   (A) nails, blood vessels, and epidermis
   (B) adrenal cortex and epidermis
   (C) neurons and epidermis
   (D) kidneys, blood vessels, and heart
   (E) tooth enamel, blood vessels, and epidermis

3. After a scientist swam in a pond with newly hatched ducklings separated from their mother, they followed him around as if he were their mother. This is an example of

   (A) discrimination
   (B) response to pheromones
   (C) imprinting
   (D) instrumental conditioning
   (E) classical conditioning

4. The best description of identical twins is

   (A) twins of the same sex
   (B) twins from a single egg
   (C) twins from two eggs that have been fertilized by the same sperm
   (D) twins from two eggs fertilized by two separate sperm
   (E) twins from a single egg fertilized by two separate sperm

5. In DNA, the amount of thymine is equal to

   (A) the amount of uracil
   (B) the amount of adenine
   (C) the amount of cytosine
   (D) the amount of guanine
   (E) the amount of guanine plus the amount of cytosine

6. Which of the following is not an organic substance?

   (A) water
   (B) glucose
   (C) phospholipid
   (D) DNA
   (E) amylase

7. Which of the following is true regarding enzymes?

   (A) enzyme activity is not affected by changes in pH
   (B) enzymes increase the activation energy
   (C) enzymes change the free energy of the products
   (D) enzyme activity is not affected by changes in temperature
   (E) enzymes form temporary enzyme-substrate complexes

8. Cystic fibrosis is caused by a mutation in a gene resulting in the incorrect folding of a transport protein. In one form of gene therapy for cystic fibrosis, the gene producing the correctly folded transport protein is inserted into a cold virus which has been modified to remove illness-causing genes. Patients with cystic fibrosis are given an inhalant containing the genetically engineered cold viruses. It is hoped that the gene will become integrated into some cells lining the lung, helping to ease the symptoms of cystic fibrosis by providing the genetic instructions for the correctly folded transport protein.

   Which of the following is false regarding this example of genetic engineering?

   (A) Cold viruses have DNA as their genetic material.
   (B) DNA ligase is used to anneal the "sticky ends" of the cold virus DNA and the cystic fibrosis gene DNA.
   (C) The cold virus is the vector by which the correct cystic fibrosis gene enters lung cells.
   (D) The cold virus DNA and the DNA containing the cystic fibrosis gene were cut with different restriction enzymes.
   (E) A cold virus was used to target lung tissue when delivering the correct cystic fibrosis gene.

9. Which cells have cilia?

   (A) cells lining the human respiratory tract
   (B) *Euglena*
   (C) cells lining the human urinary tract
   (D) human sperm cells
   (E) plant cells

10. Which of the following fixes atmospheric N_2?

    (A) lactic acid fermentation
    (B) bacteria on the roots of legumes
    (C) UV light
    (D) photosynthesis
    (E) alcohol fermentation

11. Which statement about the *lac* operon is NOT correct?

    (A) There are 3 structural genes that code for functional proteins.
    (B) There is a gene that codes for a repressor protein.
    (C) The promoter is the binding site of RNA polymerase.
    (D) The repressor protein binds to the operator, stopping gene expression.
    (E) The lac operon is found in mature red blood cells.

12. In a certain pea plant, green pea color is dominant to yellow pea color. A farmer has a field of pea plants. In this field, 34% of the peas are yellow, and 64% of the peas are green. He randomly selects 6 plants and interbreeds them. In the first generation of plants from this cross, 54% of the peas are yellow, and 46% of the peas are green. This is an example of

    (A) gradualism
    (B) punctuated equilibrium
    (C) stabilizing selection
    (D) bottleneck effect
    (E) sympatric speciation

13. A large lake evaporates slowly over time, leaving many small ponds. The large lake had been inhabited by one species of fish, but over time, each small pond is inhabited by a different species of fish, each species a descendant of the species which inhabited the large lake. This is an example of

    (A) punctuated equilibrium
    (B) disruptive selection
    (C) allopatric speciation
    (D) sympatric speciation
    (E) mutation

GO ON TO THE NEXT PAGE. ➡

14. Which of the following statements is true regarding membranes?

    (A) Membranes consist only of phospholipids.
    (B) Phospholipids within the membrane are randomly arranged.
    (C) Membranes consist only of proteins.
    (D) Proteins are not found in membranes.
    (E) Different membranes have different proteins embedded within them.

15. Active transport, unlike facilitated transport

    (A) requires energy
    (B) only takes place in animal cells
    (C) moves molecules from an area of high concentration to an area of low concentration
    (D) does not require energy
    (E) does not involve membrane proteins

16. Which statement regarding retroviruses is true?

    (A) Retroviruses use double-stranded DNA as their genetic material.
    (B) Retroviruses use single-stranded DNA as their genetic material.
    (C) Retroviruses use double-stranded RNA as their genetic material.
    (D) HIV, the virus which causes AIDS, is an example of a retrovirus.
    (E) Retroviruses use reverse transcriptase from the host cell to transcribe DNA from RNA.

17. What is the principal function of $NaHCO_3$?

    (A) $NaHCO_3$ increases the amount of CO_2 in the blood
    (B) $NaHCO_3$ inactivates bile
    (C) $NaHCO_3$ acts as a buffer
    (D) $NaHCO_3$ combines with oxygen when hemoglobin is saturated
    (E) $NaHCO_3$ combines with CO_2 in alveoli

18. A certain chemical is found to denature all enzymes in the synaptic cleft. What are the effects of this chemical on acetylcholine?

    (A) None, since acetylcholine is not released from the presynaptic membrane.
    (B) None, since acetylcholine does not bind to the postsynaptic membrane.
    (C) If the chemical is present, acetylcholine is degraded before it acts on the postsynaptic membrane.
    (D) If the chemical is present, acetylcholine is not inactivated in the synaptic cleft.
    (E) None, since acetylcholine does not function in the synaptic cleft.

19. Which of the following associations of brain structure and function is false?

    (A) hypothalamus : appetite
    (B) cerebellum : motor coordination
    (C) cerebral cortex : higher intellectual function
    (D) reticular activating system : sensory processing
    (E) medulla : basic emotional drives

20 Which statement about gastrulation is false?

    (A) In the amphibian, the initial site of gastrulation is the gray crescent.
    (B) As a consequence of gastrulation the embryo now consists of two germ layers: ectoderm and endoderm.
    (C) In the amphibian, the infolding through which blastula cells migrate during gastrulation is called the blastopore.
    (D) The primitive gut that results from gastrulation is called the archenteron.
    (E) Mesoderm develops during gastrulation.

21. Digestion of cellulose by bacteria found in a termite's gut is an example of

    (A) mutualism
    (B) parasitism
    (C) saprophytism
    (D) commensalism
    (E) autotrophism

GO ON TO THE NEXT PAGE. ➡

22. Which of the following is not a marine zone?

    (A) intertidal zone
    (B) littoral zone
    (C) pelagic zone
    (D) photic zone
    (E) crosstidal zone

23. Which of the following contains thymine?

    (A) snRNA
    (B) mRNA
    (C) tRNA
    (D) rRNA
    (E) DNA

24. Which of the following best describes the function of tumor suppressor genes?

    (A) Tumor suppresor genes are not present in normal cells.
    (B) Mutations in tumor suppressor genes may lead to cancer.
    (C) Tumor suppressor genes cause uncontrolled cell growth.
    (D) Tumor suppressor genes cause mutations.
    (E) Tumor suppressor genes can be deleted from normal cells.

25. Which of the following constitutes the secondary level of protein structure?

    (A) bonding between side chains of amino acids
    (B) sequence of amino acids joined by peptide bonds
    (C) beta pleated sheet
    (D) interactions between polypeptide subunits
    (E) alpha helix

26. In the cell cycle, DNA is replicated during which phase?

    (A) prophase
    (B) metaphase
    (C) anaphase
    (D) telophase
    (E) interphase

27. Two closely related species share the same habitat, but the sperm of one species cannot fertilize the egg of another species, and vice versa. This is an example of

    (A) gametic isolation
    (B) behavioral isolation
    (C) habitat isolation
    (D) temporal isolation
    (E) mechanical isolation

28. If a person lacks gamma globulin, a likely result would be

    (A) severe allergies
    (B) low resistance to infection
    (C) diabetes
    (D) hemophilia
    (E) color-blindness

29. Rhizomes are

    (A) underground stems with buds
    (B) stems running above and along the ground
    (C) spores
    (D) woody underground stems
    (E) the result of sexual reproduction

30. Which statement about the menstrual cycle is false?

    (A) FSH causes the development of the primary follicle.
    (B) FSH and LH are both posterior pituitary secretions.
    (C) The corpus luteum develops from the remains of the post-ovulatory graafian follicle.
    (D) As a follicle develops, it produces estrogen.
    (E) LH causes the graafian follicle to undergo ovulation.

GO ON TO THE NEXT PAGE.

31. A stable ecosystem

    (A) does not require a constant energy source
    (B) does require a living system to incorporate energy into organic compounds
    (C) is not self-sustaining
    (D) does not requires cycling of materials between the living system and the environment
    (E) does not require a stable physical environment

32. The earliest forms of life were most likely

    (A) multicellular animals
    (B) multicellular plants capable of photosynthesis
    (C) unicellular organisms with mitochondria
    (D) unicellular organisms with chloroplasts
    (E) unicellular heterotrophs which consumed organic molecules

33. Which of the following is a correct statement of a condition of Hardy-Weinberg equilibrium?

    (A) nonrandom mating
    (B) natural selection favors some phenotypes over others
    (C) large population size
    (D) migration into and out of the population
    (E) mutations which alter alleles

34. A change in pH from 3 to 4 represents

    (A) a tenfold increase in hydrogen ion concentration
    (B) a tenfold decrease in hydrogen ion concentration
    (C) a 100 fold increase in hydrogen ion concentration
    (D) a 100 fold increase in hydrogen ion concentration
    (E) none of the above

35. Cholesterol is found in the cell membranes of

    (A) *E. coli*
    (B) yeast
    (C) corn
    (D) fish
    (E) all cells

36. Which cellular organelle functions in ribosome synthesis?

    (A) the endoplasmic reticulum
    (B) the nucleolus
    (C) the nucleus
    (D) Golgi bodies
    (E) lysosomes

37. Which of the following statements is true regarding cellular respiration?

    (A) Glycolysis produces acetyl coenzyme A which enters the Krebs cycle.
    (B) The majority of ATP production results from glycolysis.
    (C) Pyruvate is converted to lactate.
    (D) NADH and $FADH_2$ function as electron carriers.
    (E) The products of cellular respiration are ATP and $C_6H_{12}O_6$.

38. Crossing over of homologous chromosomes during meiosis is important because

    (A) it decreases the genetic diversity of a population
    (B) it increases the genetic diversity of a population
    (C) it maintains the genetic diversity of a population
    (D) it insures that the next generation of organisms is identical to the previous generation of organisms
    (E) it has no effect on the next generation of organisms

GO ON TO THE NEXT PAGE. ➡

39. If a person with blood type AB mates with a person with blood type O, what are the possible blood types of their children?

    (A) AB, O
    (B) A, B, O
    (C) A, B
    (D) A, B, AB, O
    (E) O only

40. Which of the following is true regarding gene transcription and translation?

    (A) Gene transcription in prokaryotes involves intron splicing.
    (B) RNA polymerase "proofreads" RNA as it is synthesized.
    (C) Transcription in eukaryotes takes place in the cytoplasm.
    (D) A polyribosome translates several strands of RNA from one DNA strand.
    (E) A separate aminoacyl-tRNA synthetase exists for each amino acid.

41. In oogenesis, unlike spermatogenesis

    (A) meiosis is completed before fertilization
    (B) each oogonium produces four eggs
    (C) cytokinesis is unequal, producing one large ovum and three polar bodies
    (D) egg production is not hormonally regulated
    (E) oogonia are produced throughout a woman's lifetime

42. Which of the following best describes a virus?

    (A) A virus contains DNA as its genetic material, mitochondria, ribosomes, and a protein coat.
    (B) A virus contains RNA as its genetic material, mitochondria, ribosomes, and a protein coat.
    (C) A virus contains DNA as its genetic material, and a protein coat.
    (D) A virus contains RNA as its genetic material, and a protein coat.
    (E) A virus contains DNA or RNA as its genetic material, and a protein coat.

43. In *Drosophila,* normal wings are dominant to curly wings. Heterozygotes have normal wings. Pure breeding normal wing males are mated with pure breeding curly wing females; their offspring are then mated. What is the phenotypic ratio in the F_2 generation?

    (A) all offspring have curly wings
    (B) 1 normal wing offspring : 3 curly wing offspring
    (C) 1 normal wing offspring : 1 curly wing offspring
    (D) 3 normal wing offspring : 1 curly wing offspring
    (E) all offspring have normal wings

44. Energetically unfavorable reactions in cells

    (A) are coupled to protein hydrolysis
    (B) are coupled to carbohydrate hydrolysis
    (C) are coupled to adenosine triphosphate hydrolysis
    (D) are catalyzed by enzymes
    (E) cannot occur

45. When there is insufficient oxygen reaching human muscles

    (A) they convert pyruvate to ethanol
    (B) they convert pyruvate to lactate
    (C) they revert to the Calvin cycle to produce carbohydrate
    (D) they use the electron transport chain
    (E) they cannot function until aerobic conditions are restored

46. After meiosis, the number of chromosomes in a sperm cell is 6. What is the diploid number of the organism?

    (A) 3
    (B) 6
    (C) 9
    (D) 12
    (E) 18

GO ON TO THE NEXT PAGE. ➡

47. Which of the following are characteristic of prokaryotic cells only?

(A) DNA, cell membrane, mitochondria
(B) chloroplasts, cytoplasm, centrioles
(C) DNA, cytoplasm, cell membrane
(D) cell membrane, ribosomes, centrioles
(E) cell membrane, centrioles, mitochondria

48. Which of the following would be a dominant species in a climax community?

(A) mosses
(B) sun-loving trees
(C) shrubs
(D) annual grasses
(E) shade trees

49. The aquatic environment necessary for embryonic development in amphibians is replaced in reptiles by

(A) humid atmospheric conditions
(B) intrauterine development
(C) amniotic fluid
(D) use of lungs instead of gills
(E) shells that prevent against gas escaping

50. Which experiment did Oparin and Haldane perform to simulate conditions of the early Earth?

(A) They applied electric charge to a mixture of hydrogen gas, water, methane, and ozone.
(B) They applied electric charge to a mixture of hydrogen gas, water, methane, and ammonia.
(C) They added DNA to a mixture of hydrogen gas, water, methane, and ozone.
(D) They added RNA to a mixture of hydrogen gas, water, methane, and ozone
(E) They added RNA to a mixture of hydrogen gas, water, methane, and ammonia.

51. In aerobic respiration, ATP is produced via substrate-level phosphorylation in

(A) glycolysis only
(B) the Krebs cycle only
(C) the Calvin cycle only
(D) glycolysis and the Krebs cycle
(E) glycolysis and the Calvin cycle

52. Which of the following best describes the lytic cycle of lambda phage, a virus which infects bacteria?

(A) Lambda phage is an RNA virus and must use reverse transcriptase to transcribe viral RNA to DNA before entering the lytic cycle.
(B) In the lytic cycle, the phage injects its DNA into the bacterial host and uses host cell machinery to replicate viral DNA and produce viral proteins.
(C) In the lytic cycle, the viral DNA is replicated when the host bacteria replicates its DNA.
(D) Lambda phage does not enter bacterial cells in the lytic cycle.
(E) In the lytic cycle, lambda phage integrates its DNA into the bacterial host genome; once integrated into the host genome, the virus is known as a prophage.

53. Living in a close nutritional relationship with another organism where one organism benefits while the other is harmed is best defined as

(A) symbiosis
(B) mutualism
(C) saprophytism
(D) commensalism
(E) parasitism

54. Which organism is a producer?

(A) snakes
(B) owls
(C) fruits
(D) squirrels
(E) deer

GO ON TO THE NEXT PAGE. ➡

55. Which statement about the respiratory system is NOT true?

    (A) Ciliated nasal membranes warm, moisten, and filter inspired air.
    (B) Contraction of the diaphragm enlarges the thoracic cavity.
    (C) When the thoracic cavity enlarges, the pressure of air within the lungs falls.
    (D) When the pressure of air within the lungs is less than the atmospheric pressure, air will flow out of the lungs.
    (E) The respiratory process consists of inspiratory and expiratory acts following one another.

56. Prokaryotes are found in which kingdom?

    (A) Monera
    (B) Protista
    (C) Fungi
    (D) Plantae
    (E) Animalia

57. If ^{32}P is used to label both strands of double-stranded DNA, after two rounds of replication, four double-stranded molecules of DNA are produced. Of these,

    (A) one of the four double-stranded molecules of DNA will contain all the ^{32}P
    (B) two of the four double-stranded molecules of DNA will contain ^{32}P, each with half of the ^{32}P
    (C) three of the four double-stranded molecules of DNA will contain ^{32}P, each with 1/3 of the ^{32}P
    (D) all four double-stranded molecules of DNA will contain ^{32}P, each with 1/4 of the ^{32}P
    (E) none of the double-stranded molecules of DNA will contain ^{32}P

58. Which of the following organelles is thought to have originally been a free-living prokaryotic organism?

    (A) Golgi apparatus
    (B) endoplasmic reticulum
    (C) lysosomes
    (D) chloroplasts
    (E) centrioles

59. Plant cells, unlike animal cells, do not contain

    (A) cell walls
    (B) centrioles
    (C) mitochondria
    (D) a nucleus
    (E) chloroplasts

60. Which of the following hormones raises the concentration of blood calcium?

    (A) glucagon
    (B) insulin
    (C) parathyroid hormone
    (D) aldosterone
    (E) anti-diuretic hormone

61. Denitrifying bacteria

    (A) turn ammonia into NO_2
    (B) turn ammonia into N_2
    (C) turn ammonia into NO_3
    (D) do not use nitrogen in their life cycle
    (E) turn N_2 into ammonia

62. Organisms in which group lack true tissues?

    (A) Insecta
    (B) Pisces
    (C) Porifera
    (D) Arachnida
    (E) Mammalia

63. Consider the species *H. sapiens* and *H. erectus*. Which of the following statements regarding the two organisms is true?

    (A) *H. sapiens* and *H. erectus* are in the same species
    (B) *H. sapiens* and *H. erectus* are not in the same species
    (C) *H. sapiens* and *H. erectus* are not in the same phylum
    (D) *H. sapiens* and *H. erectus* are able to mate
    (E) *H. sapiens* and *H. erectus* share no evolutionary relationship

GO ON TO THE NEXT PAGE. ➡

64. It is believed that early life forms differed from modern life forms in that

    (A) early life forms used proteins both as enzymes and as genetic material
    (B) modern life forms use proteins as both enzymes and as genetic material
    (C) early life forms used proteins as genetic material while modern life forms use DNA as genetic material
    (D) early life forms used RNA both as enzymes and as genetic material
    (E) early life forms used DNA both as enzymes and as genetic material

65. Which of the following would not be involved in the production of a secretory protein?

    (A) rough endoplasmic reticulum
    (B) Golgi apparatus
    (C) ribosomes
    (D) secretory vesicles
    (E) smooth endoplasmic reticulum

66. A person with blood type AB

    (A) can donate blood to people with any blood type
    (B) can donate blood to people with blood type AB only
    (C) can receive blood of type AB only
    (D) can donate blood of type O
    (E) can donate blood to people of blood type A, B, and AB

67. In contrast to the environment today, the prebiotic environment on Earth

    (A) contained a great deal of O_2
    (B) contained a great deal of ozone
    (C) had an oxidizing atmosphere
    (D) had a reducing atmosphere
    (E) contained no carbon

68. A certain species of plant which lives in a hot, dry desert environment in Arizona developed a type of protective covering with spines which protect it from predators. An unrelated species which lives in a similar hot, dry desert environment in Africa developed a similar protective covering with spines which protect it from predators. This is an example of

    (A) homologous structures
    (B) codominance
    (C) convergent evolution
    (D) punctuated equilibrium
    (E) taxonomy

Directions: Each group of questions below consists of five lettered headings followed by a list of numbered phrases or sentences. For each numbered phrase or sentence select the one heading that is most closely related to it and fill in the corresponding oval on the answer sheet. Each heading may be used once, more than once, or not at all in each group.

Questions 69–72

    (A) Platyhelminthes
    (B) Cnidaria
    (C) Porifera
    (D) Echinodermata
    (E) Annelida

69. Organisms in this phylum have a brain and two or more nerve trunks

70. Organisms in this phylum have a central nerve ring with radial nerves

71. Organisms in this phylum have a nervous system consisting of a simple nerve net

72. Organisms in this phylum have a brain and a central nerve cord containing segmented ganglia

Questions 73–76

    (A) protostomes
    (B) deuterostomes
    (C) tracheophytes
    (D) bryophytes
    (E) coelom

73. These organisms show radial cleavage during early embryonic development, and the organism's anus develops from the blastopore.

74. These organisms show spiral cleavage during early embryonic development, and the organism's mouth develops from the blastopore.

75. These plants lack true roots, stems, and leaves.

76. These plants use vascular tissues to transport materials.

Questions 77–81

    (A) exoskeleton
    (B) cartilage
    (C) hinge joint
    (D) ball and socket joint
    (E) endoskeleton

77. The human knee is an example of this type of joint.

78. This material forms discs which are located between vertebrae in the human spine.

79. This type of skeleton contains chitin.

80. The human hip is an example of this type of joint.

81. This type of skeleton is found in the interior of the organism.

Questions 82–85

    (A) compound eye
    (B) rods
    (C) cones
    (D) iris
    (E) pupil

82. This structure regulates the amount of light entering the eye.

83. In the human eye, this type of receptor provides night vision

84. Organisms in class Insecta have this type of eye.

85. In the human eye, this type of receptor provides color vision.

GO ON TO THE NEXT PAGE. ➡

Questions 86–88 refer to the following diagram.

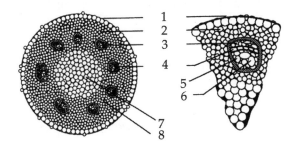

86. This vascular system is most likely from what type of plant?

    (A) monocot
    (B) dicot
    (C) either a monocot or a dicot
    (D) neither a monocot nor a dicot
    (E) cannot be determined

87. Water is transported via which structure?

    (A) 2
    (B) 3
    (C) 4
    (D) 5
    (E) 7

88. Which structure is made up of undifferentiated cells?

    (A) 2
    (B) 3
    (C) 4
    (D) 5
    (E) 7

GO ON TO THE NEXT PAGE.

Questions 89–92 refer to the following diagram.

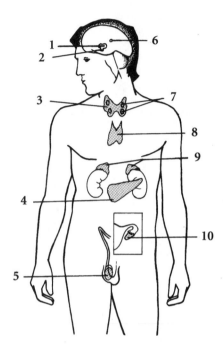

89. Gland(s) that produce(s) melatonin, which is thought to play a role in circadian rhythms.

(A) 1
(B) 2
(C) 3
(D) 6
(E) 7

90. Gland(s) that produce(s) hormones which raise blood pressure and heart rate as part of the "fight or flight" response.

(A) 1
(B) 2
(C) 4
(D) 8
(E) 9

91. Gland(s) that function(s) as part of the endocrine system as well as part of the digestive system.

(A) 3
(B) 4
(C) 7
(D) 8
(E) 9

92. Gland(s) that produce(s) estrogen and progesterone.

(A) 4
(B) 5
(C) 8
(D) 9
(E) 10

GO ON TO THE NEXT PAGE. ➡

Questions 93–96 refer to the diagrams of organic molecules below.

(A)

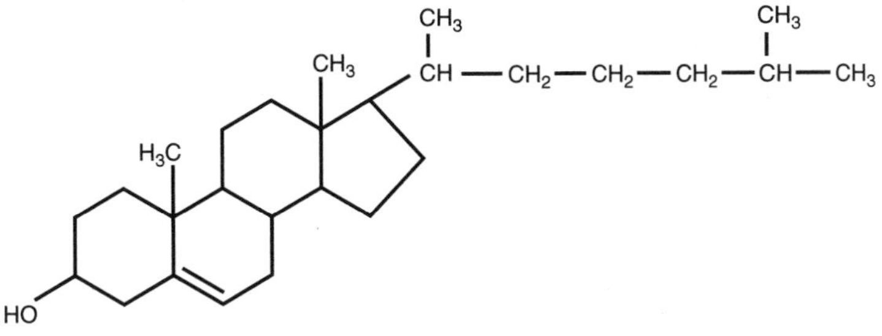

(B)

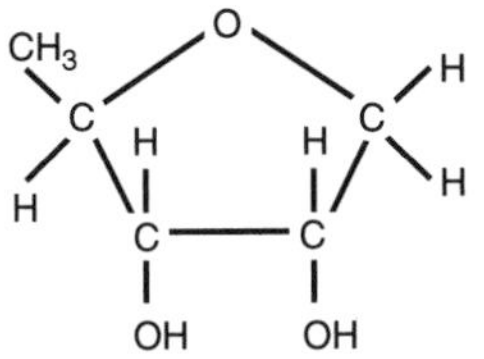

(C)

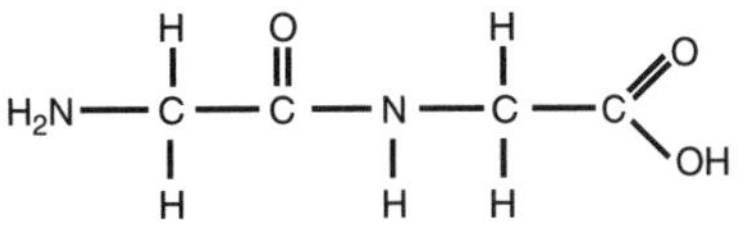

(D)

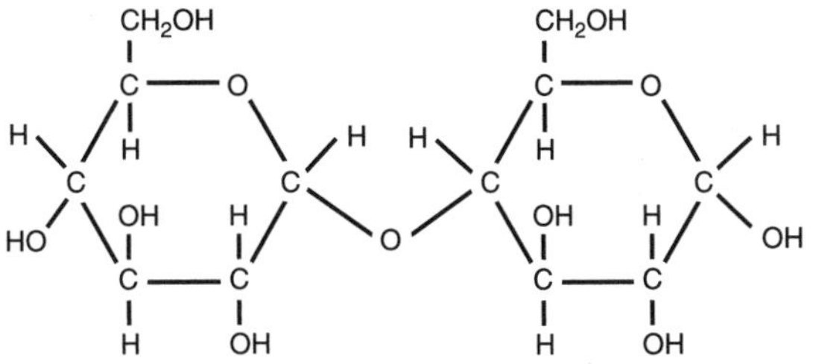

(E)

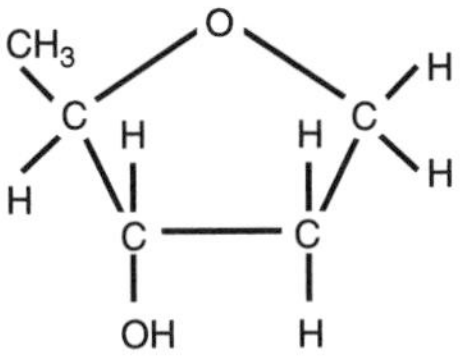

93. This molecule contains a glycosidic bond.

94. This molecule contains a peptide bond.

95. The amount of this sterol found in cell membranes is related to membrane fluidity.

96. This molecule is found in RNA.

GO ON TO THE NEXT PAGE. ➡

Directions: Each group of questions below consists of five lettered headings followed by a list of numbered phrases or sentences. For each numbered phrase or sentence select the one heading that is most closely related to it and fill in the corresponding oval on the answer sheet. Each heading may be used once, more than once, or not at all in each group.

Questions 97–99 refer to the following:

    (A) mitochondrial matrix
    (B) cytoplasm
    (C) mitochondrial intermembrane space
    (D) stroma of the chloroplast
    (E) thylakoid of the chloroplast

97. Site of glycolysis

98. Site of the Calvin cycle

99. Site of the Krebs cycle

Directions: Each group of questions below consists of five lettered headings followed by a list of numbered phrases or sentences. For each numbered phrase or sentence select the one heading that is most closely related to it and fill in the corresponding oval on the answer sheet. Each heading may be used once, more than once, or not at all in each group.

Questions 100–102 refer to the following diagram.

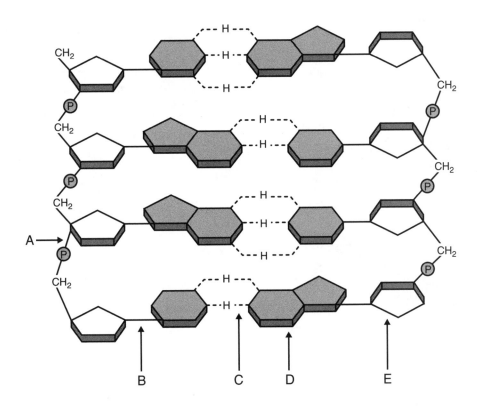

100. This molecule is a purine.

101. This molecule is part of the sugar-phosphate backbone.

102. This bond may be broken by restriction enzymes.

Questions 103–105 refer to the following diagrams.

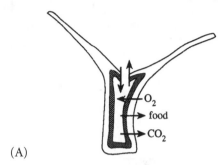

(A)

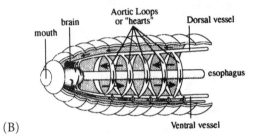

(B)

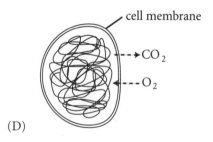

(C)

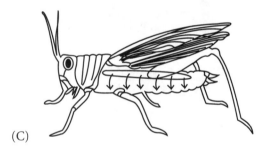

(D)

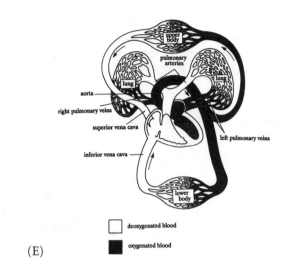

(E)

103. the circulatory system of a human

104. the circulatory system of a bacteria

105. the circulatory system of a hydra

GO ON TO THE NEXT PAGE. ➡

Questions 106–108 refer to the following information and diagram.

The following pedigree shows the pattern of inheritance of hemophilia over three generations.

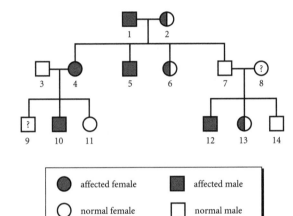

affected female   affected male

normal female   normal male

carrier female

106. Individual 8 is best described as a(n)

(A) affected male
(B) normal male
(C) carrier female
(D) normal female
(E) affected female

107. Individual 9 is best described as a(n)

(A) affected male
(B) normal male
(C) carrier female
(D) normal female
(E) affected female

108. If individual 14 marries a carrier female, what percentage of their male children will have hemophilia?

(A) 0%
(B) 25%
(C) 50%
(D) 66%
(E) 100%

GO ON TO THE NEXT PAGE. ➡

Questions 109–111 refer to the following information and graph.

The graph below shows the absorption spectrum of chlorophyll a which was extracted from spinach leaves.

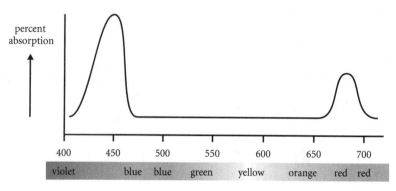

109. Which wavelength of light is best absorbed by chlorophyll a?

    (A) 450 nm
    (B) 525 nm
    (C) 550 nm
    (D) 625 nm
    (E) 675 nm

110. Which colors show greatest absorption by chlorophyll a?

    (A) red-orange
    (B) orange-yellow
    (C) yellow-green
    (D) green-blue
    (E) blue-violet

111. The fact that spinach leaves are green is best described by the fact that

    (A) green wavelengths of light are absorbed by chlorophyll a while all other colors of light are reflected
    (B) red wavelengths of light are reflected by chlorophyll a while all other colors of light are reflected
    (C) green and yellow wavelengths of light are absorbed by chlorophyll a while all other colors of light are reflected
    (D) blue-violet and red wavelengths of light are absorbed by chlorophyll a while all other colors of light are reflected
    (E) all wavelengths of light are absorbed equally well by chlorophyll a

GO ON TO THE NEXT PAGE. ➡

Questions 112–114 refer to the following diagrams and information.

The following graph represents the rate of an enzyme-catalyzed reaction as measured in a test tube:

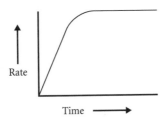

Answer choices:

(A) 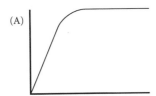 (B)

(C) 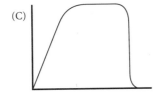 (D) 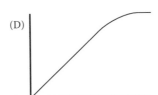 (E)

112. This graph represents the effect of adding an irreversible inhibitor, then removing excess, unbound inhibitor from the reaction tube.

113. This graph represents the effect of decreasing the concentration of substrate.

114. This graph represents the effect of increasing the concentration of substrate.

Questions 115–116 refer to the following diagram:

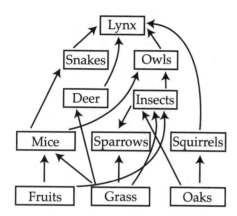

115. The organism with the greatest biomass is

(A) lynx
(B) snakes
(C) deer
(D) squirrels
(E) grass

116. The organism with the least amount of stored energy is

(A) oaks
(B) mice
(C) insects
(D) owls
(E) lynx

Questions 117–119 refer to the following information and diagram.

When a seed is fertilized, the shoot grows upward, toward the sun, while the roots grow downward, into the soil. A scientist wishes to study the hormone responsible for the growth of the shoot toward the sun. One seedling is allowed to grow normally, while the tip of the shoot of another seedling is cut off. The results are shown in the diagram below.

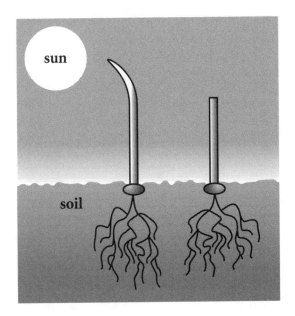

117. The growth of the shoot toward the sun is called

(A) photoperiodism
(B) gravitropism
(C) phototropism
(D) thigmotropism
(E) photosynthesis

118. The hormone involved in the growth of the shoot toward the sun is

(A) auxin
(B) ethylene
(C) cytokinin
(D) gibberellin
(E) abscisic acid

119. The hormone involved in the growth of the shoot toward the sun is most likely produced in what area of the plant?

(A) root
(B) seed
(C) stem
(D) shoot tip
(E) leaf

120. Use the following data regarding crossover frequencies to map the relative location of four chromosomes.

Genes	Frequency of Crossover
P and Q	35%
R and Q	20%
R and P	15%
T and Q	60%
P and T	25%

Which of the following represents a possible arrangement of these four genes on the chromosome relative to each other?

(A) P-Q-R-T
(B) P-R-Q-T
(C) T-P-R-Q
(D) T-P-Q-R
(E) Q-P-T-R

GO ON TO THE NEXT PAGE. ➡

AP BIOLOGY
PRACTICE TEST II

Section II

Time—1 hour and 30 minutes

Answer all questions. Number your answer as the question is numbered below.

Answers must be in essay form. Outline form is NOT acceptable. Labeled diagrams may be used to supplement discussion, but in no case will a diagram alone suffice. It is important that you read each question completely before you begin to write.

1. It is hypothesized that mitochondria and chloroplasts were originally free-living prokaryotes that formed a symbiotic relationship with eukaryotic cells.

   (A) Diagram the mitochondria and the chloroplast.
   (B) What function does each organelle perform in the cell? How does the structure of each organelle relate to its function?
   (C) State the major reactions which take place in each organelle and identify the area within the organelle in which the reactions take place.
   (D) What evidence exists to suggest that mitochondria and chloroplasts were originally free-living organisms?

2. Translation is the process by which genetic information is used to make proteins. Discuss translation in terms of the following:

   (A) How is mRNA processed before it leaves the nucleus?
   (B) What cellular constituents play a major role in translation?
   (C) How is a newly synthesized polypeptide processed before becoming a functional polypeptide?
   (D) Diagram the process of translation, beginning with pre-mRNA in the nucleus and ending with a polypeptide chain. Post-translational modifications need not be included in the diagram.

3. Bacteria have three means of transferring genetic information.

   (A) Name and briefly explain the three recombination mechanisms.
   (B) In a laboratory, a student starts with two strains of bacteria:
       -strain A: resistant to penicillin, not resistant to ampicillin
       -strain B: not resistant to penicillin, resistant to ampicillin
   Design a procedure by which the genes that code for penicillin resistance are transferred to the strain that is not penicillin resistant; the resulting strain would therefore be resistant to both penicillin and ampicillin. Identify the recombination mechanism used.
   (C) How would the student isolate the strain which is resistant to both ampicillin and penicillin?

GO ON TO THE NEXT PAGE. ➡

4. Some organisms reproduce sexually, some asexually, and some both sexually and asexually.

   (A) Give an example of:

        (I)   an organism that reproduces sexually
        (II)   an organism that reproduces asexually
        (III) an organism that reproduces both sexually and asexually

   (B) What are the advantages of sexual reproduction? What are the disadvantages of sexual reproduction?

   (C) What are the advantages of asexual reproduction? What are the disadvantages of asexual reproduction?

PRACTICE TEST II
ANSWER KEY

Section I—Multiple Choice

1. D	25. C	49. C	73. B	97. B
2. C	26. E	50 B	74. A	98. D
3. C	27. A	51. D	75. D	99. A
4. B	28. B	52. B	76. C	100. D
5. B	29. D	53. E	77. C	101. E
6. A	30. B	54. C	78. B	102. A
7. E	31. B	55. D	79. A	103. E
8. D	32. E	56. A	80. D	104. C
9. A	33. C	57. B	81. E	105. A
10. D	34. B	58. D	82. D	106. C
11. E	35. D	59. B	83. B	107. A
12. D	36. B	60. C	84. A	108. C
13. C	37. D	61. B	85. C	109. A
14. E	38. B	62. C	86. B	110. E
15. A	39. C	63. B	87. D	111. D
16. D	40. E	64. D	88. E	112. C
17. C	41. C	65. E	89. D	113. D
18. D	42. E	66. B	90. E	114. E
19. E	43. D	67. D	91. B	115. E
20. B	44. C	68. C	92. E	116. E
21. A	45. B	69. A	93. D	117. C
22. E	46. D	70. D	94. C	118. A
23. E	47. C	71. B	95. A	119. D
24. B	48. E	72. E	96. B	120. C

Answers and Explanations to Practice Test II

SECTION I: MULTIPLE-CHOICE

1. **(D)** Cell plate formation is involved in plant cell mitosis. Since plant cells have cell walls, plant cells cannot pinch along the middle to divide the cytoplasm, as animal cells do. Instead, cytokinesis in plants involves membrane vesicles which fuse together, forming a double-membraned cell plate. The new cell wall is formed between the two membranes of the cell plate.

2. **(C)** The ectodermal germ layers give rise to the epidermis of the skin and also the nervous system. The endodermis, or endodermal germ layer, gives rise to the lining of the digestive system, its associated glands and organs (such as the liver and pancreas), and the lungs. Most of the other organs and systems of the body are mesodermal, including the excretory system, the reproductive system, the muscular and skeletal systems, and the circulatory system. Choices A, B, D, and E all contain tissues or structures that are mesodermal in origin.

3. **(C)** Imprinting is a process in which environmental patterns or objects presented to a developing organism during a brief "critical period" in early life become accepted permanently as elements of their behavior environment. A duckling passes through a critical period in which it learns that the first large moving object it sees is its mother. Choice A is incorrect because discrimination involves the ability of the learning organism to differentially respond to slightly different stimuli. Choice B is incorrect because many animals secrete substances called pheromones that influence the behavior of other members of the same species. Choice D is incorrect because instrumental conditioning involves conditioning responses to stimuli with the use of reward or reinforcement. Choice E is incorrect because classical conditioning involves the learning of an association between two events after repeated pairings of the two events.

4. **(B)** Identical twins are produced when a zygote produced by one egg and one sperm splits during the four- or eight-cell stage to develop into two genetically identical organisms. Choice A is incorrect because identical twins will be the same sex but fraternal twins can also be the same sex. Choice D is incorrect because these are termed fraternal twins and are no more alike genetically than siblings. Choices C and E are impossible events.

5. **(B)** In DNA, the amount of thymine is equal to the amount of adenine. The amount of guanine is equal to the amount of cytosine. This is because adenine can pair only with thymine, and guanine can pair only with cytosine. Uracil is not found in DNA.

6. (A) Organic substances contain carbon. Glucose, phospholipid, DNA, and amylase all contain carbon; water contains only hydrogen and oxygen, so water is not an organic substance.

7. (E) Enzymes form temporary enzyme-substrate complexes. The other statements about enzymes have been corrected here: enzymes decrease the activation energy of a reaction but do not change the free energy of the products. Enzymes are affected by changes in temperature and pH.

8. (D) In genetic engineering, the host DNA and the DNA which we want to insert are both cut with the same restriction enzyme. Restriction enzymes recognize a specific sequence of bases in DNA and cut at a specific site within this sequence. If the host DNA and the DNA which we want to insert were cut with different restriction enzymes, then the "sticky ends" which result would have different base sequences, which would not necessarily complement each other. Thus the host DNA and the DNA to be inserted must be cut with the same restriction enzyme, so that the "sticky ends" have complementary base sequences. The other statements are true: cold viruses have DNA as their genetic material; DNA ligase is used to anneal the "sticky ends" of the cold virus DNA and the cystic fibrosis gene DNA; the cold virus is the vector by which the correct cystic fibrosis gene enters lung cells; and a cold virus was used to target lung tissue when delivering the correct cystic fibrosis gene.

9. (A) Cells lining the human respiratory tract have cilia, which help to keep the respiratory tract free of dust and debris. Euglena have a flagellum but no cilia; cells in the human urinary tract do not have cilia or flagella; human sperm cells have a flagellum but no cilia; plant cells do not have cilia or flagella.

10. (D) Elemental nitrogen is chemically inert and cannot be used by most organisms. Nitrogen-fixing bacteria in the roots of legumes change the nitrogen to the usable, soluble nitrates.

11. (E) Choice E is false for two reasons: mature red blood cells contain no nuclei, and the lac operon is a set of control and structural genes in E. coli (not humans) that allow the digestion of lactose. The other statements are true: there are three structural genes controlled by an operator found on another part of the genome; in the absence of lactose a repressor protein is bound to the operator, preventing RNA polymerase from binding to the DNA, thus preventing translation of the structural genes.

12. (D) The bottleneck effect occurs when a small group of organisms is separated from a larger population. The small sample of organisms may not accurately represent the gene pool of the entire population. This is what happened in the case of the pea plants. By chance, the small group of 6 individuals which the farmer pulled from the larger pea plant population has a greater concentration of alleles for yellow pea color, resulting in more offspring with yellow pea color. Gradualism is an evolutionary theory which states that evolutionary change occurs gradually over time. Punctuated equilibrium is an evolutionary theory which states that evolutionary change occurs in spurts. Stabilizing selection is a form of natural selection which favors individuals with intermediate phenotypes. An example of stabilizing selection is human birth weight, which tends to be centered around 7 pounds, 7 ounces. In sympatric speciation, a new species emerges in the midst of the parent species. For example, if there exists a population of 2n individuals and a 3n individual appears who is able to reproduce, this is an example of sympatric speciation.

13. **(C)** If a new species emerges as a result of geographical isolation from the parent species, this is known as allopatric speciation. In this example, several species of fish emerged from one parental species which had inhabited the large lake. As the lake dried up, small ponds were created which were separated from each other geographically by dry land. Disruptive selection is a type of natural selection in which extreme phenotypes are favored. In sympatric speciation, a new species emerges in the midst of the parent species. For example, if there exists a population of 2n individuals and a 3n individual appears who is able to reproduce, this is an example of sympatric speciation. Mutation is a change in the genetic information in DNA.

14. **(E)** Membranes are composed of phospholipids, which are arranged in a bilayer, with the hydrophilic heads facing the interior of the cell (or organelle) and the exterior of the cell (or organelle), and the hydrophobic tails facing each other. Membranes also contain proteins and cholesterol. Proteins function in cellular transport and recognition. Different membranes have different proteins embedded within them.

15. **(A)** Active transport requires energy while facilitated transport does not require energy. Active transport moves molecules from an area of high concentration to an area of low concentration. Both active transport and facilitated transport involve membrane proteins. Plant and animal cells both use active transport and facilitated transport to move molecules.

16. **(D)** Human Immunodeficiency Virus (HIV), the virus which causes Acquired Immune Deficiency Syndrome (AIDS), is an example of a retrovirus. Choices A and B is incorrect because retroviruses use RNA (not DNA) as their genetic material. Choice C is incorrect because retroviruses use single-stranded RNA (not double-stranded) RNA as their genetic material. Choice E is incorrect because animal cells do not produce reverse transcriptase; therefore, a virus must bring reverse transcriptase with it when it infects a cell.

17. **(C)** The principal function of $NaHCO_3$ is to serve as a buffer in the blood to maintain a slightly basic pH of 7.2 .

18. **(D)** Acetylcholine functions in the transfer of depolarization from one neuron to another. Acetylcholine is inactivated in the synaptic cleft by the enzyme acetylcholinesterase after it has acted upon the postsynaptic membrane. If a chemical denatures all enzymes in the synaptic cleft, it will inactivate acetylcholinesterase. Inactivated acetylcholinesterase will not be able to inactivate acetylcholine, and the postsynaptic membrane will be continuously depolarized.

19. **(E)** Choice E is false because the medulla monitors blood carbon dioxide levels and pH, and adjusts breathing, temperature, and heart rate. It is also the center for reflex activities such as coughing, sneezing, and swallowing. Choice A is true because the hypothalamus is the center that controls thirst, hunger, sleep, blood pressure, and water balance. Choice B is true because the cerebellum controls muscle coordination and tone, and maintains posture. Choice C is true because the cerebral cortex is the center for vision, hearing, smell, voluntary movement, and memory. Choice D is true because the reticular activating system receives and sorts sensory input.

20. **(B)** Choice B is false because three germ layers (not two) develop during gastrulation: the endoderm, mesoderm, and ectoderm. The other statements are correct: in the amphibian, the initial site of gastrulation is the gray crescent and the infolding through which blastula cells migrate during gastrulation is called the blastopore; the primitive gut that results from gastrulation is called the archenteron; and mesoderm develops during gastrulation.

21. **(A)** Mutualism is defined as a benefit to both species, a +/+ situation. As the termite ingests cellulose, it enters the digestive system. The termite actually cannot digest the cellulose but small protozoans in the digestive system can digest the cellulose. In return, they receive a home and food and water from the termite. Choice B is incorrect because parasitism is an instance where one organism gains and another organism is harmed (+/– situation) such as fleas on a dog. Choice C is incorrect because saprophytes are organisms that obtain their nutrients from dead organisms. Choice D is incorrect because commensalism is a +/0 situation in which one organism benefits and the other is not harmed such as vines on trees. Choice E is incorrect because autotropism describes self-feeders such as photosynthetic or chemosynthetic organisms.

22. **(E)** Marine biomes include: 1. Intertidal zone—a region exposed at low tides that undergoes variations in temperature and periods of dryness. Populations in the intertidal zones include algae, sponges, clams, snails, sea urchins, starfish, and crabs. 2. Littoral zone—a region on the continental shelf that contains ocean area with depths up to 600 feet, and extends several hundred miles from the shore. Populations in this zone include algae, crabs, crustacea, and many different species of fish. 3. Pelagic zone—a region typical of the open sea and divided into photic (the sunlit layer containing plankton, fish, sharks, and whales) and aphotic (the sunless zone containing the crawling and sessile organisms).

23. **(E)** Only DNA contains thymine. In RNA, uracil is substituted for thymine.

24. **(B)** The protein products of tumor suppressor genes function in the normal regulation of the cell cycle, preventing unregulated cell growth. Mutations in tumor suppressor genes prevent the regulation of cell growth, and cancer may result. Tumor suppressor genes are present in normal cells and do not cause mutations. Deleting tumor suppressor genes from normal cells may result in cancer.

25. **(C)** The secondary level of protein structure refers to the structure of a polypeptide which is generated by hydrogen bonding between amino acids at regular intervals. Some examples of the secondary level of protein structure are the alpha helix and the beta pleated sheet. The primary level of protein structure refers to the linear sequence of amino acids joined by peptide bonds. The tertiary level of protein structure refers to the three dimensional folding of a polypeptide chain brought about by the bonding of two amino acid side chains which may be far apart on the chain of amino acids. An example of the tertiary level of protein structure is the formation of disulfide bonds between two cysteines. The quaternary level of protein structure refers to the interaction between two or more polypeptide subunits.

26. **(E)** In the cell cycle, DNA is replicated during interphase.

27. **(A)** An example of gametic isolation would be two species which could not mate because the sperm of one species does not recognize the egg of another species. An example of behavioral isolation is a courtship ritual which is unique to a species: members of other species do not respond to the courtship ritual of a given species. In habitat isolation, two species cannot mate because they live in different habitats. An example of habitat isolation would be two closely related species: one which lives near mountains, another which lives near rivers; the two species would rarely encounter each other. If two species breed during different seasons, even if they are closely related, their gametes will not meet. This is a form of temporal isolation. An example of mechanical isolation would be two species which could not mate because they are anatomically incompatible.

28. **(B)** With no gamma globulins (antibodies), a person cannot produce a specific immune response to pathogens. This may result in frequent, severe, prolonged infections. Allergies are caused by the immune system's hypersensitivity to allergens. Diabetes is caused by the inability to produce insulin or the inability of body cells to respond to insulin. Hemophilia is caused by the inability of the body to produce a clotting factor. Color-blindness is caused by a deficiency of one or more types of cones. Cones are the receptors in the eye which are responsible for color vision.

29. **(D)** Rhizomes are a type of natural vegetative propagation and are characterized by woody, underground stems. At intervals, new upright stems appear. Examples of plants that utilize rhizomes are ferns and irises. Other forms of natural vegetative propagation are: 1. Bulbs such as tulips; 2. Tubers, which have underground stems with buds, such as potatoes; and 3. Runners, which have stems running above and along the ground, such as strawberries. Rhizomes are a form of asexual reproduction in plants.

30. **(B)** Choice B is false because FSH and LH are secreted by the anterior pituitary, and affect the maturation of the follicle. FSH stimulates the production of estrogen, and aids in the maturation of the primary follicle. LH causes the corpus luteum to secrete progesterone, which causes the uterine lining to thicken to prepare for implantation.

31. **(B)** A stable ecosystem is self-sustaining and will therefore remain stable if there is a relatively stable physical environment (abiotic factors) and a relatively stable biotic community. A stable ecosystem requires: 1. A constant energy source; 2. A living system incorporating this energy into organic compounds; and 3. A cycling of materials between the living system and the environment.

32. **(E)** The earliest forms of life were most likely unicellular heterotrophs which consumed organic molecules. Choices A and B are incorrect because unicellular organisms developed before multicellular organisms. Choices C and D are incorrect because mitochondria and chloroplasts are thought to have originally been free living bacteria which developed a symbiotic relationship with heterotrophic cells.

33. **(C)** One of the conditions of Hardy-Weinberg equilibrium is large population size. The other conditions of Hardy-Weinberg equilibrium have been corrected here: random mating, no natural selection, no migration into or out of the population, no mutation.

34. **(B)** A change of one digit in pH indicates a tenfold change in hydrogen ion concentration. Lower numbers have greater concentrations of hydrogen ion. An acid with a pH of 3 has a hydrogen ion concentration which is ten times greater than an acid with a pH of 4.

35. **(D)** Cholesterol is found in the cell membranes of animals. Cholesterol is not found in the cell membranes of bacteria, yeast, or plant cells.

36. **(B)** The nucleolus synthesizes ribosomes. The rough endoplasmic reticulum functions in protein synthesis and the smooth endoplasmic reticulum functions in lipid synthesis, along with other functions. The nucleus contains chromosomes and is important in directing mitosis. The Golgi bodies modify new proteins and package them for distribution. Lysosomes function in intracellular digestion of proteins, polysaccharides, fats, and nucleic acids, as well as obsolete organelles.

37. **(D)** In aerobic cellular respiration, NADH and $FADH_2$ function as electron carriers. The other statements are false and have been corrected here: glycolysis produces pyruvate which is converted to acetyl coenzyme A before entering the Krebs cycle; the majority of ATP is produced by the electron transport chain and oxidative phosphorylation; pyruvate is converted to lactate in anaerobic respiration; the products of cellular aerobic respiration are ATP, CO_2, and H_2O.

38. **(B)** Crossing over of homologous chromosomes during meiosis is important because it increases the genetic diversity of a population. New phenotypes may emerge which make an organism better equipped to survive in a changing environment.

39. **(C)** A person with blood type AB has two alleles: one which codes for the A surface marker, and another which codes for the B surface marker. Let's call the allele which codes for the A surface marker I^A, and the allele which codes for the B surface marker I^B. A person with blood type AB has a genotype of I^AI^B. A person with blood type O codes for neither the A surface marker nor the B surface marker. Let's call the allele which codes for no surface markers i. A person with blood type O has a genotype of ii. If we look at the Punnett square , their children can have the genotypes I^Ai and I^Bi only; therefore, their children can be type A or type B only.

	I^A	I^B
i	I^Ai	I^Bi
i	I^Ai	I^Bi

40. **(E)** Each amino acid has a separate aminoacyl-tRNA synthetase which is responsible for attaching the correct amino acid to a tRNA molecule with the correct anticodon. Choice A is incorrect because gene transcription in eukaryotes (not prokaryotes) involves intron splicing. Choice B is incorrect because DNA polymerase (not RNA polymerase) proofreads DNA (not RNA) as it is synthesized. Choice C is incorrect because transcription in eukaryotes takes place in the nucleus (not the cytoplasm). Choice D is incorrect because a polyribosome translates several amino acid chains (not strands of RNA) from one RNA (not DNA) strand.

41. **(C)** Unlike spermatogenesis, in oogenesis, cytokinesis is unequal, producing one large ovum and three polar bodies. Choice A is false since in oogenesis, meiosis is completed only at fertilization. Choice B is false because each oogonium produces one ovum and three polar bodies. Choice D is false because egg production is hormonally regulated, the hormones involved being estrogen and progesterone. Choice E is false because a woman's ovaries do not produce new oogonia after birth.

42. **(E)** Viruses contain DNA or RNA as their genetic material, and a protein coat. Viruses do not contain organelles such as mitochondria and ribosomes, so choices A and B are incorrect. Choices C and D are incorrect because viruses contain either DNA or RNA as their genetic material.

43. **(D)** We are told that normal wings are dominant to curly wings, and heterozygotes have normal wings. Let's call the allele for normal wings W and the allele for curly wings w. Pure breeding normal males would have the genotype WW and pure breeding curly wing females would have the genotype ww. If WW males are mated with ww females, the offspring in the F_1 generation will all be Ww. If Ww flies are mated with Ww flies (here the sex of the flies does not matter since curly wing is not a sex-linked trait), the offspring in the F_2 generation will have the following genotypes (see Punnett square) — 1 WW : 2 Ww : 1 ww. Since heterozygotes have normal wings, the phenotypic ratio in the F_2 generation is 3 normal wing offspring : 1 curly wing offspring.

	W	*w*
W	*WW*	*Ww*
w	*Ww*	*ww*

44. **(C)** In cells, reactions which are energetically unfavorable (which require an input of energy) are coupled to the hydrolysis of adenosine triphosphate (ATP), which is energetically favorable (the hydrolysis of ATP releases energy).

45. **(B)** When there is insufficient oxygen reaching human muscles, they utilize lactic acid fermentation to regenerate NAD^+ from NADH. (A buildup of lactic acid causes soreness after a workout.)

46. **(D)** The number of chromosomes in a sperm cell is half the diploid number. If half the diploid number is 6, then the diploid number of the organism is 12.

47. **(C)** Prokaryotic cells contain DNA, cytoplasm, ribosomes, and a cell membrane, but no membrane bound organelles. Prokaryotic cells do not contain mitochondria, chloroplasts, or centrioles.

48. **(E)** The summary of ecological succession in a rocky barren area to a final climax community looks like this: lichen–mosses–annual grasses–perennial grasses–shrubs–sun-loving trees–thick shade trees such as hemlock and beech.

49. **(C)** The amniotic fluid surrounds the egg with a liquid environment and protects it from shock. Surrounding the amniotic fluid is the amnion on the membrane of the egg. The embryonic membranes include: 1. The chorion—lines the inside of the shell and permits gas exchange through the egg shell; 2. The allantois—a saclike structure developed from the digestive tract. Functions include respiration and excretion, particularly the exchange of gases with the external environment; 3. Amnion—encloses the amniotic fluid, which provides a watery environment for the embryo to develop in and provides protection against shock; 4. Yolk sac—encloses the yolk and transfers food to the developing embryo.

50. **(B)** To simulate conditions of the early Earth, Oparin and Haldane mixed hydrogen gas, water, methane, and ammonia, then applied electric charge to simulate lightning. Choice A is incorrect because the environment of the early Earth had no ozone. Choice C is incorrect because Oparin and Haldane did not use DNA, and the environment of the early Earth had no ozone. Choice D is incorrect because Oparin and Haldane did not use RNA , and the environment of the early Earth had no ozone. Choice E is incorrect because Oparin and Haldane did not use RNA.

51. **(D)** In aerobic respiration, ATP is produced via substrate-level phosphorylation in both glycolysis and the Krebs cycle. The Calvin cycle is part of photosynthesis.

52. **(B)** Lambda phage is a DNA virus which infects bacteria. In the lytic cycle, the phage injects its DNA into the bacterial host and uses host cell machinery to replicate viral DNA and produce viral proteins. Choice A is incorrect because lambda phage is a DNA (not RNA) virus, and therefore does not need reverse transcriptase. Choice C is incorrect because in the lytic cycle, the phage does not integrate into the host genome, and therefore viral DNA is not replicated when the host bacteria replicates its DNA. Choice D is incorrect because lambda phage does enter bacterial cells in both the lytic and lysogenic cycles. Choice E is incorrect because lambda phage integrates its DNA into the bacterial host during the lysogenic cycle, not the lytic cycle.

53. **(E)** Parasitism is a relationship in which one organism gains while the other loses; it is also known as a "+/-" relationship. An example is the relationship between a tapeworm and its human host. The tapeworm receives a secure home, nutrients, and water while the host loses nutrients and water. Choice A is incorrect; symbiosis describes close nutritional relationships of all types, including mutualism, commensalism, and parasitism. It is a more general term than the others. Choice B is incorrect because mutualism is a relationship in which both species benefit, also known as a "+/+" relationship. An example is the relationship between the algal and fungal organisms that live in the organism known as a lichen. Choice C is incorrect because a saprophyte is an organism such as a mushroom, that gains nutrients by digesting dead plant or animal material. Choice D is incorrect because in a commensal relationship, which is a form of symbiotic relationship, two organisms live in close association with each other, and one benefits from this association, while the other is neither harmed nor benefited. This is also sometimes described as a "+/0" relationship. An example of a commensal relationship is epiphyte plants, which live on the branches of rainforest trees. These plants gain the advantage of being closer to sunlight by being on the branches.

54. **(C)** In a food pyramid, a producer gets energy from the sun (or in some cases, organic molecules) and uses this energy to produce carbohydrates, lipids, and proteins from simple molecules. Of the choices given, the fruits are the producers in this food pyramid.

55. **(D)** Choice D is false: when the pressure of air within the lungs is less than the atmospheric pressure, air enters the lungs. The other statements are true: ciliated nasal membranes warm, moisten, and filter inspired air; contraction of the diaphragm enlarges the thoracic cavity; when the thoracic cavity enlarges, the pressure of air within the lungs falls; and the respiratory process consists of inspiratory and expiratory acts following one another.

56. **(A)** Prokaryotic organisms (bacteria), which lack a nucleus and membrane-bound organelles, are found in kingdom Monera. Organisms in the kingdoms Protista, Fungi, Plantae, and Animalia are all eukaryotic.

57. **(B)** If ^{32}P is used to label a strand of DNA, after one round of replication, there will be two double-stranded molecules of DNA, each containing half of the ^{32}P. DNA replication is semi-conservative, meaning that the strands split apart, and each half is used as a template for the synthesis of the complementary strand. So after one round of replication, we have two double-stranded DNA molecules, each containing one strand labeled with ^{32}P. During the second round of replication, the DNA strands split, and each strand is used as a template for the synthesis of the

complementary strand. Therefore, after two rounds of replication, we have four double-stranded DNA molecules. Two of the double-stranded molecules contain a strand of DNA labeled with ^{32}P; the other two double-stranded molecules contain no ^{32}P.

58. **(D)** Chloroplasts (and mitochondria) are thought to be descended from free-living prokaryotic organisms which entered heterotrophic host cells.

59. **(B)** Plant cells contain a nucleus, mitochondria, chloroplasts, and are enclosed by a cell wall. Plant cells do not contain centrioles.

60. **(C)** The parathyroid glands are four small pea-shaped structures embedded in the posterior surface of the thyroid. These glands synthesize and secrete parathyroid hormone, which raises the concentration of blood calcium. Parathyroid hormone raises the calcium concentration in the blood by increasing bone resorption and decreasing calcium excretion in the kidneys. Glucagon stimulates the conversion of glycogen to glucose in the liver and therefore increases blood glucose. Insulin lowers blood glucose and increases storage of glycogen. Aldosterone regulates plasma levels of sodium and potassium and consequently the total extracellular water volume. It causes the active reabsorption of sodium and passive reabsorption of water in the nephron. Anti-diuretic hormone stimulates water reabsorption by the kidneys by increasing the nephron's permeability to water.

61. **(B)** Denitrifying bacteria break down NH_3 into N_2. Nitrifying bacteria turn ammonia into NO_2 (nitrites) while nitrogen fixing bacteria turn N_2 into NO_3 (nitrates).

62. **(C)** Organisms in the phylum Porifera (sponges) lack true tissues. Organisms in the class Insecta (insects), the class Pisces (fish), the class Arachnida (spiders), and the class Mammalia (for example, humans) all have true tissues, which are formed from the germ layers which develop during early embryonic development (the endoderm, mesoderm, and ectoderm).

63. **(B)** *H. sapiens* and *H. erectus* are not in the same species. *H. sapiens* and *H. erectus* are two different species; thus choice A is incorrect. *H. sapiens* represents the genus and species names of an organism. *H. sapiens* and *H. erectus* are in the same genus, so they also share the same higher taxonomic information: kingdom, phylum, class, order, and family. Choice C is incorrect because *H. sapiens* and *H. erectus* are in the same phylum. Choice D is incorrect because *H. sapiens* and *H. erectus* are not in the same species and therefore are unable to mate. Choice E is incorrect because *H. sapiens* and *H. erectus* are related in terms of evolution; if they were not related, they would not be in the same genus.

64. **(D)** It is thought that early life forms used RNA both as enzymes and as genetic material. Choice A is incorrect because early life forms did not use proteins as enzymes or genetic material. Choice B is incorrect because modern life forms use DNA as genetic material and proteins as enzymes. Choice C is incorrect because early life forms used RNA as genetic material. Choice E is incorrect because early life forms did not use DNA as genetic material or as enzymes.

65. **(E)** Smooth endoplasmic reticulum functions in lipid synthesis, carbohydrate synthesis, and the detoxification of toxins, but not in the production of proteins. Secretory proteins are manufactured on ribosomes, which are located on the rough endoplasmic reticulum. From the endoplasmic reticulum, the proteins enter the Golgi apparatus where they are modified and packaged. From the Golgi apparatus, the proteins are packaged into secretory vesicles. When the secretory vesicles fuse with the plasma membrane, the proteins are released outside the cell.

66. **(B)** Human blood type is based on the presence or absence of markers on a person's red blood cells. A person with type O blood has no surface markers on his or her red blood cells. A person with type AB blood has both type A surface markers and type B surface markers. A person with type A blood has type A surface markers, and a person with type B blood has type B surface markers. A person with blood type O makes antibodies against type A surface markers and type B surface markers, and therefore cannot receive blood of types A, B, or AB. A person with type AB blood does not make antibodies against either type A markers or type B markers, and therefore can receive blood of type A, B, or AB, or O. A person with type A blood makes antibodies against type B surface markers, and therefore cannot receive blood of types AB or B. A person with type B blood makes antibodies against type A surface markers, and therefore cannot receive blood of types AB or A. If a person receives a transfusion of blood which they are not compatible with, antibodies in their blood will react with the transfused blood.

    A person with blood type AB can only donate blood to people with blood type AB. This eliminates choice A. Choice C is false because a person with blood type AB can receive blood of type A, B, AB, or O. Choice D is false because a person with blood type AB cannot donate blood of type O. Choice E is false because a person with blood type AB can only donate to people with blood type AB.

67. **(D)** The prebiotic environment on Earth had a reducing atmosphere, with very little if any free O_2, and no ozone layer to protect against the UV rays of the sun. The prebiotic environment contained carbon in the form of methane (CH_4).

68. **(C)** The two plants developed similar structures because they adapted to similar environments, not because of common ancestry, so this is an example of convergent evolution. Homologous structures are structures which developed in organisms with similar ancestry. An example would be the arm of a human and the leg of a dog. Codominance is a pattern of heredity in which one allele is not dominant over another allele; instead, the products of both alleles result in a unique phenotype. An example of codominance is the human blood type AB. The presence of both the A allele and the B allele means that the individual with type AB blood has both type A markers and type B markers on his or her red blood cells. Punctuated equilibrium is an evolutionary theory which states that evolutionary change occurs in spurts. Taxonomy is a system of naming and classifying organisms.

69. **(A)** Organisms in phylum Platyhelminthes (flatworms) have a brain and two or more nerve trunks

70. **(D)** Organisms in phylum Echinodermata (starfish, for example) have a central nerve ring with radial nerves

71. **(B)** Organisms in phylum Cnidaria (*Hydra*, for example) have a nervous system consisting of a simple nerve net

72. **(E)** Organisms in phylum Annelida (segmented worms) have a brain and a central nerve cord containing segmented ganglia

73. **(B)** Deuterostomes show radial cleavage during early embryonic development, and the anus of deuterostomes develops from the blastopore.

74. **(A)** Protostomes show spiral cleavage during early embryonic development, and the mouth of protostomes develops from the blastopore.

75. **(D)** Bryophytes lack true roots, stems, and leaves. An example of a bryophyte is a moss.

76. **(C)** Tracheophytes use vascular tissues to transport materials. Tracheophytes include gymnosperms, an example of which is the pine, and angiosperms, or flowering plants.

77. **(C)** The human knee is an example of a hinge joint. This type of joint allows movement in one direction only; thus it is possible to bend the knee from front to back, but not side to side.

78. **(B)** Cartilage forms discs which are located between vertebrae in the human spine. These discs function to cushion the spine.

79. **(A)** Exoskeletons, which enclose an organism, such as the shells of lobsters and crabs, contain chitin.

80. **(D)** The human hip is an example of a ball and socket joint. A ball and socket joint allows 360 degree motion.

81. **(E)** Endoskeletons, such as those in humans, are found in the interior of the organism.

82. **(D)** By changing size, the iris regulates the amount of light entering the eye.

83. **(B)** In the human eye, rods are the receptors which provide night vision.

84. **(A)** Organisms in class Insecta have compound eyes, consisting of up to several thousand ommatidia, each with its own cornea and lens.

85. **(C)** In the human eye, cones are the receptors which provide color vision.

86. **(B)** This vascular system is most likely from a dicot. The vascular bundles of dicots are arranged in a ring, while the vascular bundles of monocots are complexly arranged, with vascular bundles throughout the ground tissue.

87. **(D)** The structure indicated by 5 is xylem, which transports water through the plant. The structure indicated by 2 is the cortex, which is not involved with water transport. The structure indicated by 3 is the phloem, which transports nutrients through the plant. The structure indicated by 4 is the cambium, which is undifferentiated and may develop into xylem or phloem. The structure indicated by 7 is the pith, which consists of non-specialized, undifferentiated parenchyma cells.

88. **(E)** The structure indicated by 7 is the pith, which consists of non-specialized, undifferentiated parenchyma cells. The structure indicated by 2 is the cortex, which is not involved with water transport. The structure indicated by 3 is the phloem, which transports nutrients through the plant. The structure indicated by 4 is the cambium, which is undifferentiated and may develop into xylem or phloem. The structure indicated by 5 is xylem, which transports water through the plant.

89. **(D)** The structure indicated by 6 is the pineal gland. The pineal gland secretes melatonin, which is thought to play a role in circadian rhythms. The structure indicated by 1 is the hypothalamus, which produces hormones (one example: antidiuretic hormone) which are released by the anterior pituitary, as well as releasing hormones which regulate the anterior pituitary. The structure indicated by 2 is the pituitary gland. The pituitary gland secretes some hormones manufactured by the hypothalamus, as well as hormones produced by the anterior pituitary such as growth hormone, follicle-stimulating hormone, luteinizing hormone, thyroid-stimulating hormone, and adrenocorticotropin. The structure indicated by 3 is the thyroid gland, which

produces hormones involved in the regulation of metabolism and blood calcium levels. The structures indicated by 7 are the parathyroid glands, which release parathyroid hormone. Parathyroid hormone acts to raise plasma calcium levels.

90. **(E)** The structures indicated by 9 are the adrenal glands. The adrenal glands release epinephrine and norepinephrine, which are involved in the "fight or flight" response to perceived danger. Epinephrine and norepinephrine raise heart rate and blood pressure. The structure indicated by 1 is the hypothalamus, which produces hormones (one example: antidiuretic hormone) which are released by the anterior pituitary, as well as releasing hormones which regulate the anterior pituitary. The structure indicated by 2 is the pituitary gland. The pituitary gland secretes some hormones manufactured by the hypothalamus, as well as hormones produced by the anterior pituitary such as growth hormone, follicle-stimulating hormone, luteinizing hormone, thyroid-stimulating hormone, and adrenocorticotropin. The structure indicated by 4 is the pancreas, which produces insulin and glucagon, which regulate blood sugar levels, as well as pancreatic amylase, which breaks down disaccharides. The structure indicated by 8 is the thymus, which is involved in T-lymphocyte development.

91. **(B)** The structure indicated by 4 is the pancreas, which produces insulin and glucagon as part of the endocrine system, and pancreatic amylase as part of the digestive system. Insulin and glucagon regulate blood sugar levels while pancreatic amylase breaks down disaccharides. The structure indicated by 3 is the thyroid gland, which produces hormones involved in the regulation of metabolism and blood calcium levels. The structures indicated by 7 are the parathyroid glands, which release parathyroid hormone. Parathyroid hormone acts to raise plasma calcium levels. The structure indicated by 8 is the thymus, which is involved in T-lymphocyte development. The structures indicated by 9 are the adrenal glands. The adrenal glands release epinephrine and norepinephrine, which are involved in the "fight or flight" response to perceived danger. Epinephrine and norepinephrine raise heart rate and blood pressure.

92. **(E)** The gland indicated by 10 is the ovary, which is found in females. The ovary produces estrogen and progesterone. Estrogen initiates and maintains female sex characteristics, while both estrogen and progesterone function in the development of the uterine lining in preparation for pregnancy. The structure indicated by 4 is the pancreas, which produces insulin and glucagon as part of the endocrine system, and pancreatic amylase as part of the digestive system. Insulin and glucagon regulate blood sugar levels while pancreatic amylase breaks down disaccharides. The structure indicated by 5 is the testis, which is found in males. The testis produce androgens, which initiate and maintain male sex characteristics and support spermatogenesis. The structure indicated by 8 is the thymus, which is involved in T-lymphocyte development. The structures indicated by 9 are the adrenal glands. The adrenal glands release epinephrine and norepinephrine, which are involved in the "fight or flight" response to perceived danger. Epinephrine and norepinephrine raise heart rate and blood pressure.

93. **(D)** Two glucose molecules are joined by a glycosidic bond to form maltose.

94. **(C)** Two amino acids are joined by a peptide bond, forming a polypeptide chain.

95. **(A)** This molecule is cholesterol. The amount of cholesterol in a cell membrane is related to its fluidity.

96. **(B)**   Ribose is found in RNA. Ribose differs from deoxyribose in the placement of one -OH group.

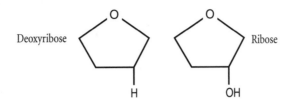

97. **(B)**   Glycolysis takes place in the cytoplasm.

98. **(D)**   The Calvin cycle takes place in the stroma of the chloroplast.

99. **(A)**   The Krebs cycle takes place in the mitochondrial matrix.

100. **(D)**   Purines are two-ring molecules, while pyrimidines are one-ring molecules. Cytosine and thymine are the pyrimidines found in DNA (uracil is a pyrimidine found in DNA); adenine and guanine are the purines found in DNA and RNA. (Here's a mnemonic device: purines have a shorter name but a bigger 2-ring shape, while pyrimidines have a longer name but a smaller 1-ring shape.)

101. **(E)**   Deoxyribose, the molecule indicated by E, is the sugar which forms the sugar-phosphate backbone of DNA.

102. **(A)**   Restriction enzymes cut DNA between bases, so the bond indicated by A may be broken by restriction enzymes.

103. **(E)**   The circulatory system of humans has a four chambered heart which pumps blood. Gas exchange takes place in the lungs.

104. **(C)**   Bacteria exchange materials with the environment directly through the plasma membrane.

105. **(A)**   Hydra have a gastrovascular cavity which functions to exchange materials with the environment.

106. **(C)**   We are told that hemophilia is a sex-linked genetic disorder. Let's denote the normal X chromosome as X, and the X chromosome containing the gene for hemophilia X^+. An affected male would have the genotype X^+Y, since males have one X chromosome and one Y chromosome. A carrier female would have the genotype XX^+, since she carries the gene on one of her X chromosomes but her other X chromosome is normal. Individual 8 mates with Individual 7, who does not carry the gene for hemophilia on his X chromosome. Individual 7 cannot pass the gene for hemophilia on to his children. Therefore, Individual 8 must have at least one copy of the gene for hemophilia, since one of her male children is affected, and her female child is a carrier. Since her other male child is not affected, she must pass a normal X chromosome to him, since he receives a Y chromosome from his father. Therefore, individual 8 is XX^+, or a carrier female.

107. **(A)**   Since individual 9's mother has hemophilia, her genotype is X^+X^+. His father does not have hemophilia, so his father's genotype is XY. Individual 9 gets a Y chromosome from his father, and an X^+ chromosome from his mother. Therefore, individual 9's genotype is X^+Y, and he is an affected male.

108. **(C)** Individual 14 is a normal male, so his genotype is XY. A carrier female has a genotype of XX^+. If we work through the Punnett square, their children have the following possible genotypes: XX^+, XX, X^+Y, and XY. Therefore, 50% of their male children will have hemophilia.

	X	Y
X^+	X^+X	X^+Y
X	XX	XY

109. **(A)** From the graph, the highest peak occurs at 450 nm, indicating that chlorophyll a absorbed the most light with a wavelength of 450 nm.

110. **(E)** From the graph, maximum absorption occurs at 450 nm, which corresponds to blue-violet.

111. **(D)** From the graph, chlorophyll a absorbs wavelengths of light corresponding to blue-violet and red. Thus, orange, yellow, and green light is reflected.

112. **(C)** If we add an irreversible inhibitor, the enzyme is unable to function, and the rate of reaction decreases until there is no more reaction taking place. The rate of reaction does not increase again, since the inhibition is irreversible and the inhibitor remains attached to the enzyme, even though the excess, unbound inhibitor is removed from the reaction tube. Graph B shows the result of adding a reversible inhibitor: the rate of reaction decreases initially, but increases once the excess, unbound inhibitor is removed, as the inhibitor detaches from the enzyme and the enzyme is allowed to function.

113. **(D)** If we decrease the concentration of substrate, the reaction rate does not increase as rapidly than the reaction rate at the original substrate concentration.

114. **(E)** If we increase the concentration of substrate, the reaction rate increases more rapidly than the reaction rate at the original substrate concentration.

115. **(E)** Organisms at the bottom of the food pyramid have the greatest biomass. Of the choices given, grass has the lowest position on the food pyramid, so it has the greatest biomass.

116. **(E)** Organisms at the top of the food pyramid have the least stored energy. Of the choices given, the lynx has the lowest position on the food pyramid, so it has the least amount of stored energy.

117. **(C)** The response of a plant to light is called phototropism. Some plants require a certain amount of light and/or darkness to flower. This is an example of photoperiodism, in which a physiological response depends on the amount of light and darkness. In thigmotropism, a plant is responsive to touch. Photosynthesis is the process by which plants use the energy of the sun to synthesize carbohydrates.

118. **(A)** The response of a plant to light is called phototropism. The hormone auxin regulates the growth of the shoot toward the sun. Ethylene is involved in the ripening of fruit. Cytokinin functions to stimulate cell division and germination. Gibberellin functions in stem and leaf growth, and flower and fruit development. Abscisic acid inhibits growth.

119. **(D)** The hormone involved in the growth of the shoot toward the sun is most likely produced in the shoot tip, since the plant in which the shoot tip was cut off did not grow toward the sun.

120. **(C)** In this problem we are given the data regarding crossover frequencies of pairs of genes. Genes which are closer together on a chromosome will have a lower frequency of crossover. If we think of the percentages as "crossover units", we can create a "map" of the four genes and the distances between them. The distance between P and Q would be 35 units, the distance between R and Q would be 20 units, the distance between R and P would be 15 units, the distance between T and Q would be 60 units, and the distance between P and T would be 25 units. Let's start with the genes which are closest together, P and R. Genes P and R have 15 units between them, so the arrangement of genes so far is P-R. Next let's take a look at gene Q. The distance between Q and P is 35. We have two choices for the placement of Q: to the right of gene P or to the left of gene P. If gene Q is to the right of gene P, the distance between gene R and gene Q would be equal to the distance between gene R and gene P (15) plus the distance between P and Q (35), for a total of 50. However, no two genes listed have a crossover frequency of 50%. Let's try putting gene Q to the left of gene P. If this is the case, then the distance between gene P and gene Q would be equal to the distance between gene R and gene P (15) plus the distance between gene Q and gene R (20), for a total of 35. This matches the distance between gene Q and gene P that we are given (35), so gene Q is to the left of gene P. Now let's look at gene T. Gene T is 60 units away from gene Q. There are two options for the placement of gene T: to the left of gene Q or to the right of gene Q. If gene T is to the left of gene Q, then the distance between gene Q and gene T is equal to the distance we are given, which is 60 units. However, we are also given that the distance between gene P and gene T is 25 units. If gene T is to the left of gene Q, then the distance between gene P and gene T is equal to the distance between gene T and gene Q (60) plus the distance between gene Q and gene P (35), for a total of 100. This does not match the data we are given, so gene T must lie to the right of gene Q. The distance between gene P and gene T is therefore equal to the distance between gene T and gene Q (65) minus the distance between gene Q and gene P (35), or 25 units. This matches the data we are given, so the arrangement of genes is Q-R-P-T.

SECTION II

It's important to note that essays will differ, and these sample essay responses are not necessarily the only responses that would earn credit for a particular essay question.

1. **(A)**

 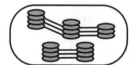

mitochondria   **chloroplast**

  (B) The mitochondria is the power source of the cell; it is the site of cellular respiration, which produces ATP for cellular use. The chloroplast converts solar energy to chemical energy; it is the site of photosynthesis. Each organelle consists of an outer membrane and a convoluted inner membrane. The folding of the inner membrane increases the surface area available to cellular reactions.

  (C) Major reactions of the mitochondria: Glycolysis takes place in the cytoplasm. The pyruvate from glycolysis is transported into the mitochondrion, where it enters the Krebs cycle in the intermembrane space of the mitochondria. The electron transport chain completes cellular respiration; this occurs in the inner mitochondrial membrane. ATP is released into the mitochondrial matrix.

       Major reactions of chloroplasts: The light reactions in photosystem I and photosystem II use solar energy to produce ATP, an energy source, and NADPH, a source of reducing power for the Calvin cycle. In the Calvin cycle, hydrogen atoms derived from water and NADPH from the light reactions are used to fix CO_2 from the environment into glucose. The light reactions take place in the thylakoid membranes of the choloroplast, while the Calvin cycle takes place in the stroma.

  (D) There exists some evidence which suggests to scientists that mitochondria and chloroplasts were originally free-living organisms (the endosymbiont hypothesis). The DNA of mitochondria and choloroplasts is similar to prokaryotic DNA, and their ribosomes are similar. In addition, the enzymes produced by mitochondria and chloroplasts are similar to those produced by prokaryotes.

2. **(A)** DNA is the genetic code which contains information for the synthesis of polypeptides. DNA is transcribed to mRNA in the nucleus. Before it leaves the nucleus, it is processed in several ways. To the 5′ end of the newly synthesized polypeptide, a modified guanine polypeptide is added. To the 3′ end of the newly synthesized polypeptide, a poly-A tail made up of 150 to 200 adenine nucleotides is added. The introns, or intervening sequences of DNA which are not transcribed, are removed from the mRNA.

(B) Once outside the nucleus, the mRNA is translated into a polypeptide chain. Translation involves mRNA, tRNA, amino acids, small and large ribosomal subunits, enzymes, and other proteins. Each tRNA molecule is specific for a single amino acid. Thus, a correct pairing of amino acid and aminoacyl tRNA is essential for accurate translation.

(C) Before becoming a functional polypeptide, a newly synthesized amino acid chain may undergo several post-translational modifications; some possibilities include methylation, phosphorylation, and carboxylation.

(D)

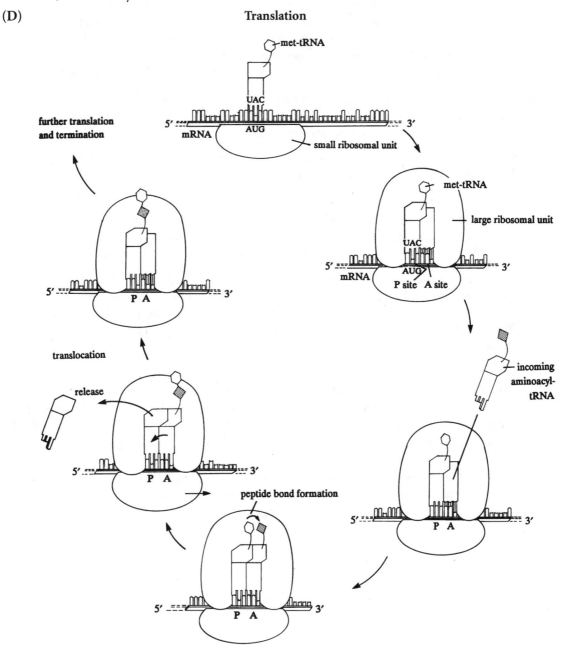

3. **(A)** The three mechanisms by which bacteria transfer genetic information are transformation, transduction, and conjugation. In transformation, bacterial cells pick up genetic information directly from the environment. For example, if a bacterium resides in the intestinal tract, and DNA from a recently lysed cell is nearby, the DNA may enter the bacterium directly. Note that the source of the DNA need not be another bacterium. Transduction is the transfer of DNA from one bacterial cell to another by phages. A phage is a virus that infects bacteria. When the phage replicates ard repackages itself into new phages, occasionally a piece of bacterial DNA is incorporated into the phage genome. When the phage infects another bacterium, the bacterial DNA may be incorporated into the genome with the phage DNA. Conjugation involves two bacterial cells: one must contain the F plasmid; this cell is designated F^+. The other bacterial cell does not contain the F plasmid; this cell is designated F^-. The F^+ cell has the ability to form sex pili which attach to the F- cell. The F^+ cell transfers DNA to the F^- cell through the sex pili.

   (B) The student could infect strain A with a phage, then incubate strains A and B together. Strain A DNA would be incorporated into phage DNA. When the phage infects strain B, some of the DNA from strain A would be incorporated into the genome of strain B.

   (C) After infecting strain A with the phage and then incubating strains A and B together, the student could plate the culture on media containing both penicillin and ampicillin. Any bacteria which grow on the plate must have genes for resistance to both penicillin and ampicillin.

4. **(A)** An example of an organism which reproduces sexually is *Homo sapiens*. An example of an organism which reproduces asexually is the starfish. An example of an organism which reproduces both sexually and asexually is the segmented worm (phylum Annelida).

   (B) The main advantage of sexual reproduction is that the offspring are not genetically identical to the parents. Sexual reproduction introduces genetic variation into the population, which is an advantage in a changing environment. If a genetic recombination results in offspring which are better able to survive in the changed environment, the species is more likely to survive. A disadvantage of sexual reproduction is that two individuals of different sexes are usually required for sexual reproduction.

   (C) An advantage of asexual reproduction is that the offspring are almost identical to the parent in terms of genetic material (a possible source of genetic change is mutation). Asexual reproduction does not introduce variation in the offspring. This is an advantage if the parent is well adapted to an unchanging environment, since the offspring will be equally well adapted. Also, asexual reproduction does not require two individuals of different sexes. The fact that asexual reproduction does not introduce genetic variation in the offspring is also a disadvantage of asexual reproduction. In a changing environment, the organism may be less able to survive; without genetic recombination, the offspring are genetically identical to the parents and therefore also less able to survive.

Glossary

abiotic. Nonliving, as in the physical environment

absorption. The process by which water and dissolved substances pass through a membrane

acetylcholine. A transmitter substance released from the axons of nerve cells at the synapse

active immunity. Protective immunity to a disease in which the individual produces antibodies as a result of previous exposure to the antigen

adaptation. A behavioral or biological change that enables an organism to adjust to its environment

adaptive radiation. The production of a number of different species from a single ancestral species

adenosine phosphate. Adenosine diphosphate (ADP) and adenosine triphoshate (ATP), which are energy storage molecules

ADH (vasopressin). A hormone that regulates water reabsorption

adipose. Fatty tissue, fat-storing tissue, or fat within cells

adrenal cortex. The outer part of the adrenal gland that secretes many hormones, including cortisone and aldosterone

adrenal medulla. The inner part of the adrenal gland that secretes adrenalin

adrenaline (epinephrine). An "emergency" hormone stimulated by anger or fear; increases blood pressure and heart rate in order to supply the emergency needs of the muscles

adrenocorticotrophic hormone. Usually referred to as ACTH and secreted by the anterior lobe of the pituitary gland; stimulates the adrenal cortex to produce its characteristic hormones

aerobe. An organism that requires oxygen for respiration and can live only in the presence of oxygen

aerobic. Requiring free oxygen from the atmosphere for normal activity and respiration

aldosterone. Hormone active in osmoregulation; a mineral corticoid produced by the adrenal cortex; stimulates reabsorption of Na^+ and secretion of K^+

allantois. The extraembryonic membrane of birds, reptiles, and mammals that serves as an area of gaseous exchange and as a site for the storage of noxious excretion products

allele. One of two or more types of genes, each representing a particular trait; many alleles exist for a specific gene locus

alternation of generations. The description of a plant life cycle that consists of a diploid, asexual, sporophyte generation and a haploid, sexual, gametophyte generation

alveolus. An air sac in the lung; the site of respiratory exchange, involving diffusion of oxygen and carbon dioxide between the air in the alveolus and the blood in the capillaries (plural = *alveoli*)

ameboid movement. Movement involving the flowing of cytoplasm into pseudopods, as in amoeba

amnion. The extraembryonic membrane in birds, reptiles, and mammals that surrounds the embryo, forming an amniotic sac

anaerobe. An organism that does not require free oxygen in order to respire

anaerobic. Living or active in the absence of free oxygen; pertaining to respiration that is independent of oxygen

analogous. Describes structures that have similar function but different evolutionary origins; e.g., a bird's wing and a moth's wing

anaphase. The stage in mitosis that is characterized by the migration of chromatids to opposite ends of the cell; the stage in meiosis during which homologous pairs migrate (Anaphase I), and the stage in meiosis during which chromatids migrate to different ends of the cell (Anaphase II)

androgen. A male sex hormone (e.g., testosterone)

angiosperm. A flowering plant; a plant of the class Angiospermae that produces seeds enclosed in an ovary and is characterized by the possession of fruits and flowers

animal pole. The end of the egg cell that contains the least amount of cytoplasm

Annelida. The phylum to which segmented worms belong

anther. The part of the male reproductive organ (the stamen) that produces and stores pollen

antheridium. The organ of plants that produces sperm

antibiotic. An antipathogenic substance (e.g., penicillin)

antibody. Globular proteins produced by tissues that destroy or inactivate antigens

antigen. A foreign protein that stimulates the production of antibodies when introduced into the body of an organism

aorta. The largest artery; carries blood from the left ventricle

aortic arch. Blood vessels located between ascending and descending aortas that deliver blood to most of upper body

appendage. A structure that extends from the trunk of an organism and is capable of active movements

aqueous humor. Fluid in the eye, found between the cornea and the lens

archegonium. The organ of plants that produces eggs

archenteron. The central cavity of the gastrula, which is lined by endoderm and gives rise to the adult digestive cavity

artery. A blood vessel that carries blood away from the heart

Arthropoda. The phylum to which jointed-legged invertebrates belong, including insects, arachnids, and crustaceans

asexual reproduction. The production of daughter cells by means other than the sexual union of gametes (as in budding and binary fission)

atrium. The thin-walled anterior chamber of the heart (also called the auricle)

autonomic nervous system. The part of the nervous system that regulates the involuntary muscles, such as the walls of the alimentary canal; includes the parasympathetic and sympathetic nervous systems

autosome. Any chromosome that is not a sex chromosome

autotroph. An organism that utilizes the energy of inorganic materials such as water and carbon dioxide or the sun to manufacture organic materials; examples of autotrophs include plants

auxin. A plant growth hormone

axon. A nerve fiber

bacillus. Bacteria that are rod shaped

bacteriophage. A type of virus that can destroy bacteria by infecting, parasitizing, and eventually killing them

bile. An emulsifying agent secreted by the liver

bile salts. Compounds in bile that aid in emulsification

binary fission. Asexual reproduction; in this process, the parent organism splits into two equal daughter cells

binomial nomenclature. The system of naming an organism by its genus and species names

biome. A habitat zone, such as desert, grassland, or tundra

biotic. Living, as in living organisms in the environment

blastopore. The outer opening of the archenteron that develops during gastrulation, becoming the anus in deuterostomia and the mouth in protostomia

blastula. A stage of embryonic development in which the embryo consists of a hollow ball of cells

Bowman's capsule. Part of the nephron in the kidney; involved in excretion

bryophyte. A plant phylum that incorporates mosses and liverworts

bud. In plants, an area of undifferentiated tissue covered by embryonic leaves

budding. A process of asexual reproduction in which the offspring develop from an outgrowth of the plant or animal

buffer. A substance that prevents appreciable changes in pH in solutions to which small quantities of acids or bases are added

calorie. A unit of heat; the amount of heat required to raise the temperature of one gram of water by one degree centigrade (Note: a large Calorie (food calorie) = 1,000 calories)

Calvin cycle. Cycle in photosynthesis that reduces fixed carbon to carbohydrates through the addition of electrons (also known as the "dark cycle")

cambium. Undifferentiated tissue in the stem of a plant that aids growth in width

capillary. A tube one cell thick that carries blood from artery to vein; the site of material exchange between the blood and tissues of the body

carbohydrate. An organic compound to which hydrogen and oxygen are attached; the hydrogen and oxygen are in a 2:1 ratio; examples include sugars, starches, and cellulose

carbon cycle. The recycling of carbon from decaying organisms for use in future generations

carnivore. A flesh-eating animal; a holotrophic animal that subsists on other animals or parts of animals

carotene. An orange plant pigment that is the precursor of Vitamin A

cation. An ion with a positive charge, or an ion that migrates towards the cathode (negative electrode) in an electric field

cell wall. A wall composed of cellulose that is external to the cell membrane in plants; it is primarily involved in support and in the maintenance of proper internal pressure

cell wall plate. In mitosis of higher plants, the structure that forms between the divided nuclei of the two daughter cells and eventually becomes the cell wall

central nervous system (CNS). Encompasses the brain and the spinal cord

centriole. The small granular body within the centrosome to which the spindle fibers attach

centromere. The place of attachment of the mitotic fiber to the chromosome

centrosome. A structure in animal cells containing centrioles from which the spindle fibers develop

cerebellum. The hindbrain region that controls equilibrium and muscular coordination

cerebral cortex. The outer layer of cerebral hemispheres in the forebrain, consisting of gray matter

cerebral hemisphere. One of the paired lateral divisions of the forebrain

cerebrum. The largest portion of the human brain; it is believed to be the center of intelligence, conscious thought, and sensation

chemosynthesis. The process by which carbohydrates are formed through chemical energy; found in bacteria

chemotropism. The orientation of cells or organisms in relation to chemical stimuli; the growth or movement response of organisms to chemical stimuli

chitin. A white or colorless, amorphous, horny substance that forms part of the outer integument of insects, crustaceans, and some other invertebrates; it also occurs in certain fungi

chloroplast. A plastid containing chlorophyll

chlorophyll. A green pigment that performs essential functions as an electron donor and light "entrapper" in photosynthesis

chlorophyte. A green alga; a member of the phylum Chlorophyta

Chondrichthyes. Fish that have a cartilage skeleton; a class of vertebrates that includes sharks, skates, rays, and related types

Chordata. An animal phylum in which all members have a notochord, dorsal nerve cord, and pharyngeal gill slits at some embryonic stage; includes the Cephalochordata and the Vertebrates

chorion. The outermost, extra-embryonic membrane of reptiles and birds

chromatid. One of the two strands that constitute a chromosome; chromatids are held together by the centromere

chromatin. A nuclear protein of chromosomes that stains readily

chromosome. A short, stubby rod consisting of chromatin that is found in the nucleus of cells; contains the genetic or hereditary component of cells (in the form of genes)

chromosome map. The distribution of genes on a chromosome, derived from crossover frequency experiments

chyme. Partially digested food in the stomach

circadian rhythms. Daily cycles of behavior

cleavage. The division in animal cell cytoplasm caused by the pinching in of the cell membrane

climax community. The stable, biotic part of the ecosystem in which populations exist in balance with each other and with the environment

clotting. The coagulation of blood caused by the rupture of platelets and the interaction of fibrin, fibrinogen, thrombin, prothrombin, and calcium ions

cloaca. The chamber in the alimentary canal of certain vertebrates located below the large intestine, into which the ureter and reproductive organs empty (as in frogs)

cochlea. The sensory organ of the inner ear of mammals; it is coiled and contains the organ of Corti

codominant. The state in which two genetic traits are fully expressed and neither dominates

Coelenterata. An invertebrate animal phylum in which animals possess a single alimentary opening and tentacles with stinging cells; examples are jellyfish, corals, sea anemones, and hydra

coelom. The space between the mesodermal layers that forms the body cavity of some animal phyla

coenzyme. An organic cofactor required for enzyme activity

colon. The large intestine

commensal. Describes an organism that lives symbiotically with a host; this host neither benefits nor suffers from the association

conditioning. The association of a physical, visceral response with an environmental stimulus with which it is not naturally associated; a learned response

cone. A cell in the retina that is sensitive to colors and is responsible for color vision

consumer. Organism that consumes food from outside itself instead of producing it (primary, secondary, and tertiary)

contractile vacuole. A specialized structure that controls osmotic pressure by removing water from the cell; exists in protozoans

cornea. The outer, transparent layer of the eye

corpus callosum. A tract of nerve fibers connecting the two cerebral hemispheres

corpus luteum. A remnant of follicle after ovulation that secretes the hormone progesterone

cortex. In plants, the tissue between the epidermis and the vascular cylinder in the roots and stems of plants; in animals, the outer tissue of some organs

cortisone. A hormonal secretion of the adrenal cortex

cotyledon. A "seed leaf;" responsible for food digestion and storage in a plant embryo

crossing over. The exchange of parts of homologous chromosomes during meiosis

cross-pollination. The pollination of the pistil of one flower with pollen from the stamen of a different flower of the same species

Crustacea. Crustaceans; a large class of arthropods, including crabs and lobsters

cuticle. A waxy protective layer secreted by the outer surface of plants, insects, etcetera

cytochrome. A hydrogen carrier containing iron that functions in many cellular processes, including respiration

cytokinesis. A process by which the cytoplasm and the organelles of the cell divide; the final stage of mitosis

cytoplasm. The living matter of a cell, located between the cell membrane and the nucleus

cytoskeleton. The organelle that provides mechanical support and carries out motility functions for the cell

cytosine. A nitrogen base that is present in nucleotides and nucleic acids; it is paired with guanine

deamination. The removal of an amino group from an organism, particularly from an amino acid

deletion. The loss of all or part of a chromosome

deme. A small, local population

dendrite. The part of the neuron that transmits impulses to the cell body

deoxyribose. A five carbon sugar that has one oxygen atom less than ribose; a component of DNA (deoxyribose nucleic acid)

diastole. The passive, rhythmical expansion or dilation of the cavities of the heart (atria or ventricles) that allows these organs to fill with blood; preceded and followed by systole (contraction)

dicotyledon. A plant that has two seed leaves or cotyledons

diencephalon. The hind portion of the forebrain of vertebrates

differentiation. A progressive change from which a permanently more mature or advanced state results; for example, a relatively unspecialized cell's development into a more specialized one

diffusion. The movement of particles from one place to another as a result of their random motion

digestion. The process of breaking down large organic molecules into smaller ones

dihybrid. An organism that is heterozygous for two different traits

dimorphism. The instance of polymorphism in which there is a difference of form between two members of a species, as between males and females

diploid. Describes cells that have a double set of chromosomes in homologous pairs ($2n$)

disaccharide. A sugar composed of two combined monosaccharides (e.g., sucrose, lactose)

disjunction. The separation of homologous pairs of chromosomes following meiotic synapsis

DNA. Deoxyribonucleic acid; found in the cell nucleus, its basic unit is the nucleotide; contains coded genetic information; can replicate on the basis of heredity

dominance. A dominant allele suppresses the expression of the other member of an allele pair when both members are present; a dominant gene exerts its full effect regardless of the effect of its allelic partner

dorsal root. The sensory branch of each spinal nerve

duodenum. The most anterior portion of the small intestine of vertebrates, adjacent to the stomach; the continuation of the stomach into which the bile duct and pancreatic duct empty

Echinodermata. The phylum of spiny-skinned animals that includes starfish and sea urchins

ecological succession. The orderly process by which one biotic community replaces another until a climax community is established

ecology. The study of organisms in relation to their environment

ectoderm. The outermost embryonic germ layer that gives rise to the epidermis and the nervous system

egg (ovum). The female gamete; it is nonmotile, large in comparison to male gametes, and stores nutrients

electron transport chain. A complex carrier mechanism located on the inside of the inner mitochondrial membrane of the cell; releases energy, and is used to form ATP

embolus. A blood clot that is formed within a blood vessel

endemic. Pertaining to a restricted locality; ecologically, occurring only in one particular region

endocrine gland. A ductless gland that secretes hormones directly into the bloodstream

endocytosis. A process by which the cell membrane is invaginated to form a vesicle which contains extracellular medium

endoderm. The innermost embryonic germ layer that gives rise to the lining of the alimentary canal and to the digestive and respiratory organs

endoplasmic reticulum. A network of membrane-enclosed spaces connected with the nuclear membrane; transports materials through the cell; can be soft or rough

enzyme. An organic catalyst and protein

endoplasm. The inner portion of the cytoplasm of a cell or the portion that surrounds the nucleus

endosperm. The triploid tissue in some seeds that contains stored food and is formed by the union of one sperm nucleus with two nuclei of the female's gametophyte

epidermis. The outermost surface of an organism

epididymis. The coiled part of the sperm duct, adjacent to the testes in mammals

epiglottis. In mammals, a flap of tissue above the glottis; it folds back over the glottis in swallowing to close the air passages of the lungs; contains elastic cartilage

epicotyl. The portion of seed plant embryo above the cotyledon

epinephrine see *adrenalin*

epithelium. The cellular layer that covers external and internal surfaces

epiphyte. A plant that lives on another plant commensalistically

erythrocyte. An anucleate red blood cell that contains hemoglobin

esophagus. The portion of alimentary canal connecting the pharynx and the stomach

estrogen. A female sex hormone secreted by the follicle

ethanol fermentation. A form of anaerobic respiration found in yeast and bacteria

ethylene. A hormone that ripens fruit and induces aging

eukaryote. Multicellular organism

excretion. The elimination of metabolic waste matter

exocrine. Pertaining to a type of gland that releases its secretion though a duct; e.g., the salivary gland, the liver

exocytosis. A process by which the vesicle in the cell fuses with the cell membrane and releases its contents to the outside

exoskeleton. Describes arthropods and other animals whose skeletal or supporting structures areoutside the skin

eye. A sensory organ capable of detecting light

F_1. The first filial generation (first offspring)

F_2. The second filial generation; offspring resulting from the crossing of individuals of the F_1 generation

fallopian tube. The mammalian oviduct that leads from the ovaries to the uterus

feedback mechanism. The process by which a certain function is regulated by the amount of the substance it produces

femur. The thigh bone of vertebrates

fermentation. Anaerobic respiration that yields 2 molecules of ATP, lactic acid, ethyl alcohol and carbon dioxide, or some similar compound via the glycolytic pathway

fertilization. The fusion of sperm and the egg to produce a zygote

fibrin. Protein threads that form in the blood during clotting

fibrinogen. Blood protein that is transformed to fibrin upon clotting

fitness. The ability of an organism to contribute its alleles and therefore its phenotypic traits to future generations

flagellate. An organism that possesses one or more whiplike appendages called flagella

flagellum. A microscopic, whiplike filament that serves as a locomotor structure in flagellate cells

follicle. The sac in the ovary in which the egg develops

food vacuole. A vacuole in the cytoplasm in which digestion takes place (in protozoans)

frame shift mutation. A mutation involving the addition or loss of nucleotides

fruit. A mature ovary

FSH. An anterior pituitary hormone that stimulates the follicles in females and the function of the seminiferous tubules in males

functional groups. Chemical groups attached to carbon skeletons that give compounds their functionality

gall bladder. An organ that stores bile

gamete. A sex or reproductive cell that must fuse with another of the opposite type to form a zygote, which subsequently develops into a new organism

gametophyte. The haploid, sexual stage in the life cycle of plants (alternation of generations)

ganglion. A grouping of neuron cell bodies that acts as a coordinating center

gastrula. A stage of embryonic development characterized by the differentiation of the cells into the ectoderm and endoderm germ layers and by the formation of the archenteron; can be two-layer or three-layer

gene. The portion of a DNA molecule that serves as a unit of heredity; found on the chromosome

gene frequency. A decimal fraction that represents the presence of an allele for all members of a population that have a particular gene locus

genetic code. A four-letter code made up of the DNA nitrogen bases A, T, G, and C; each chromosome is made up of thousands of these bases

genetic drift. Random evolutionary changes in the genetic makeup of a (usually small) population

genotype. The genetic makeup of an organism without regard to its physical appearance; a homozygous dominant and a heterozygous organism may have the same appearance but different genotypes

genus. In taxonomy, a classification between species and family; a group of very closely related species, e.g., *Homo*, *Felis*

geographical barrier. Any physical feature that prevents the ecological niches of different organisms (not necessarily different species) from overlapping

geotropism. Any movement or growth of a living organism in response to the force of gravity

germ cell. A reproductive cell

germ layer. One of the primary tissues of the embryo

gibberellin. A hormone that stimulates plant stem elongation

gill slit. A perforation leading from the pharynx to the outside environment that is a characteristic of chordates at one stage of their development

glomerulus. A network of capillaries in the Bowman's capsules of the kidney

glottis. In mammals, the slitlike opening formed by the vocal folds in the larynx

glycogen. A starch form in animals; glucose is converted to glycogen in the liver

glycolysis. The anaerobic respiration of carbohydrates

goiter (simple). An enlargement of the thyroid gland due to lack of iodine

Golgi apparatus. Membranous organelles involved in the storage and modification of secretory products

gonads. The reproductive organ that produces sex cells (e.g., ovary, testes)

Graffian follicle. The cavity in the mammalian ovary in which the egg ripens

granum. The smallest particle that is capable of carrying out photosynthesis; the functional unit of a chloroplast

gray matter. A portion of the CNS consisting of cytons (cell bodies), their dendrites, and synaptic connections

guanine. A purine (nitrogenous base) component of nucleotides and nucleic acids; it links up with cytosine in DNA

guard cell. One of a pair of kidney-shaped cells that surround a stomate and regulate the size of the stomate in a leaf

gymnosperm. A plant that belongs to the class of seed plants in which the seeds are not enclosed in an ovary; includes the conifers

haploid. Describes cells (gametes) that have half the chromosome number typical of the species (n chromosome number)

hemoglobin. A protein compound containing iron that is found in red blood cells; hemoglobin pigment combines with oxygen and gives the red blood cells their respiratory function

hepatic portal system. The veins that carry blood from the digestive organs to the liver

herbivore. A plant-eating animal

hermaphrodite. An organism that possesses both the male and the female reproductive organs

heterotroph. An organism that must get its inorganic and organic raw materials from the environment; a consumer

heterozygous. Describes an individual that possesses two contrasting alleles for a given trait (Tt)

homeotherm. An animal with a constant body temperature

homologous. Describes two or more structures that have similar forms, positions, and origins despite the differences between their current functions; examples are the arm of a human, the flipper of a dolphin, and the foreleg of a horse

homozygous. Describes an individual that has the same gene for the same trait on each homologous chromosome (TT or tt)

hormone. A chemical messenger that is secreted by one part of the body and carried by the blood to affect another part of the body, usually a muscle or gland

host. Any organism that is the victim of a parasite

humerus. A bone of the upper arm

hybrid. An offspring that is heterozygous for one or more gene pairs

hydrostatic skeleton. Fluid skeleton of annelids

hyperthyroidism. An oversecretion of thyroid that leads to high metabolism and exophthalmia goiter

hypertonic. Describes a fluid that has a higher osmotic pressure than another fluid it is compared to; it exerts greater osmotic pull than the fluid on the other side of a semipermeable membrane; hence, it possesses a greater concentration of particles, and acquires water during osmosis

hypocotyl. The portion of the embryonic seed plant below the point of attachment of the cotyledon; forms the root

hypothalamus. A section of the posterior forebrain associated with the pituitary gland

hypotonic. Describes a fluid that has a lower osmotic pressure than a fluid it is compared to; it exerts lesser osmotic pull than the fluid on the other side of a semipermeable membrane; hence, it possesses a lesser concentration of particles, and loses water during osmosis

immunity. A resistance to disease developed through immune system

imprinting. The process by which environmental patterns or objects presented to a developing organism during a "critical period" of its growth is accepted as a permanent element of its behavior

incomplete dominance. Genetic blending; each allele exerts some influence on the phenotype (for example, red and white parents may yield pink offspring)

independent assortment. The law by which genes on different chromosomes are inherited independently of each other

ingestion. The intake of food from the environment into the alimentary canal

inner ear. A fluid-filled sensory apparatus that aids in balance and hearing

insulin. A hormone produced by the Islets of Langerhans in the pancreas; regulates blood sugar concentration by converting glucose to glycogen (in the process lowering glucose level)

integument. Refers to protective covering, such as the covering of an ovule, that develops into the seed coat, or an animal's skin

interphase. A metabolic stage between mitoses in which genetic material is reproduced

interstitial cells. Cells which in the female are located between the ovarian follicles, and in the male are located between the seminiferous tubules of the testes; in both cases, these cells produce male sex hormones

inversion. Occurs when a segment of genetic material on a chromosome becomes reversed

iris. The colored part of the eye that is capable of contracting and regulating the size of the pupils

irritability. The ability to respond to a stimulus

isolation. The separation of some members of a population from the rest of their species; prevents interbreeding and may lead to the development of a new species

isomer. One of a group of compounds that is identical in atomic composition, but different in structure or arrangement

isotonic. Describes a fluid that has the same osmotic pressure as a fluid it is compared to; it exerts the same osmotic pull as the fluid on the other side of a semipermeable membrane; hence it neither gains nor loses net water during osmosis, and possesses the same concentration of particles before and after osmosis occurs

Krebs cycle. Process of aerobic respiration that fully harvests the energy of glucose; also known as the citric acid cycle

lactase. The enzyme that acts upon lactose

lacteal. A lymph tubule located in the villus that absorbs fatty acids

lactid acid fermentation. A type of anaerobic respiration found in fungi, bacteria, and human muscle cells

larva. A period in the development of animals between the embryo and adult stages; starts at hatching and ends at metamorphosis

legume. A flowering plant with simple dry fruit, characterized by nodes on their roots that contain nitrogen-fixing bacteria (e.g., beans, clover)

lens. A structure of the eye that focuses images on the retina by changing its convexity

levels of structure. Different relationships that are formed in proteins between the original sequence of amino acids and more complex three-dimensional compounds

lichen. An association between an algae and a fungus that is symbiotic and mutualistic in nature

linkage. Occurs when different traits are inherited together more often than they would have been by chance alone; it is assumed that these traits are linked on the same chromosome

lipase. A fat-digesting hormone

lipid. A fat or oil

littoral zone. A marine biome; a region on the continental shelf that contains an ocean area with depths of up to 600 ft

Loop of Henle. The thin, bent part of the renal tubule that is the site of the counter-current flow and the sodium gradient

luteinizing hormone (LH). Secreted by the anterior pituitary gland, this hormone stimulates the conversion of a follicle into the corpus luteum and the secretion of progesterone by the corpus luteum; it also stimulates the secretion of sex hormones by the testes

lymph. A body fluid that flows in its own circulatory fluid in lymphatic vessels separate from blood circulation

lymph capillary. One of many tubules that absorb tissue fluid and return it to the bloodstream via the lymphatic system

lymphocyte. A kind of white blood cell in vertebrates that is characterized by a rounded nucleus; involved in the immune response

lysosome. An organelle that contains enzymes that aid in intracellular digestion

macula. A sensory hair structure in the utriculus and the sacculus of the inner ear; orients the head with respect to gravity

malleus. The outermost bone of the middle ear (hammer)

malpighian tubules. Tubules that excrete metabolic wastes into the hindgut in arthropods

maltase. An enzyme that acts upon maltose and converts it into glucose

maltose. A 12-carbon sugar that is formed by the union of two glucose units (a disaccharide)

marsupial. A pouched mammal, such as the kangaroo or opossum

medulla. The inner layer of an organ surrounded by the cortex

medulla oblongata. The posterior part of the brain that controls the rate of breathing and other autonomic functions

medusa. A jellyfish; the bell-shaped, free-swimming stage in the life cycle of coelenterates

meiosis. A process of cell division whereby each daughter cell receives only one set of chromosomes; the formation of gametes

Mendelian laws. Laws of classical genetics established through Mendel's experiments with peas

meristem. An undifferentiated, growing region of a plant that is constantly undergoing cell division and differentiation

mesoderm. The primary germ layer, developed from the lip of the blastopore, that gives rise to the skeleton, the circulatory system, and many organs and tissues between the epidermis and the epithelium

metabolism. A group of life-maintaining processes that includes nutrition, respiration (the production of usable energy), and the synthesis and degradation of biochemical substances

metamorphosis. The transformation of an immature animal into an adult; a change in the form of an organ or structure

metaphase. A stage of mitosis; chromosomes line up at the equator of the cell

microbodies. Organelles that serve as specialized containers for metabolic reactions

micron (micrometer). One-thousandth of a millimeter; a unit of microscopic length

mitochondria. Cytoplasmic organelles that serve as sites of respiration; a rod-shaped body in the cytoplasm known to be the center of cellular respiration

mitosis. A type of nuclear division that is characterized by complex chromosomal movement and the exact duplication of chromosomes; occurs in somatic cells

monocotyledon. A plant that has a single cotyledon or seed-leaf

monohybrid. An individual that is heterozygous for only one trait

monosaccharide. A simple sugar; a 5- or 6-carbon sugar (e.g., ribose or glucose)

morphology. The study of form and structure

morula. The solid ball of cells that results from cleavage of an egg; a solid blastula that precedes the blastula stage

mucosa. A mucus-secreting membrane, such as the inner intestinal lining

mutagenic agent. Agent that induces mutations; typically carcinogenic

mutation. Changes in genes that are inherited

mutualism. A symbiotic relationship from which both organisms involved derive some benefit

myelin sheath. A fatty sheath surrounding the axon of a neuron that aids in stimulus transmission; it is secreted by the Schwann cells

NAD. An abbreviation of nicotinamide-adenine-dinucleotide, also called DPN; a respiratory oxidation-reduction molecule

NADP. An abbreviation of nicotinamide-adenine-dinucleotide-phosphate, also called TPN; an organic compound that serves as an oxidation-reduction molecule

nephron. Functional urinary tubules responsible for excretion in the kidney of vertebrates

nerve. A bundle of nerve axons

nerve cord. A compact linear organization of nerve tissues with ganglia in the CNS

nerve net. A multidirectional sensory system of lower animals such as the hydra, consisting of nerve fibers spread throughout the ectoderm

neural tube. An embryonic structure that gives rise to the central nervous system

neuron. A nerve cell

niche. The functional role and position of an organism in an ecosystem; embodies every aspect of the organism's existence

nictitating membrane. A thin, transparent, eyelidlike membrane that opens and closes laterally across the cornea of many vertebrates (the third eyelid)

nitrogen cycle. The recycling of nitrogen from decaying organisms for use in future generations

nondisjunction. The failure of some homologous pairs of chromosomes to separate following meiotic synapsis

notochord. A flexible, supportive rod running longitudinally through the dorsum ventral to the nerve cord; found in lower chordates and in the embryos of vertebrates

nuclear membrane. A membrane that envelopes the nucleus and separates it from the cytoplasm; present in eukaryotes

nucleolus. A dark-staining small body within the nucleus; composed of RNA

nucleotide. An organic molecule consisting of joined phosphate, 5-carbon sugar (deoxyribose or ribose), and a purine or a pyrimidine (adenine, guanine, uracil, thymine, or cytosine)

nucleus. An organelle that regulates cell functions and contains the genetic material of the cell

olfactory. Related to the sense of smell

oogenesis. A process of formation of ova

organelle. A specialized structure that carries out particular functions for eukaryotic cells; examples include the plasma membrane, the nucleus, and ribosomes

osmoregulation. The ways in which organisms regulate their supply of water

osmosis. The diffusion of water through a semipermeable membrane, from an area of greater concentration to an area of lesser concentration

ovary. The female gonad in animals; the base of the pistil in plants

oviduct. A tube connecting the ovaries and the uterus

oxidation. The removal of hydrogen or electrons from a compound or addition of oxygen; half of a redox (oxidation or reduction) process

pairing (synapsis). An association of homologous chromosomes during the first meiotic division

parasitism. A relationship in which one organism benefits at the expense of another

parasympathetic. Pertaining to a subdivision of the autonomic nervous system of vertebrates

parathyroid. An endocrine gland of vertebrates, usually paired, and located near or within the thyroid; it secretes parathormone, which controls the metabolism of calcium

parenchyma. Plant tissue consisting of large thin-walled cells for storage

passive immunity. A resistance to disease produced through the injection of antibodies

parthenogenesis. A form of asexual reproduction in which the egg develops in the absence of sperm

pathogen. A disease-causing organism (*pathogenic* = disease inducing)

pedigree. A family tree depicting the inheritance of a particular genetic trait over several generations

pelagic zone. A marine biome typical of the open seas

pepsin. A stomach enzyme that partially digests proteins

peptide. The kind of bond formed when two amino acid units are jointed end to end; a double unit is called a dipeptide; the joining of many amino acid units into a chain results in a polypeptide that is the structural unit of a protein molecule

peripheral nervous system. Comprises somatic and autonomic nervous systems; consists of cranial nerves and spinal nerves

peristalsis. Waves of contraction and relaxation passing along a tubular structure, such as the digestive tube

permeability. Degree of penetrability, as in membranes that allow given substances to pass through; the ability to penetrate

pH. A symbol that denotes the relative concentration of hydrogen ions in a solution: the lower the pH, the more acidic a solution; the higher the pH, the more basic is a solution; pH is equal to $-\log (H^+)$

phagocyte. Any cell capable of ingesting another cell

pharynx. The part of the alimentary canal between the mouth and the esophagus

phenotype. The physical appearance or makeup of an individual, as opposed to its genetic makeup

phloem. The vascular tissue of a plant that transports organic materials (photosynthetic products) from the leaves to other parts of the plant

photolysis. A process of photosynthesis in which water is split into H^+ and OH^-; the hydrogen ion is then joined to NADP

photoperiodism. A response by an organism to the duration and timing of light and dark conditions

photosynthesis. The process by which light energy and chlorophyll are used to manufacture carbohydrates out of carbon dioxide and water; an autotrophic process using light energy

phototropism. Plant growth stimulated by light (stem: +, towards light; root: –, away from light)

phylogeny. The study of the evolutionary descent and interrelations of groups of organisms

phylum. A category of taxonomic classification that is ranked above class; kingdoms are divided into phyla

physiology. The study of all living processes, activities, and functions

pineal body. A structure found between the cerebral hemispheres of vertebrates; secretes melatonin, which may help regulate the pituitary by regulating hypothalamic releasing factors

pinocytosis. The intake of fluid droplets into a cell

pistil. The part of the flower that bears the female gametophyte

pith. The central tissue of a stem, used for food storage

pituitary. A gland composed of two parts, anterior and posterior, each with its own secretions; called the "master gland" because its hormones stimulate secretion by other glands

placenta. A structure formed by the wall of uterus and the chorion of embryo; serves as the area in which the embryo obtains nutrition from the parent

planaria. The class of free-living flatworms

plankton. Passively floating or drifting flora and fauna of a body of water; consists mainly of microscopic organisms

plasma. The liquid part of blood

plasma membrane. The cell membrane

plasmodium. A motile, multinucleate mass of protoplasm resulting from fusion of uninuclear amoeboid cells; an organism consisting of such a structure, e.g., a slime mold

plastid. Cytoplasmic bodies within a plant cell that are often pigmented (e.g., chloroplasts)

platelet. Small, disc-shaped bodies in the blood that play a chief role in coagulation

pleural cavity. The cavity between the lungs and the wall of the chest

point mutation. A mutation in which a single nucleotide base is substituted for another nucleotide base

polar body. Nonfunctional haploid cells created during meiosis in females; they have very little cytoplasm—most has gone into the functional egg cell

pollen. The microspore of a seed plant

pollination. The transfer of pollen to the micropyle or to a receptive surface that is associated with an ovule (such as a stigma)

polymer. A large molecule that is composed of many similar molecular units (e.g., starch)

polymorphism. The individual differences of form among the members of a species

polyp. A typical coelenterate individual with a hollow tubular body whose outer ectoderm is separated from its inner endoderm by mesoglea

polyploidy. A condition in which an organism may have a multiple of the normal number of chromosomes ($4n$, $6n$, etcetera)

polysaccharide. A carbohydrate that is composed of many monosaccharide units joined together, such as glycogen, starch, and cellulose

pons. The part of the hindbrain located in the brain stem

population. All the members of a given species inhabiting a certain locale

Porifera. The phylum of sponges

primary oocyte. A cell that divides to form the polar body and the secondary oocyte

primary spermatocyte. A cell that divides to form two secondary spermatocytes

producer. Organism that produces its own food; first stage in the food chain

progesterone. The hormone secreted by the corpus luteum of vertebrates and the placenta of mammals; its function is to maintain the endometrium

prokaryote. Unicellular organism with simple cell structure

prophase. A mitotic or meiotic stage in which the chromosomes become visible and during which the spindle fibers form; synapsis takes place during the first meiotic prophase

protein. One of a class of organic compounds that is composed of many amino acids; contains C, H, O, and N

prothrombin. A constituent of the plasma of the blood of vertebrates; it is converted to thrombin by thrombokinase in the presence of calcium ions, thus contributing to the clotting of blood

Protista. A kingdom of unicellular living organisms that are neither animals nor plants; includes some groups of algae, slime molds, and protozoa

pulmonary. Relating to the lung

pupil. An opening in the eye whose size is regulated by the iris

purine. A nitrogenous base such as adenine or guanine; when joined with sugar and phosphate, a component of nucleotides and nucleic acids

pyrimidine. A nitrogen base such as cytosine, thymine, and uracil; when joined with sugar and phosphate, a component of nucleotides and nucleic acids

pyloric valve. A muscular valve regulating the flow of food from the stomach to the small intestine

recessive. Pertains to a gene or characteristic that is masked when a dominant allele is present

recombinant DNA technology. Technology that allows for manipulation of genetic material

reduction. A change from a diploid nucleus to a haploid nucleus, as in meiosis

regeneration. The ability of certain animals to regrow missing body parts

respiration. A chemical action that releases energy from glucose to form ATP

respiratory center. The area of medulla that regulates the rate of breathing

reticulum. A network or mesh of fibrils, fibers, or filaments, as in the endoplasmic reticulum

retina. The innermost tissue layer of the eyeball that contains light-sensitive receptor cells

rhizome. An underground stem

ribosome. An organelle in the cytoplasm that contains RNA; serves as the site of protein synthesis

rhodopsin. The pigment in rod cells that causes light sensitivity

RNA. An abbreviation of ribonucleic acid, a nucleic acid in which the sugar is ribose; a product of DNA transcription that serves to control certain cell activities; acts as a template for protein translation; types include mRNA (messenger), tRNA (transfer), and rRNA (ribosomal)

rod. A cell in the retina that is sensitive to weak light

root hair. Outgrowths of a root's epidermal cells that allow for greater surface area for absorption of nutrients and water

saprophyte. An organism that obtains its nutrients from dead organisms

secondary tissue. Tissue formed by the differentiation of cambium that causes a growth in width of a plant stem

selective breeding. The creation of certain strains of specific traits through control of breeding

self-pollination. The transfer of pollen from the stamen to the pistil of the same flower

semicircular canals. Fluid-filled structures in the inner ear that are associated with the sense of balance

seminal fluid. Semen

seminiferous tubules. Structures in the testes that produce sperm and semen

sensory neuron. A neuron that picks up impulses from receptors and transmits them to the spinal cord

serum. The fluid that remains after fibrinogen is removed from the blood plasma of vertebrates

sex chromosome. There are two kinds of sex chromosomes, *X* and *Y*; *XX* signifies a female and *XY* signifies a male; there are fewer genes on the *Y* chromosome than on the *X* chromosome

sex linkage. Occurs when certain traits are determined by genes on the sex chromosomes

sinus. A space in the body (e.g., blood sinus or maxillary sinus)

small intestine. The site of most digestion of nutrients and absorption of digested nutrients (e.g. the wall of the alimentary canal)

smooth muscle. Involuntary muscle (e.g., the wall of the alimentary canal)

somatic cell. Any cell that is not a reproductive cell

species. A group of populations that can interbreed

spermatogenesis. The process of forming the sperm cells from primary spermatocytes

spindle. A structure that arises during mitosis and helps separate the chromosomes; composed of tubulin

spiracle. The external opening of the trachea in insects, opening into respiratory system

sphincter. A ring-shaped muscle that is capable of closing a tubular opening by constriction; one example is the orbicularis oris muscle around the mouth

spore. A reproductive cell that is capable of developing directly into an adult

sporophyte. An organism that produces spores; a phase in the diploid-haploid life cycle that alternates with a gametophyte phase

stamen. The part of the flower that produces pollen

steroid. One of a class of organic compounds that contains a molecular skeleton of four fused rings of carbon; includes cholesterol, sex hormones, adrenocortical hormones, and Vitamin D

stigma. The uppermost portion of pistil upon which pollen grains alight

stoma (stomate). A microscopic opening located in the epidermis of a leaf and formed by a pair of guard cells; the guard cells interact physically and regulate the passage of gas between the internal cells and the external environment

stomach. The portion of alimentary canal in which some protein digestion occurs; its muscular walls of stomach churn food so that it is more easily digested; its low pH environment activates certain protein-digesting enzymes

stroma. A dense fluid within the chloroplast; the site at which CO_2 is converted into sugars in photosynthesis

style. A stalklike or elongated body part, usually pointed at one end; part of the pistil of the flower

substrate. A substance that is acted upon by an enzyme

sucrase. An enzyme that acts upon sucrose

symbiosis. The living together of two organisms in an intimate relationship; includes commensalism, mutualism, and parasitism

sympathetic. Pertaining to a subdivision of the autonomic nervous system

synapse. The junction or gap between the axon terminal of one neuron and the dendrites of another neuron

synergistic. Describes organisms that are cooperative in action, such as hormones or other growth factors that reinforce each other's activity

synaptic terminal. The swelling at the end of an axon

synapsis. The pairing of homologous chromosomes during meiosis

systole. The contraction of the atria or ventricles of the heart

taiga. A terrestrial habitat zone that is characterized by large tracts of coniferous forests, long and cold winters, and short summers; bounded by tundra in the north and found particularly in Canada, northern Europe, and Siberia

taxonomy. The science of classification of living things

telophase. A mitotic stage in which nuclei reform and nuclear membrane reappears

test cross. The breeding of an organism with a homozygous recessive in order to determine whether an organism is homozygous dominant or heterozygous dominant for a given trait

testes. The male gonads that produce sperm and male hormones

tetrad. A pair of chromosome pairs present during the first metaphase of meiosis

thalamus. A lateral region of the forebrain

thermoregulation. The ways in which organisms regulate their internal heat

thoracic duct. A major lymphatic that empties lymph into a vein in the neck

thorax. The part of the body of an animal that is between the neck or head and the abdomen

thrombin. A substance that participates in the clotting of blood in vertebrates; formed from prothrombin, it converts fibrinogen into fibrin

thrombokinase. The enzyme released from the blood platelets in vertebrates during clotting; transforms prothrombin into thrombin in the presence of calcium ions; also known as thromboplastin

thymine. A pyrimidine component of nucleic acids and nucleotides; pairs with adenine in DNA

thymus. A ductless gland in upper chest region concerned with immunity and the maturation of lymphocytes

thyroid. An endocrine gland located in the neck that produces thyroxin

thryoxin. A hormone of the thyroid that regulates basal metabolism

tissue. A mass of cells that have similar structures and perform similar functions

trachea. An air-conducting tube, e.g., the windpipe of mammals or the respiratory tubes of insects

transcription. The first stage of protein synthesis, in which the information coded in the DNA base is transcribed onto a strand of mRNA

translation. The final stages of protein synthesis in which the genetic code of nucleotide sequences is translated into a sequence of amino acids

translocation. The transfer of a piece of chromosome to another chromosome

transpiration. The evaporation of water from leaves or other exposed surfaces of plants

trypsin. An enzyme from the pancreas that digests proteins in the small intestine

tundra. The biome located between the polar region and the taiga; characterized by a short growing season, no trees, and frozen ground

turgor pressure. The pressure exerted by the contents of a cell against the cell membrane or cell wall

umbilicus. The navel; the former site of connection between the embryo and the umbilical cord

ungulate. A hoofed animal

uracil. A pyrimidine found in RNA (but not in DNA); pairs with DNA adenine

urea. An excretory product of protein metabolism

ureter. A duct that carries urine from the kidneys to the bladder

urethra. A duct through which the urine passes from the bladder to the outside

urinary bladder. An organ that stores urine temporarily before it is excreted

urine. Fluid excreted by the kidney containing urea, water, salts, etcetera

uterus. The womb in which the fetus develops

vacuole. A space in the cytoplasm of a cell that contains fluid

vagus nerve. The tenth cranial nerve that innervates digestive organs, heart, and other areas

vegetal pole. The end of the egg cell containing the most yolk; undergoes less division than the animal pole

vein. A blood vessel that carries blood back to the heart from the capillaries

ventral root. The basal branch of each spinal nerve; carries motor neurons

ventricle. The more muscular chamber(s) of the heart that pump blood to the lungs and to the rest of the body

vestigial organ. An organ that is not functional in an organism, but was functional at some period in its evolution

villus. A small projection in the walls of the small intestine that increases the surface area available for absorption *(plural: villi)*

vitamin. An organic nutrient required by organisms in small amounts to aid in proper metabolic processes; may be used as an enzymatic cofactor; since it is not synthesized, it must be obtained prefabricated in the diet

white matter. An accumulation of axons within the CNS that is white because of its fatty, myelin sheath

wood. Xylem that is no longer being used; gives structural support to the plant

xylem. Vascular tissue of the plant that aids in support and carries water

yolk sac. A specialized structure that leads to the digestive tract of a developing organism and provides it with food during early development

zygote. A cell resulting from the fusion of gametes

Index

NOTES

NOTES

NOTES

NOTES

How Did We Do? Grade Us.

Thank you for choosing a Kaplan book. Your comments and suggestions are very useful to us. Please answer the following questions to assist us in our continued development of high-quality resources to meet your needs.

The Kaplan book I read was: _____

My name is: _____

My address is: _____

_____

My e-mail address is: _____

What overall grade would you give this book?   (A) (B) (C) (D) (F)

How relevant was the information to your goals?   (A) (B) (C) (D) (F)

How comprehensive was the information in this book?   (A) (B) (C) (D) (F)

How accurate was the information in this book?   (A) (B) (C) (D) (F)

How easy was the book to use?   (A) (B) (C) (D) (F)

How appealing was the book's design?   (A) (B) (C) (D) (F)

What were the book's strong points? _____

How could this book be improved? _____

Is there anything that we left out that you wanted to know more about?

Would you recommend this book to others?   ☐ YES   ☐ NO

Other comments: _____

_____

_____

Do we have permission to quote you?   ☐ YES   ☐ NO

Thank you for your help. Please tear out this page and mail it to:

Dave Chipps, Managing Editor
Kaplan Educational Centers
888 Seventh Avenue
New York, NY 10106

Or, you can answer these questions online at www.kaplan.com/talkback.

Thanks!

SIXTY · YEARS · OF · BUILDING · FUTURES
KAPLAN
60

Just in case the rock star thing doesn't work out.

Kaplan gets you in.

For over 60 years, Kaplan has been helping students get into college. Whether you're facing the SAT, PSAT or ACT, take Kaplan and get the score you need to get into the schools you want.

1-800-KAP-TEST

kaptest.com AOL keyword: kaplan

*Test names are registered trademarks of their respective owners.

KAPLAN

World Leader in Test Prep

About

Educational Centers

K aplan Educational Centers is one of the nation's leading providers of education and career services. Kaplan is a wholly owned subsidiary of The Washington Post Company.

TEST PREPARATION & ADMISSIONS

Kaplan's nationally recognized test prep courses cover more than 20 standardized tests, including secondary school, college and graduate school entrance exams and foreign language and professional licensing exams. In addition, Kaplan offers private tutoring and comprehensive, one-to-one admissions and application advice for students applying to college and graduate programs. Kaplan also provides information and guidance on the financial aid process. Students can enroll in online test prep courses and admissions consulting services at www.kaptest.com

SCORE! EDUCATIONAL CENTERS

SCORE! after-school learning centers help K-9 students build confidence, academic and goal-setting skills in a motivating, sports-oriented environment. Its cutting-edge, interactive curriculum continually assesses and adapts to each child's academic needs and learning style. Enthusiastic Academic Coaches serve as positive role models, creating a high-energy atmosphere where learning is exciting and fun. SCORE! Prep provides in-home, one-on-one tutoring for high school academic subjects and standardized tests. www.eSCORE.com provides customized online educational resources and services for parents and kids ages 0 to 18. eSCORE.com creates a deep, evolving profile for each child based on his or her age, interests and skills. Parents can access personalized information and resources designed to help their children realize their full potential.

KAPLAN LEARNING SERVICES

Kaplan Learning Services provides customized assessment, education and professional development programs to K-12 schools and universities.

KAPLAN INTERNATIONAL PROGRAMS

Kaplan services international students and professionals in the U.S. through a series of intensive English language and test preparation programs. These programs are offered campus-based centers across the USA. Kaplan offers specialized services including housing, placement at top American universities, fellowship management, academic monitoring and reporting, and financial administration.

KAPLAN PUBLISHING

Kaplan Publishing produces books and software. Kaplan Books, joint imprint with Simon & Schuster, publishes titles in test preparation, admissions, education, career development and life skills Kaplan and Newsweek jointly publish guides on getting into college finding the right career, and helping your child succeed in school Through an alliance with Knowledge Adventure, Kaplan published educational software for the K-12 retail and school markets.

KAPLAN PROFESSIONAL

Kaplan Professional provides assessment, training, and certification services for corporate clients and individuals seeking to advance their careers. Member units include Dearborn, a leading supplier of licensing training and continuing education for securities, real estate, and insurance professionals; Perfect Access/CRN, which delivers software education and consultation for law firms and businesses; and Kaplan Professional Call Center Services, a total provider of services for the call center industry.

DISTANCE LEARNING DIVISION

Kaplan's distance learning programs include Concord School of Law, the nation's first online law school; and Kaplan College, leading provider of degree and certificate programs in criminal justice and paralegal studies.

COMMUNITY OUTREACH

Kaplan provides educational career resources to thousands of financially disadvantaged students annually, working closely with educational institutions, not-for-profit groups, government agencies and other grass roots organizations on a variety of national and local support programs. Kaplan enriches local communities by employing high school, college and graduate students, creating valuable work experiences for vast numbers of young people each year.

BRASSRING

BrassRing Inc., headquartered in New York and San Mateo, CA, is the first network that combines recruiting, career development and hiring management services to serve employers and employees at every step. Through its units BrassRing.com and HireSystems BrassRing provides an array of on- and off-line resources that help employers simplify and accelerate the hiring process, and help individuals to build skills and find a better job. Kaplan is BrassRing's majority shareholder.

Want more information about our services, products, or the nearest Kaplan center?

 Call our nationwide toll-free numbers:

1-800-KAP-TEST for information on our courses, private tutoring and admissions consulting

1-800-KAP-ITEM for information on our books and software

1-888-KAP-LOAN* for information on student loans

 Connect with us in cyberspace:

On AOL, keyword:"Kaplan"

On the World Wide Web, go to:

1. www.kaplan.com
2. www.kaptest.com
3. www.eSCORE.com
4. www.dearborn.com
5. www.BrassRing.com
6. www.concord.kaplan.edu
7. www.kaplancollege.com

Via e-mail: info@kaplan.com

 **Write to:**

 Kaplan Educational Centers
888 Seventh Avenue
New York, NY 10106

Kaplan is a registered trademark of Kaplan Educational Centers. All rights reserved. *Kaplan is not a lender and does not participate in determinations of loan eligibility.

Paying for college just got easier...

The Kaplan/American Express Student Loan Information Program.

Get free information on the financial aid process before you apply.

When you request your student loan applications through The Kaplan/American Express Student Loan Information Program, we'll send you our free Financial Aid Handbook. Then, you'll have access to some of the least expensive educational loans available.

- The Federal Stafford Loan—Eligible students can borrow various amounts depending on their year in college. Loan amounts range from $2,625–$5,500 for dependent students and $6,625–$10,500 for independent students.

- The Federal Parent Loan for Undergraduate Students (PLUS)—Eligible parents may borrow up to the total cost of education, less other financial aid received.

Return your request form today!

www.kaploan.com

Educational Financing

1-888-KAP-LOAN

*Kaplan is not a lender and does not participate in the determination of loan eligibility.
Telephone inquiries to 1-888-KAP-LOAN will be answered by a representative of a provider of federal and certain private educational loans.

Plan Ahead!

When you return this form to Kaplan, you and your family will receive a copy of Kaplan's free Financial Aid Handbook. Then, we'll send you your Federal Stafford and Federal PLUS loan applications.

It's Easy!

Yes! Please send us our Free Financial Aid Handbook, as well as our educational loan kit with applications for the Federal Stafford (student) and Federal PLUS (parent) loans.

Last Name

First Name

Parent Last Name

Parent First Name

Permanent Address

City

State

Zip Code

Phone (Daytime)

Phone (Evening)

Email Address

Current GPA

Top School Choice(s)

Earliest Start Date (month/year)

Please return this form to Kaplan.

You may fax it toll free any time to 1-800-844-7458, or mail to:
Att: KapLoan, 888 Seventh Avenue 22nd Floor, New York, NY. 10106

www.kaploan.com

Educational
Financing

1-888-KAP-LOAN

KAP–34